Don Pollard

About the Author

DOUGLAS E. SCHOEN has been a Democratic campaign consultant for more than thirty years with his firm Penn, Schoen, and Berland Associates. He lives in New York City.

Also by Douglas E. Schoen

On the Campaign Trail:
The Long Road of Presidential Politics 1860–2004

Enoch Powell and the Powellites

Pat: A Biography of Daniel P. Moynihan

THE POWER OF THE VOTE

Electing Presidents, Overthrowing Dictators, and Promoting Democracy Around the World

DOUGLAS E. SCHOEN

HARPER

NEW YORK · LONDON · TORONTO · SYDNEY

HARPER

A hardcover edition of this book was published in 2007 by William Morrow, an imprint of HarperCollins Publishers.

HarperCollins books may be purchased for educational, business, or sales promotional use. For information please write: Special Markets Department, HarperCollins Publishers, 10 East 53rd Street, New York, NY 10022.

FIRST HARPER PAPERBACK PUBLISHED 2008.

Designed by Publications Development Company of Texas

The Library of Congress has catalogued the hardcover edition as follows:

Schoen, Douglas E., 1953–
The power of the vote / Doug Schoen.—1st ed.
    p. cm.
Includes index.
ISBN-13: 978-0-06-123188-9
ISBN-10: 0-06-123188-6
1. Schoen, Douglas E., 1953–   2. Campaign management—United States.   3. Political consultants—United States—Biography.   I. Title.
JK2281.S36 2007
324.7092—dc22
[B]
                                                                            2006052877

ISBN: 978-0-06-144080-9 (pbk.)

08   09   10   11   12   PDC/RRD   10   9   8   7   6   5   4   3   2   1

To Josh

CONTENTS

PART THREE   THE BIG TIME

PART FOUR   AN UNCERTAIN FUTURE

INTRODUCTION

S ome kids went to Europe for the summer. I took the Coney
Island Express.

My earliest memory in politics is the sensation of speed—
sprinting through the D Train station with a thick stack of campaign
literature, plastering the walls with "Bob Low for City Council President" stickers, all the while doing my best to avoid the police. I was
a sixteen-year-old kid, a junior at the Horace Mann School, a prestigious private school in the Riverside section of the Bronx. Most of
my classmates weren't interested in politics, and they certainly
weren't interested in the working-class neighborhoods served by the
D Train. But for me, New York City in 1969 was a mysterious and
magical place, and politics was my ticket to explore. Moving car-to-
car down the clattering trains, you saw it all: graffiti artists tagging
cars; proud blue-collar immigrants eyeing young hippies with distaste; young black men mimicking the style of the burgeoning Black
Power movement; the middle-class metropolis of the 1950s changing into the menacing city of the 1970s.

As the D Train crossed the East River into Brooklyn and
headed toward the shore, I saw a city in the midst of a momentous
political transformation. FDR's New York, the citadel of New Deal

liberalism, was experiencing a backlash born of racial and ethnic turmoil combined with a growing alienation from mainstream liberalism. What was stirring in neighborhoods like Sheepshead Bay was not Barry Goldwater–Ronald Reagan-style conservatism but rather a growing skepticism about liberal policies (be they the policies of Democratic President Lyndon B. Johnson or of Republican Mayor John Lindsay) and a mounting suspicion that government was either unable or unwilling to help people like them. It was a mood that Manhattan Democrats seemed not to have noticed, but as I walked the boardwalk in Coney Island (an expense account at Nathan's was the payoff for my subway work), I could see the anger and resentment brewing among voters who had once been Democratic stalwarts. Understanding these voters' new attitude would become a major preoccupation for me as a political consultant. It would soon pose a major challenge for my party: in an increasingly conservative climate, how can Democrats win?

That day in the subway, however, I had a more immediate challenge—the transit police. Strictly speaking, putting up advertising of any sort in the subways was vandalism. As a result, I performed my job as quickly as possible, always looking out for the police. But after successfully avoiding arrest for the entire journey from Manhattan, I stepped off the D Train in Coney Island—and into the arms of a beefy transit cop. My first encounter with the law taught me that its concerns were practical and immediate ones.

"Hey kid," he said, "how much are you getting paid to do this?"

"Twenty-five dollars," I mumbled (not knowing how to qualify the value of my concessions at Nathan's).

He tapped his head. "Not very smart kid," he said in a tone that indicated he was sad to find himself standing before a simpleton. "The fine for illegally posting bills is fifty dollars."

"So I could lose money?" I asked, playing along.

He nodded. "Who got you to do this?"

"Um, a big kid," I said. With that, I was free—and off to Nathan's.

Communicating with different constituencies was what my job was all about.

✳   ✳   ✳

My polling partner Mark Penn and I got our first chance to build a winning coalition in 1977 when we helped elect a liberal Democrat-turned-centrist—Ed Koch—mayor of New York City. We did so at two levels: first, with a message that emphasized fiscal discipline, independence, and competence, as well as a healthy dose of middle-class values; second, with a remarkable technological breakthrough—a kit-built personal computer that enabled us to gauge voters' reactions to advertisements and endorsements overnight. This combination—a centrist message married to innovative tactics—would soon become a hallmark of our business, which, following the addition of Mike Berland in 1987, became Penn, Schoen & Berland. Instant overnight polling; the first sophisticated use of message polling that allowed candidates to pretest messages before putting them in the field; analyses of voters that paid as much attention to their attitudes and lifestyle choices (psychographic characteristics) as to their demographic characteristics and political preferences; the application of political techniques to corporate campaigns (and corporate crises); the use of exit polls as checks on authoritarian political leaders abroad; the compilation of a massive database with sophisticated psychological portraits of voters in an entire city; using the internet to poll specific target audiences—all were Penn, Schoen & Berland innovations. Each changed the outcome of an election and ultimately the practice of politics, both here in the United States and around the world. Some changed the course of history.

Of course, Mark and I have hardly been political communications' sole pioneers. At a time when we were just beginning our careers, political consultants were already emerging as campaign celebrities. In 1976, pollster Pat Caddell, one of the architects of Jimmy Carter's unlikely rise to the presidency, had become the first wunderkind pollster. Joseph Napolitan was working on foreign campaigns when we were still in college. The Sawyer/Miller Group did for corporate communications what we were doing for polling. Harris Diamond and Jack Leslie, respectively the CEO and chairman of Weber Shandwick Worldwide, were pioneers in the field of corporate crisis management. Bob Squier pioneered the massive advertising campaigns that now fuel political campaigns. These were among the many accomplished strategists who were plotting political campaigns when we were still fledgling players in the industry. At important moments in history, these men shaped the fate of the Democratic Party and the practice of politics more generally. Few, however, have matched the thirty-year run that Mark and I have enjoyed. Our long record of innovation and success has given us an unusually expansive vantage point from which to comment on the changing American political landscape and the landscape of communications in general.

Innovative thinking and great tactics have been central to our success, but winning also requires the right strategy. In the decades following our first big campaign, we've found ways to win time and time again—in 1980 and again in 1984 with Jay Rockefeller in West Virginia, in 1982 with Frank Lautenberg in New Jersey, in 1986 with then-Democrat Richard Shelby in Alabama, in 1988 in Indiana with Evan Bayh, in 1996 with Bill Clinton, in 2000 and 2005 with Jon Corzine, and most recently with Democrat-turned-Republican Michael Bloomberg in 2001 and then again in 2005. As a firm, we've also found ways to bring campaign techniques to the corporate marketplace, winning important victories for companies like Texaco, AT&T, AOL, Eli Lilly, and Microsoft and challenging long-held ideas about how corporate marketing should work. Although our tools varied, in campaign after campaign, our fundamental strategy has been the same: neutralize the issues that hurt our client (be they guns, taxes, values, or new products that threaten market share) and find a set of issues on which our candidates can build a decisive, centrist majority (or dominant market share).

Yet in the political sphere, despite our successes, our message has still not been heard. Some Democrats today seem to believe that the only thing standing between them and the White House are better tactics—faster responses, sharper language, a tougher electoral playbook. As a political consultant, I'd be the last person to downplay the importance of tactics. In the 2001 New York mayoral race, an election morning exit poll detected that Giuliani supporters weren't turning out in large enough numbers to vote for Michael Bloomberg. Faced with this potential catastrophe, our automated call system went into action, flooding voters we'd identified as likely Bloomberg supporters with prerecorded messages from Giuliani and Bloomberg urging them to go to the polls and vote. They did, providing the margin of victory that made Mike Bloomberg mayor. Clearly, effective tactics can often be the difference between victory and defeat.

But good tactics and the latest technology ultimately are not enough. To win, candidates have to be where the voters are. Unfortunately, many Democrats are still too far to the Left. History could not be clearer on this point: Democrats who win at the national level are the ones who are tough on security, fiscally conservative, and responsive to people of faith. In 1960, John F. Kennedy (who was generally tougher on the Soviet Union and security issues than Richard Nixon) made a campaign stop in Houston where he candidly addressed religious issues before an audience of Southern Baptist ministers, a stirring performance that helped settled concerns about his Catholicism. Every subsequent Democrat elected president—Lyndon Johnson, Jimmy Carter, and, of course, Bill Clinton—has likewise spoken the language of faith. All of these Democratic presidents were tough-minded interventionists—Kennedy and Johnson in Vietnam, Jimmy Carter in the Persian Gulf (proclaiming the United States considered the Persian Gulf an area of vital strategic national interest), and Bill Clinton in Bosnia. By the standards of today's politicians, these Democrats also were all were fiscal conservatives who balanced budgets in ways no Republican president since Dwight Eisenhower has. (Even Johnson insisted on leaving his successor with a small surplus.)

There should be no dispute about how Democrats win presidential elections and elections in swing states. Yet the commonsense observation that Democrats need to move to the center is often greeted with thunderous denunciations of "treason!" or "sell out!" by those claiming to represent "the Democratic wing of the Democratic Party." Returning to 1970s-style liberalism isn't courageous; it's self-defeating madness. That kind of politics didn't play well in 1970s New York City, and it certainly won't work today. The fact that fellow Democrats like Howard Dean are still channeling Bella Abzug thirty-plus years after she flamed out with New York's overwhelmingly Democratic electorate is a depressing reminder that old habits die hard. In reality, Democrats like Harold Ford—not Howard Dean—should be chairmen of the Democratic National Committee.

One of my hopes for this book is that it will demonstrate the need for Democrats to aggressively reject the policies of the Far Left in favor of a more moderate, centrist approach. Nowhere is it written that the Democrats must espouse a left-wing agenda that emphasizes the redistribution of wealth, liberal social and cultural values, and an isolationist foreign policy. Such a vision is in fact myopic in the extreme. The Democratic Party has traditionally been a broad coalition of disparate interests. By standing for the broad interests of the whole at home and championing American interests abroad, it can once again be the dominant party in American political life, as the political landslide of 2006 compellingly proved. Recent successes should not conceal the fact that any political party that moves away from fiscal discipline, traditional values, and an assertive foreign policy dooms itself to a marginal role in American political life.

I came of age at a propitious time: New York City's Tammany Hall and the old urban machines were dying; the old system of votes-for-patronage was giving way to a new politics where voters favored the candidates who were most responsive to their concerns. In this brave new world, information—the ability to understand voters' preferences—was the decisive advantage. Yet the ability to gather such information was virtually nonexistent. Madison Avenue had turned away from politics after a brief dalliance in the 1950s. Political consultants were few; pollsters who tried to specialize in politics were rare and often marginalized. (Lou Harris, John F. Kennedy's pollster, was famously cut off by the candidate with a dismissive, "Just the numbers, Lou," when he tried to advise Senator Kennedy on campaign strategy during the primary season.) As immigration and suburbanization changed populations, the old bosses were losing touch with their voters. Being an obedient servant of the machine was no longer a reliable path to higher office; new tools were needed. The practice of politics was ripe for transformation. It was my good fortune to launch my career in politics at the precise moment when candidates were open to—indeed, desperate for—fresh answers.

I was also fortunate to find colleagues like Mark Penn, my first partner and the inspired cofounder of our firm, and Michael

Berland, architect of so much of our corporate business. Whereas I tended to be interested in people and political strategy, Mark's initial forte was polling and statistical analysis. Two years behind me at Horace Mann, Mark had conducted his first polls in high school. Like me, he had then gone on to Harvard, where he'd quickly become the resident pollster of the *Harvard Crimson,* the student newspaper. He had spent his summers doing research for NBC and indeed plotting marketing strategy for the network—that is, until the new head of the research department figured out that he was actually just a summer intern. I'd known him slightly from the *Crimson* and Horace Mann, and I knew him to be a gifted pollster and a sharp analyst. So when, after graduating from college, I found myself in command of a phone bank of gubernatorial candidate Hugh Carey, I instantly thought of Mark. Evidently, he was thinking of me too. I'll never forget how, when I got Mark on the phone, he said, "Hey, Doug, I expected to hear from you sooner. What took you so long?"

Jerry Bruno, an old Kennedy advance man, was ostensibly our boss, but he had no idea what could be done with a bank of 110 telephones. Mark and I did. We decided to use it to call voters, particularly members of minority groups, and try to determine which endorsements were meaningful and most effective. We then passed the results up the chain of command. The Carey campaign was thrilled. No one had done quick polls of this sort before. Indeed, as far as I know, no one had thought to use a phone bank for such precise purposes before. It was the beginning of our effort to innovate. Yet while the technology has advanced, the underlying emphasis was always the same: speed, proper targeting, and quick responses.

Despite the political success that polls have generated, polling has a mixed, even negative reputation, with some political observers doubting both the accuracy and importance of gauging public opinion—not to mention the overall financial cost to the campaign. In 1977, Mark and I got our first big break working on the campaign of a then-obscure member of Congress who wanted to be mayor, Ed Koch. David Garth, the legendary campaign consultant, was working on Koch's campaign, and though Garth liked us because we were two smart young kids, the real reason he kept us was that we were dirt cheap. Candidate Koch didn't have much money; we were about all he could afford. That affordable price tag did not last for long. As public opinion research became more integral to campaigns, its cost greatly increased, and along with it, there was an increase in the debate over the actual function that political consultants should play in a campaign.

"Some of my best friends are consultants," writes *Time* magazine columnist Joe Klein. "They tend to be the most entertaining people in the political community: eccentric, fanatic, creative, violently verbal and deeply hilarious." But while he enjoys our company, Klein argues that our impact on politics has been "perverse."

"Rather than make the game more interesting, they have drained a good deal of the life from our democracy," argues Klein. "They have become specialists in caution, literal reactionaries—they react to the results of their polling and focus groups; they fear anything they haven't tested." Others have described us as "mercenaries" who somehow induce politicians to give up their personal beliefs. Such feverish fears stem in part from the secrecy in which political consultants operate. Consultants who attract top billing are a distraction from their candidates: no example is more illustrative of this than the tragic flameout of my one-time colleague in the Clinton campaign, Dick Morris. Most of the best political consultants operate under the radar. We tend to guard our trade secrets.

This book breaks with that habit. By exploring how the process really works, how elections are really won, how decisions are really made, I hope to put to rest some of the hoary myths surrounding our profession. For in fact, far from disemboweling representative democracy, polling has made elected officials more responsive to their constituents than ever before. Political consultants rarely tell candidates what to say. More often, we start with a candidate's preexisting views and help him or her shape them to the attitude of the electorate. In doing so, we typically begin a dialogue between the candidate and the voters.

In fledgling democracies, the potential benefits of public opinion research are in some ways even more striking. When a nation is struggling to establish or protect democracy, the presence of credible, outside polling—particularly exit polls on Election Day—can mean the difference between rule by the people and autocracy. As this book will show, in Mexico and in Serbia in 2000—and again in the Dominican Republic and in the Ukraine in 2004—exit polls played a critical role in ensuring the triumph of democracy, often proving the difference between unnecessary bloodshed and a peaceful transfer of power.

Yet despite the United States' explicit goal of advancing democracy across the world, our country's leaders, both Democratic and Republican, have been slow to recognize the importance of public opinion research techniques and the related tools of democratization as an instrument of American foreign policy. Accordingly, they have been strangely hesitant to use them, even when vital American interests were at stake. In 1992, the people of Serbia voted to oust dictator Slobodan Milosevic, handing an electoral victory to his opponent, Serbian American entrepreneur Milan Panic. But even in the face of exit polls that pointed to a Milosevic defeat, the West stayed quiet and allowed Milosevic to lie about the tally and steal the election. There was no condemnation, no effort to throw its weight behind Panic. It was a decision that would cost the Balkans tens of thousands of lives, unleash war on Bosnia-Herzegovina and Kosovo, and ultimately necessitate NATO's first-ever offensive attack. Similarly, during the 2004 election in Venezuela, former U.S. President Jimmy Carter, the Organization of American States, and the Bush administration turned away from compelling evidence that incumbent President Hugo Chavez had committed election fraud. This gross oversight ensured that an increasingly authoritarian, avowedly anti-American leader, who openly modeled himself on Cuba's Fidel Castro, would continue to control the largest oil supply in the Western hemisphere. In Zimbabwe, the governments of Africa have likewise stood by while dictator Robert Mugabe impoverished his country, persecuted the opposition leaders, and stole one election after another. It is well past time to end such a shortsighted policy of indifference to electoral fraud. Exit polls are now a time-proven technique; they should be a mainstay of closely contested elections both here and abroad.

Nor should the United States limit itself to funding exit polls. It is critically important that we pay closer attention to public opinion abroad and take advantage of the opportunities it sometimes presents.

Some of the most pressing problems facing the United States today can be attributed, at least in part, to its failure to take public opinion seriously. Consider the so-called axis of evil—Iran, North Korea, and Saddam Hussein's Iraq. It is striking indeed that in two of these countries, both Democratic and Republican administrations failed to support democratic forces at critical moments. Take Iran. In 1997, a Western-oriented reformer, Mohammad Khatami, became president of Iran, setting the stage for a showdown with the mullahs. His success would have transformed Iran and benefited the entire world, but the West allowed Khatami to fail.

Today, Iran is ruled by Mahmoud Ahmadinejad, a theocratic, fundamentalist demagogue who seems intent on securing nuclear weapons and whose popularity, buoyed by seventy dollar barrels of oil, is high. Yet now figures like former Secretary of State James Baker and former Representative Lee Hamilton are arguing that the United States must engage in dialogue with Iran. They may well be right, but U.S. policymakers should also note that internal opposition to Ahmadinejad is already emerging and has already had some success at the ballot box in recently concluded local elections. The United States should support that internal opposition, which is at least as promising as a policy of encouraging negotiations, all the while continuing with threats and possible sanctions. Think of how much better off we would be if it had been tried a decade ago.

The United States has displayed a similarly tone-deaf approach to South Korea. In 2002, the United States needlessly inflamed South Korean public opinion during a pivotal presidential race by failing to apologize fully for a terrible but clearly accidental homicide involving U.S. soldiers. It was a gaffe that would tilt a country away from a pro-American leader who was tough on North Korea, and toward an anti-American candidate whose policies seemed at times to border on appeasement. If the Bush administration had taken public opinion in these countries more seriously, the axis of evil today might well be a much smaller one.

There's a great deal of talk today about the need to enhance human intelligence. That need is certainly real, but human intelligence should be only one part of our approach. Monitoring public opinion and understanding the attitudes of elites and the mass public is arguably equally important. In Iraq, for instance, where even a rudimentary understanding of Islamic insurgents and the various sectarian factions eludes the United States, surveys and public opinion research should have been an integral part of the allied efforts, as journalists like George Packer have pointed out.

This is not to say that the techniques of political communications and democratization are a panacea. The example of countries such as Venezuela shows that they are not. Indeed, the story of Venezuela shows how the techniques of democracy can be *misused* by clever authoritarian regimes to enhance their own legitimacy. Frighteningly enough, there is even evidence that the Chavez government may be trying to penetrate the U.S. electoral system by having the Venezuelan-owned company that was at the center of the 2004 referendum controversy bid on contracts for voting machine systems in Chicago and elsewhere. But used properly and transparently, these techniques can promote American interests while peacefully preserving and enhancing the values we hold dear.

However, not everything about the current era is good. Political communications today often sound canned; some politicians have become hesitant to act on their own initiative. This is not to say that authenticity has disappeared from American politics. Indeed, I continue to witness politicians rejecting poll-driven policy time and time again in favor of strongly held personal beliefs. In general, though, our domestic political system has become more dysfunctional than at any point in recent history. The problem is not political consultants but a deep and growing dissatisfaction with the two-party system. Democrats' big wins during the 2006 midterm election cycle should not obscure the fact that most voters believe the Democrats have yet to offer a compelling alternative to the discredited policies of President George W. Bush. Unless Democrats do so soon, the party runs the real risk that victory could give way to defeat, just as Newt Gingrich's 1994 Republican "revolution" gave way to Bill Clinton's triumphant reelection in 1996.

In truth, the narrow dogmatism of both parties and their inability to cooperate in a bipartisan fashion could give rise to something different—a new political party or perhaps even an independent, centrist presidential candidate at some point in the future. Of course, the rules of politics in this country make it very difficult for third parties or other challenges to the status quo to emerge. Nevertheless, the desire for something different is there. It would be a profound misreading of American public opinion not to see that voters are deeply skeptical about the system itself—and that skepticism is growing every day.

The displacement of the old political machines by new information technologies, the ways in which innovation has transformed the practice of politics, the role political consultants play in campaigns, the keys to victory—these are fascinating and important topics to anyone interested in understanding the practice of politics today. Ultimately, however, politics are still about people. To truly understand the history of this political transformation, it's necessary to understand the people who make it, heroes like Yitzhak Rabin and Shimon Peres, villains like Slobodan Milošević and Hugo Chavez, business visionaries like Steve Case, Bob Pittman, and Bill Gates, and gifted, complex, and all-too-human politicians like President Bill Clinton.

Like most political professionals, I chose to make my career in politics because of my fascination with people—their relationships, their histories, their strengths, and, yes, their personal weaknesses. My most treasured memories are of the remarkable people I've met and worked with and the incredible experiences I've had. Working late into the night with Mark and a yeshiva student in 1977 to churn out overnight poll results at a time when such a feat was almost inconceivable; participating in the development of the New Democrat philosophy as young men in the early 1980s; dodging the Serbian secret police in 1992 while designing a campaign to topple Serbian dictator Slobodan Milošević; trying to help Shimon Peres turn his vision of peace in the Middle East into a winning electoral coalition in Israel; working with Steve Case and Bob Pittman to resuscitate AOL; coming to the assistance of Bill Clinton at the nadir of his political fortunes and helping to reelect him decisively in 1996; devising strategies to elect successful businessmen like Jon Corzine and Mike Bloomberg to public office—these are my most cherished memories and ultimately the heart of this book.

PART ONE

BIRTH OF A POLLSTER

CHAPTER ONE

CANVASSING, THE KLAN, AND THE PLOT TO TAKE OVER NEW YORK CITY

It was one of those classic, miserable, late-autumn afternoons in New York City—rainy, windy, and raw—when no sensible person lingers outside. Yet there I was, a sixteen-year-old high school senior sitting on a park bench at Eighty-First Street and Columbus Avenue next to a recent Columbia graduate—and newly elected district leader—named Jerry Nadler. Today, Jerry represents the Upper West Side and part of Brooklyn in Congress, but that afternoon he was merely an operative—a large, effusive operative with a message so important, so consequential that he couldn't risk divulging it within the earshot of another person.

Jerry's message concerned the powers that controlled the Upper West Side. Three years earlier, supporters of presidential candidate Eugene McCarthy had begun to wrest control of the neighborhood away from the reform Democrats who only a few years before had dispatched the last remnants of the old Tammany Hall machine. But now, Nadler whispered, the reformers who had ousted Tammany were themselves finished. A new power structure was forming, he exclaimed breathlessly, led by a secret new boss who was, moreover, a political genius—a reclusive figure who was wise beyond his years, a master political strategist who could analyze political trends "down to the individual apartment building."

"He will be one of the top strategists in America very soon—if he isn't already," Nadler declared. He went on to describe how this mysterious figure and his followers (of whom Jerry was one)—a group that had come to be known as the West Side Kids—were systematically canvassing and organizing the Upper West Side block by block, building by building, with the goal of controlling first that bastion of liberalism and then, ultimately, all of New York City. In effect, Nadler was describing in embryonic detail a new world of campaigning—a world where understanding public opinion and the issues that motivated voters was the true currency of political power. In 1969, this was a thrilling and novel idea. More thrilling still, the unnamed mastermind of this effort had chosen me to be one of his minions—if I could prove myself worthy.

As cold and wet as I was, I was intrigued. All my life, I had been obsessed with politics. My decision to spend my junior year playing football for the Horace Mann School rather than engaging in anti-war activities had given rise to intense guilt. Politics was in my family: my uncle Jack Bronston was an influential state senator in Queens, and my fondness for statistics—baseball, at that point—was a telltale indicator of my future as a pollster. Nadler was now offering me a chance to participate in the political life of New York City. Exactly how I would be participating was not at all clear, but I was interested. I told Nadler that I definitely wanted to meet this figure. Jerry was pleased.

"We'll be in touch," he told me, and then he was gone.

The phone call came several days later. "Doug? Yes? Good. You talked to Jerry? Yes? Good. I want you to come over for dinner on Tuesday. You can come? Good. See you at seven." The voice was brusque, the conversation fast, but I did manage to catch my interlocutor's name—Dick Morris. Yes, *that* Dick Morris.

For any student of modern politics, the name Dick Morris is familiar and, some would say, infamous. To me, Dick is a discredited and tragic figure, a man whose personal and professional excesses during President Clinton's 1996 campaign almost destroyed his career and marriage. But in 1969, Morris was a fresh face—a recent Columbia graduate and the secret mastermind of the West Side Kids.

I arrived at Morris's building at Riverside Drive and Ninety-fifth Street in a state of some anxiety. I knew from Nadler that Morris was a fanatical canvasser. The fact that he was willing to take time off to have dinner with me was a signal honor, and I wanted to make a good impression. I was also nervous about Morris's building. With its peeling paint and broken elevator, it didn't exactly look like the emerging center of power in New York City. When the door opened, I didn't feel much better. Standing before me was an ordinary looking twenty-two-year-old in a button-down shirt holding a glass of orange juice. The apartment looked forlorn, like a seldom-used Bolshevik safe house, furnished only with a card table and folding metal chairs.

Generally, when you go to someone's apartment for dinner, there are pleasantries involved. You make small talk, discuss current movies, what's in the news, and so on. That was not the case with Dick and his then-wife Gita. Spending time with them, there was no discussion of personal lives. It was all about politics. As Gita struggled with a take-out chicken, the talk came fast, almost breathlessly. We're building a cadre of skilled activists, Dick told me, among them Gita ("I married her because of her organizing ability," Dick later confided) and Dick Dresner, a one-time professional bowler whom Morris had convinced to give up the lanes for block associations. Everyone was hard working, dedicated to the cause, single-minded, and utterly committed to the art of canvassing. I would be part of this elite crew.

"I know things about political organizing that no one else does," he said. "Sign on, and you'll learn them too."

It was like an indoctrination ritual for a cult. Gita listened rapturously, making sure that Morris's glass of orange juice was never empty.

"But I want to go to Harvard," I told him.

"Give it up," Dick said flatly. "Go to Columbia and work for me as a canvasser. I'll give you a block to organize," he continued. "You will canvass it; you will know every building's issues; you will own it. It will be yours."

The work would be hard, he warned, but the rewards would be great. Morris promised that, in a year or two, I would be a district leader. Two or three years later, I might be an assemblyman or even a state senator. After that, who knew where I could end up?

By the end of the evening, I was half-intrigued and half-scared out of my wits. I knew I wasn't going to give up Harvard to work for Dick Morris. Even then, I knew he was only really interested in one thing—Dick Morris. However, I was dazzled by the opportunity. In the end, the chance to work on a nascent political movement and involve myself firsthand in the day-to-day world of urban politics was irresistible. I told Morris that I was ready to sign on—as long as he understood that, if admitted, I would be going on to Harvard. Reluctantly, he agreed to my terms.

Soon Nadler and I were regularly spending our evenings crammed into Morris's tiny apartment, using two phone lines—at the time, Morris's only indulgence—to canvass West Side voters by telephone. Our techniques were crude. If I were about to call someone with an old-fashioned name, Selma Brownstein say, I'd prepare a speech about senior citizen's issues. I was beginning to understand how the coupling of exhaustive information dispersal and skillful and strategic messaging could bring extraordinary political benefits. Today, we take such tactics for granted. In fact, they are the essence of modern political campaigning. But in 1970, they were practically unheard of.

I was immediately thrown into the West Side Kids' next big campaign on behalf of an ambitious, young politician named Dick Gottfried. Gottfried, who wasn't much older than I was, was running for an Assembly seat on the Upper West Side—a seat he still holds today. The West Side Kids wanted to use Gottfried's seat on the Assembly as a stepping-stone to take political power across the entire West Side. They would then elect Gottfried as City Council President—a step they saw as a surefire path to true political power.

Morris, Gottfried, and Nadler brought a shrewd insight to the West Side; namely, that it wasn't enough to simply rail against the incumbents. They would have to become machine politicians in their own right, but they were pragmatists, not ideologues. They understood that if they didn't take care of constituent needs, their political life wouldn't last very long. Nadler, for example, became an expert on landlord-tenant issues and gained a reputation for helping residents understand the intricacies of New York's rent laws. In the turbulent era of the 1960s, these new reformers also grasped the importance of taking the major issues of the day down to the local level. This meant everything from organizing peace committees against the Vietnam War to picketing merchants who sold scab grapes. The goal of the West Side Kids—power—may have been a fairly conventional one (even at that idealistic time). However, the tactics were anything but conventional.

Above all else, Morris and Nadler brought to the table a macro view of the electorate. They understood that political power came from a more complete understanding of voter preferences and attitudes. In many ways, the approach was revolutionary. Morris sought to gather as much information as possible about the voters of the Upper West Side and then use that information to tailor messages that would appeal to individual voters. Even then I saw this organizing approach as something new and different—a sort of reincarnation of Tammany Hall that revolved around shared issues and ideologies (or at least the appearance of shared ideologies), rather than just around the provision of jobs and services.

Today, voting groups are generally categorized by broad characteristics such as age, wealth, and gender. Back then, it was a bit less sophisticated. We broke the West Side down by individual apartment buildings and developed a message appropriate to each. For example, if there was a building that was a stronghold of radical liberals, we would craft a message ideological enough to please even Trotsky. If there had been a rent strike in another building, we would talk about tenants' rights. No one had ever done such focused and such flexible messaging before. This was microstrategic messaging on a neighborhood-wide scale.

Despite its sophistication at the time, the operation itself was fairly bare bones. There was no direct mail, no television, no radio ads, no leafleting at subway stops. Considering the central role of money in President Clinton's reelection in 1996 and the campaigns I worked on for multimillionaire candidates like Senators Frank Lautenberg and Jon Corzine—not to mention billionaire candidate Mike Bloomberg—it is perhaps ironic that in my first campaign, money was something of an afterthought. In fact, I remember Morris saying at one point that he didn't even care about money. The only thing he believed in was canvassing. And that's what he wanted from me.

I quickly learned the ins and outs of canvassing, because in New York City, you can't just go door-to-door; you have to find a way to sneak into the apartment buildings. I would start off by breezing into a building lobby and announce that I was headed up to visit, say, Mr. and Mrs. Levy on the sixteenth floor. These were contacts that we had developed who would let us into the building and, if need be, provide us with a hideout if we were pursued by an especially determined doorman. Once I made it to the elevator, I would head straight to the top floor. It was crucial to start at the top of a building and work your way down—that way you wouldn't get tired. In those unusual circumstances when we encountered an overzealous doorman, we would start at the top and then move down to the middle to throw them off the scent.

Our systematic nature was critical to our canvassing methods. We were basically transforming an activity, usually done in the most haphazard way, into a virtual science. We developed profiles of different buildings and sent out canvassers with the relevant political expertise. Once we were in the building, we didn't just knock on doors at random. If you came to an unfriendly apartment, it was only a matter of time before the doorman was alerted and the chase would begin anew. Instead, Morris had segmented the population so thoroughly that we only needed to talk to about seven or eight people in each building—those most likely to vote. Why waste your time trying to convince someone who probably wasn't going to the polls anyway?

In those days, voters weren't interested in simply getting a piece of campaign literature; they were gunning for a political debate. So it was quite common to be invited inside to defend the merits of your candidates and their positions. I remember once knocking on the door of Gerard Piel, who was the president and publisher of *Scientific American,* and being asked to join his family for dinner. Not surprisingly, Morris and Nadler advised us to minimize such collegiality. In fact, Morris wouldn't even allow me to canvass Lincoln Towers, a massive complex of buildings between West Sixty-Sixth and West Seventieth Streets, which was home to many radical retired teachers. He felt I wasn't "mature" enough to handle the formidable debaters I might encounter.

All the information we gathered was put to good use on Election Day. The West Side Kids had amassed a list of twelve thousand people in the West Side who were going to vote. Those names were checked against the lists at the polls, and the names of people who had not yet voted were brought back to a group of callers, who would then phone them with a reminder to get to the polls. This was all done with about one hundred to one hundred fifty volunteer workers. On the final day of the campaign, Nadler even had me calling transients at the West Side YMCA (we had developed a custom-tailored campaign message for the homeless as well, complete with the promise of food at the polls). I'm not sure if that was what put us over the top, but Gottfried won the race handily.

With Dick Morris, though, nothing was exactly as it appeared. What truly clinched the race for Gottfried was a deal that Morris made with a fourth-generation Tammany Hall political leader named Jim McManus. McManus came from a legendary political family that wrote the book on machine politics in New York City, and more important, controlled about 20 percent to 25 percent of the district. For a West Side leftist to make a deal with a Tammany

Hall politician in 1970 was akin to Al Sharpton making a political deal with Trent Lott. It was simply unheard of. But this was the real precursor to Morris's later work: for Dick, winning, as opposed to being right, was the ultimate prize.

At the time, I didn't truly understand what motivated McManus to throw his weight behind Gottfried, although I supposed that like anyone else he put his finger in the air and figured out which way the political winds were blowing. Only later did I learn that Gottfried and Morris went to McManus and asked for his support. While some might perceive such a step as akin to selling out, it's worth remembering that Dick Gottfried has been in the State Assembly for thirty-five years and he's championed many causes near and dear to the heart of progressives.

Dick's skills weren't limited to making political deals. He also understood how to take care of his people. As a show of appreciation for my work on the campaign, he landed me a summer job as a surveyor on the Board of Water Supply. At the age of seventeen, I was a political appointee making about three hundred dollars a week. It was patronage politics at its best.

As a surveyor for the Water Board, I passed my work days (which often began with a hearty breakfast and ended four hours later) out in the Bronx's Van Cortland Park or on Randall's Island holding a pole and wearing an orange vest, surveying reservoirs, I suppose. To this day, I'm somewhat unclear on what exactly we were supposed to be doing, but whatever it was, we didn't do much of it. If it was raining, we didn't work at all.

In the afternoons, I would come back to the city and work for a fellow named Tony Olivieri, who was running for New York State Assembly. Olivieri, a fundraiser for Harvard, had made me an enticing offer: "Work for my campaign, and I'll do everything I can to get you into Harvard." It was a proposal that Olivieri made good on when he later called someone named Rufus Peebles on my behalf. Peebles, a Harvard functionary with a delightful old Boston name, supposedly told Olivieri that my admissions had been taken care of. Olivieri called me to share the good news, and two weeks later I received a letter of admission from the school that confirmed he was as good as his word. It wasn't that I didn't have the credentials to get in, but for someone who had spent his life in the protective cocoon of the Upper East Side, I was quickly learning that there were ways to work most systems—getting what politicians call "an insurance policy." Even Harvard, it seemed, had quid pro quos. As for Olivieri, he ended up winning the Assembly seat over his Republican opponent by a mere 495 votes.[1]

At the age of seventeen, this was a heady introduction to a world well beyond what I had learned about in school. This was the way politics really worked—intensive canvassing and backroom deals, hard work and political patronage, sacrifice and back-scratching. However, I quickly understood that Dick Morris and the West Side Kids had showed me a very narrow slice of the world. If I were going to make my way in politics, I would need to broaden my experiences and look elsewhere for education. I would soon stumble into just such a place—Mississippi.

In the end, Harvard did admit me, and in the summer after my freshman year, I landed an internship for a New York City councilman named Carter Burden. Burden, who owned the *Village Voice* as well as a stake in *New York Magazine,* had a reputation in 1971 as a young politician who could well prove to be the next John F. Kennedy.

Burden certainly seemed to have everything going for him. He was the charming, wealthy heir to the vast Vanderbilt fortune and was married to the stepdaughter of CBS founder William Paley. A *New York Times Magazine* article described him as "tall, lean, blond, deeply tanned, and patricianly handsome in a way that somewhat reminds one of a young John V. Lindsay."[2] Moreover, he was building a name for himself as a formidable politician in his own right—tirelessly advocating and working on behalf of his low- and middle-income constituents. The whispers around town were that Burden, the new star of New York City politics, was setting his sights on the White House.

My job with Burden started in July. That left the month of June free. So I decided to do what hundreds of other brave northeastern college students had been doing for years—travel to Mississippi and fight for civil rights. My goal was to sign on as a volunteer for the long-shot gubernatorial campaign of Charles Evers, the brother of the slain civil rights leader Medgar Evers who had been assassinated in 1963.

Charles Evers was a remarkable figure. A former bootlegger, dishwasher, disc jockey, and nightclub operator in Chicago (and, some said, panderer and numbers-runner for organized crime elements in that city), Evers returned to Mississippi after his brother's brutal assassination and took over his work as the National Association for the Advancement of Colored People's (NAACP) state field director. In 1969, he became Mississippi's first elected black mayor since Reconstruction. By 1970, he had set his sights even higher, deciding to run for governor. When I arrived, an Evers staffer took one look at my political experience and said "Doug, how would you like to be our research director?"

For an eighteen-year-old college freshman who had never before traveled south of the Mason-Dixon Line, campaigning in Mississippi was an eye-opening experience. I found a bedroom in a group house for black kids with special needs run by the Reverend Eddie McBride in a black neighborhood in Jackson near the governor's mansion. The fact that Reverend McBride felt compelled to carry a pistol told me something about the political atmosphere. Still, I tried to make the best of it. Every Sunday morning, I would leave the home and walk down Pearl Street toward the State Capitol to grab the local newspapers, like the *Clarion-Ledger,* the Memphis *Commercial Appeal* and, on a good weekend, the Sunday *New York Times*. I still remember the stares I used to get. It was easy to understand the locals' confusion: I was not only white but also about the only person in town who wasn't on the way to church. With so many papers, more than a few residents thought I was the new paperboy.

Besides sideways glances, I had another, if more daunting, problem: I really wasn't quite sure what a research director was supposed to do. Luckily, during a trip to the statehouse I stumbled upon Bill

Minor, the legendary correspondent for the New Orleans *Times-Picayune*. Minor had been covering Mississippi politics for more than two decades, and he was outraged by the racism and dirty politics that characterized the state. Unsolicited, he began offering his invaluable assistance in helping me put together policy papers for the campaign. Each morning I would come to his tiny, newspaper-strewn office in the State Capitol for my tutorial. One day it would be property taxes; the next might be the State Sovereignty Commission. He would tell me who to talk to, where to look, sometimes even what to write, and off I would go. Since a Jew from the northeast hanging around the state archives would tend to raise eyebrows, I began posing as a graduate student from Tulane University.

At first, the Evers operation made for a dismaying contrast with the perfectionism of the Upper West Side. It was hopelessly disorganized with barely concealed tensions between the white volunteers and black Mississippians. Evers's white press secretary, Jason Berry, and I were both frustrated with the Evers's operation and occasionally had heated exchanges with other staff members. I was also sometimes frustrated by Evers's passivity, by his unwillingness to publicly acknowledge that we were right and that other members of the campaign should follow our directions.

One morning, tensions at his Jackson headquarters came to a head. In the middle of an intense debate, Evers drew himself up. He was a very big man. If that wasn't enough to command attention, there was also his alleged criminal background to consider. So when Evers started talking, I listened. The point he made was not about the strategy, or how it was being implemented, much less about the campaign structure. The point was that he—a black man in Mississippi—was the man in charge.

"When I worked for Bobby Kennedy, I did what Bob Kennedy told me," Evers rumbled. "Now, this is my campaign. You're working for me. So you're going to do things my way."

Despite (or perhaps because of) his familiarity with the underworld, Evers was extraordinarily affable and, to his credit, quite realistic about his chances. About three days into the campaign, I realized we had absolutely no chance of winning—and everyone knew it. Since becoming a political consultant, I have tended to avoid races where I thought there was no chance of winning. What I learned with Evers was that when it comes to Mississippi politics, election results only tell half the story. One of the last events I attended before going home to New York was a rally with Evers in Copiah County, Mississippi. It was an extraordinary moment. Our motorcade set out from Jackson, escorted by a retinue of white Mississippi police officers. In 1971, in Mississippi, this was practically unimaginable. Two hours later, we arrived at the rally. It was in a ramshackle, white wooden church in the Delta. There was a palpable sense of fear in the steamy room. Nevertheless, Evers again approached the situation from a much higher level. He was like a Baptist preacher, delivering a sermon-like address. He cajoled, harangued, and pleaded with the crowd to get involved, "What the hell are you doing if you don't get yourself registered?" he said. The mood in the room was electric.

To say the least, this was not the kind of thing one was used to seeing on New York's Upper West Side. It was amazing stuff and it made a lasting impression on me. Mississippi taught me something very important about politics—something that Dick Morris couldn't teach me—namely, that winning isn't always the point. Evers, I realized, wasn't running to get elected so much as he was running to empower black Mississippians and to show them that after years of being beaten down and mistreated by the white majority, the time had come for them to start playing a role in the political process. On Election Day, he may have lost quite badly, but when it came to empowering a community and building a political outlet for black Mississippians, the campaign was an enormous success. To bastardize Vince Lombardi's famous aphorism, "Winning is everything, but it's *not* the only thing."

In July, I returned to New York. Carter Burden, patrician heir to the Vanderbilt fortune and newly elected city councilman, immediately put me to work deciphering New York City's nearly inscrutable finances. The intricacies of the budget didn't particularly interest me; and luckily, I soon got a far more welcome and surprising assignment—taking on one of the most powerful politicians in Manhattan in one of the toughest slums in America—East Harlem.

I knew there would be risks to running a campaign in this part of town. It was a time when some black and Hispanic neighborhoods in New York were turning into urban combat zones. I expected to deal with gangs, guns, and drugs. What I did not expect was to be doing battle with the Mob.

The man Carter Burden aimed to take down was one of the most powerful Democratic politicians in New York City—Assemblyman Frank Rossetti. Rossetti had been a fixture in city and state politics for nearly three decades. Since 1967, he had served as county leader for New York City Democrats. The root of his political power was the position he held as a Democratic district leader in East Harlem. If Rossetti lost control of his district leadership, his dominant position in the Democratic Party would crumble. Simply put, a victory over Rossetti would establish Carter Burden as the new kingmaker of Manhattan (and probably New York City) politics.

From a demographic perspective, Burden sensed an opening. East Harlem had once been home to generations of Italian Americans. But over the years, their numbers had dwindled, and blacks and Puerto Ricans had moved in. Burden was determined to capitalize on the opportunity this presented. He spent as much as fifty thousand dollars of his own money and leveraged his political capital in support of Rossetti's opponent—a young lawyer and neighboring district leader named Eugene Nardelli, who had boldly moved into the neighborhood to challenge its long-time boss.

Nardelli had some potent supporters. Herman Badillo, a local congressman who was weighing a bid for mayor in the upcoming 1973 race, was on board. So was Father Louis Gigante, a prominent Roman Catholic priest, social activist, and local politician who had recently run for Congress and won overwhelming support in the neighborhood. Despite these allies, Rossetti was hardly outgunned. He had the support of all the city's party bosses, and his campaign manager was Buddy Beame, the son of Abe Beame, city comptroller and an old school clubhouse politician, who ultimately was elected mayor in 1973.

What would ordinarily have been a rather innocuous race for district leadership had become a proxy fight for the 1973 mayoral race. And when Burden asked me to serve as Nardelli's campaign manager, I found myself—at the age of eighteen—in the midst of the fray.

We knew from the beginning that ousting Rossetti was a long shot, but Carter Burden's money and charisma gave us much-needed credibility. I decided to model Nardelli's campaign strategy on Dick Gottfried's successful campaign on the West Side. This was truly retail politics 101 and we were out—selling our candidate, knocking on tenement doors in the Italian areas around 116th Street and First Avenue—introducing Nardelli to the largely elderly voters living in the community, which had become increasingly black and Hispanic in the past several years. We distributed leaflets in the projects that were geared toward Hispanic and black voters. And we sent sound trucks driving through the neighborhood blasting our message.

However, I soon discovered that East Harlem was not the Upper West Side. Early in the campaign, I made the near fatal mistake of bringing a salsa band to a rally at Lincoln Houses, a low-income housing project at 132nd Street between Fifth and Park Avenues. The residents were largely African American and the music was not exactly a hit. Before we knew what was happening, we were being pelted from the rooftops with rocks and other debris. We beat a hasty retreat.

But racial tensions were only the beginning of the story. Rossetti's support didn't just come from his Tammany buddies. He had a few "friends" of a very different sort as well, and these friends were accustomed to making offers that people couldn't refuse. Not long after the campaign began, I asked Father Gigante to send out a mass mailing to neighborhood residents in support of Nardelli.

"Doug, I don't think that will do much good," he told me with a sigh.

"Father, don't be modest," I replied. "You got something like 90 percent of the vote in this area. These people will listen to you."

"Those voters weren't for me," Gigante replied. "They were for my friends in the neighborhood who were supporting me."

Nardelli essentially confirmed my mounting suspicions about the neighborhood. When I first asked him what he thought of Rossetti, he told me that whenever he'd see him, Rossetti would say, "Hey kid, here's a couple hundred bucks, buy yourself a new suit." For Nardelli, this was the ultimate insult. First of all, he wasn't interested in being bought off—and especially not for such a paltry sum.

Second, two hundred dollars was certainly not going to pay for a good suit. Third, and most important, Nardelli prided himself on wearing better suits than Rossetti. I had met Rossetti once before. I interviewed him briefly for a paper I was writing on Adam Clayton Powell during my freshmen year at Harvard. He was symbolic of everything that I found objectionable about politics. He was over-dressed, obnoxious, and he talked out of both sides of his month. I didn't believe, or take seriously, a word that he said.

Gradually it dawned on me that the nominal head of the Gen-ovese family—"Fat" Tony Salerno—oversaw his family's businesses from the Palma Boys Social Club on 115th Street, between First and Pleasant Avenues. The reputed overlord for my part of East Harlem was Antonio Ferro, known on the street simply as "Buckaloo." I'll never forget him. He spent his days sitting impassively in a beach chair outside a social club at 116th Street and Second Avenue, with his pants pulled up absurdly high and an unlit cigar dangling from his mouth. On the surface, he didn't seem too frightening, but after a few weeks in East Harlem, we quickly came to understand that this was not a man with whom one trifled. (Indeed, he later became the acting *consigliere* of the Genovese crime family.)

While we canvassed this area, we quickly decided that our best bet was to simply avoid these "knock-around" guys, since it was pretty clear they weren't buying what we were selling. In the pro-cess, we did everything that good canvassers were supposed to do. We were aggressive. We found the old people in the neighborhood and introduced them to Nardelli. We got high-profile endorsements from former Mayor Robert Wagner as well as Father Gigante. The *New York Times* even came out with a ringing endorsement on our behalf saying, "Nardelli . . . would open doors and windows, let the smoke out and bring an end to stifling backroom deals."[3] We ran a picture-perfect campaign. But in that neighborhood, people didn't read the *New York Times* and when Buckaloo put the word out that Rossetti was his guy, all bets were off.

I still remember the mood of the neighborhood when news fil-tered out that Buckaloo had come out in support of Rossetti. As far as I could tell, he didn't actually do anything to support his candi-date: he simply put the word out. All of a sudden, people in the district who had taken our posters and our literature didn't want to have anything to do with us. I remember going into a candy store between First and Pleasant Avenues where only days earlier the people had been quite friendly. Overnight, however, "Anything you want, anything we can do to help," evolved into, "Please don't come in here. It can only cause us problems." You could literally feel the mood on the street change.

To make matters worse, Burden was MIA. Sure he was bankrolling the campaign, but it didn't take too long for me to notice that his heart wasn't really in it. The more time I spent with Burden, the more fragile and tragic a figure he seemed. He came across as a profoundly lonely man who didn't seem to like people very much. I remember he used to come to work every day around 11:00 AM or 11:30 AM and lock himself in his office, leafing through shopping catalogs.

One of my most resonant images of Burden came early on in the campaign when I was invited to his apartment for breakfast. Burden lived in River House, which still today is one of the most exclusive buildings in all of New York City. There I was, a freshman at Harvard in this beautiful apartment with Nardelli, his running mate Elba Diaz, and Angelo Guerrero, Elba's Italian working-class husband, seated around an ornate dining room table. First, Burden's wife Amanda came down—"Ba," Burden called her. Her makeup and outfit were perfect—the picture of Upper East Side decorum—and quite the opposite of our rather shabby appearance. Then Burden entered—wearing a bathrobe. As the meeting went on, I stole glances at him. It was fairly clear that the last place in the world he wanted to be was in this meeting. He seemed bored to tears. Even then, I was old enough to realize that Burden would never be a serious politician. He simply didn't have the passion for it.

I got considerably more support from a fellow Harvard student who was one year ahead of me—E.J. Dionne. Today, Dionne is a nationally syndicated columnist for the *Washington Post* and a fellow at the Brookings Institution. Back then, he was a sophomore and a star writer for the *Harvard Crimson*. I had met Dionne when he asked to borrow my notes for a class that he had attended only infrequently. Dionne got an A-minus on the final exam. I got a B-plus.

Not withstanding this rather frustrating injustice, we became fast friends and he came down to help on the campaign.

Despite the setbacks we faced, by the end of the campaign, I was surprisingly hopeful. However, on Election Day, it was clear that Rossetti had outsmarted us, in part by taking a page out of the Morris playbook. He too had analyzed every voter. He had put election district captains on each street, like an old Tammany Hall machine. In short, he had his people on every block and in each neighborhood who could deliver for him. For example, there was a captain on 110th Street by the name of Joey Verdiccio, who had close ties to Rossetti, but not to Rossetti's running mate Wilma Sena. Verdiccio was close to Elba Diaz, who was running with Nardelli. Verdiccio did his part for Rossetti, giving him 103 votes to Nardelli's ten. As it turned out, it was the same margin of victory for Diaz over Sena. It wasn't simply Rossetti's people who came out. Every district leader, from Meade Esposito in Brooklyn to Patrick Cunningham in the Bronx understood that if Rossetti lost, they could be next.

In addition, it didn't hurt that some of the people on Rossetti's team were also overseeing the election. Even the chairman of the Board of Elections was on Rossetti's team; in fact, he was Rossetti's captain for 119th Street between First and Second Avenue. I remember when we went to the Board of Elections, he said to Nardelli, "I like you. You're a nice kid. I'm going to give you a big break: I'm going to be fair to you."

However, on Election Day, I discovered that Rossetti's people were actually campaigning inside the polling booths on 120th Street. This was blatantly illegal. I quickly went to find a police officer and told him what was going on. He peered at me for a moment and dismissively said, "Kid, mind your business." I knew there wasn't going to be a level playing field.

Nardelli lost the election by a margin of nearly three to one. A *New York Times* article, which included a kind description of me as a "young genius" and "the Carmine de Sapio of the New Left," also saw fit to mention that on election night, when the first results came in, I turned, "Whiter than a bed sheet that had been washed in Axion, with Clorox added."[4] In six years' time, I would help a little-known congressman become mayor of New York. Dionne would go on to be a distinguished political correspondent for the *New York Times* and then the *Washington Post*. But on that night, we got our butts kicked by an old-school, Tammany ward boss who had never gone to college.

After his loss, the long knives were out for Nardelli. We later heard that Rossetti told a colleague that Nardelli was "deader than Kelly's nuts," referring to a racehorse that had recently been castrated. Amazingly, however, Nardelli survived and went on to become a prominent and distinguished judge in the appellate division of New York's State court system.

The Nardelli campaign was a sobering wake-up call. My early experiences had taught me that if I understood the mood of the electorate, crafted the best messaging, and canvassed well, I could ensure victory. But the Nardelli race showed me another, equally important, lesson: if you weren't able to establish legitimate roots and ties to a neighborhood, you simply weren't going to win. You can have great literature; you can go door-to-door; you can have the most effectively targeted messaging; but if you're running against someone with an entrenched political organization, you're in trouble. That's as true for state and national politicians as it is for local ones. East Harlem taught me that the political formulas I learned on the West Side had their limits. While quite clearly the tools and tactics of the West Side reformers were the wave of the future, one could never lose sight of the fact that "All politics is local." The right message was important, but to be truly successful, it would take much more—like being able to crawl inside the heads of the voters and take a long look around.

The Nardelli campaign also imparted a more personal lesson: campaigning door-to-door—particularly in East Harlem—was not the life I wanted. We had canvassed in some of the worst projects in America, where people with knives and guns sold drugs on the corner. One of my most resonant memories was canvassing door-to-door with Nardelli on 115th Street, on the block where Fat Tony had his headquarters. We were walking up the stairs of one fairly Italian building when we encountered a junkie—who was clearly wired—walking down the stairs and banging a stick against the ground. I wasn't that comfortable being on Fat Tony's block to begin with and so the approaching bang-bang-bang sounds, frankly, had me scared to death. Matters only got worse when Nardelli bent down and pulled a gun out of what was clearly a hidden ankle holster and said, "That's why I carry this." That was the moment I decided canvassing wasn't the life for me.

I had learned about as much as I could about grass roots organizing from Dick Morris and the West Side Kids. It was clear there were real limits to how far I could go and what I could do in East Harlem politics for a variety of ethical and practical reasons. Moreover, I sensed from my experience in Mississippi that there was a larger world I needed to draw on that had a broader scope and context than the D Train to Coney Island. I wanted a career in politics, but first I needed to get away from the dangers of Pleasant Avenue.

CHAPTER TWO

HARVARD, IVY LEAGUE RADICAL CHIC, AND THE CRISIS OF THE DEMOCRATIC PARTY

In the fall of 1972, I came to my first revelation about the role I would play in politics. I was taking a practical tutorial course on how to poll with an assistant professor named Bill Schneider. Today, Schneider is a prominent CNN analyst, but back then, he was merely a young professor in the Department of Government at Harvard with time on his hands.

Political polling was something of a novelty when I started at Harvard. While George Gallup and Bud Roper had been doing "scientific" political polls since before World War II, their polls were after-the-fact samples of public opinion. Moreover, given the lead-time involved with polling, surveys were frequently out of date by the time the results had been tabulated (hence the results of polling that showed Thomas E. Dewey with an insurmountable lead over Harry Truman in 1948). Polling was also dogged by methodological questions that gave rise to a general skepticism about polls' conclusions.

No one believed polls should be an important part of day-to-day political campaigning, much less orchestrate every aspect of a campaign. Consequently, polls played only a minor role in most campaigns. Indeed, politicians such as George McGovern probably made less use of polls than John F. Kennedy had twelve years earlier. Campaigns might begin with a "benchmark" poll that identified issues of concern to the electorate; however, once these issues had been incorporated into the campaign message, a pollster's job was more or less over. The notion of using polls to monitor a race or determine tactics in midstream was simply unheard of.

Schneider saw another possibility. He realized that polls and other quantitative techniques offered politicians and journalists a valuable yet underutilized way to understand public opinion and how elections were actually decided. He was interested in moving the discipline out of the academy, where polls were detailed but often out of date, and into the realm of journalism and retail politics. Schneider had long been frustrated by the vast divide between academic polling, which tended to be comprehensive but deathly slow, and media polls, which were quick but shoddy. Moreover, Schneider was convinced that these techniques could help people approach politics in a systematic way. In the place of the wily machine boss, Schneider hoped to help create a new type of expert—a cadre of political analysts with practical as well as analytic skills.

Schneider's class was a hands-on affair. We spent most of the semester writing questionnaires, fielding polls in Cambridge and around Boston, collecting data, and analyzing our responses using sophisticated and, at the time, novel programming tools like Statistical Package for the Social Sciences (SPSS). For me, it was an exciting and eye-opening experience. I could see that behind the techniques and the statistics, polls were about gathering data in much the same way that I had gathered information as a canvasser working for Dick Morris on the West Side. The difference was that polling was more systematic, more scientific, and certainly much, *much* safer, with no racists to dodge, no drug addicts to avoid, and no organized crime figures to waylay carefully designed political strategies. Looking back on it, I doubt that in his wildest dreams Buckaloo ever thought he would have this much impact on the education of a Harvard boy, much less on American politics.

As I was learning the intricacies of rudimentary polling, Harvard was quickly succumbing to the overwhelming forces of social change. In 1969, students protesting the Vietnam War had paralyzed the university for two weeks, disrupting classes, and seizing control of University Hall. Eventually, President Nathan Pusey called in the police to evict students, a decision for which he was bitterly criticized. Subsequent attempts to appease the student body were largely unsuccessful. While the Vietnam War dragged on and figures like Martin Luther King Jr. and Robert Kennedy died at the hands of assassins, the hopeful idealism of the 1960s' civil rights movement had morphed into the more militant radicalism of the Black Panthers, Students for a Democratic Society, and the Weathermen. By the time I arrived at Harvard in the fall of 1970, many of my classmates were becoming disillusioned with conventional politics. While I remained firmly committed to the system, I nonetheless managed to get a taste of the new militancy soon after arriving on campus.

The occasion was a counter teach-in, a *pro*-Vietnam War rally hosted by the conservative group Young Americans for Freedom at Saunders's Theater, the famous Harvard auditorium built after the Civil War in honor of the Union dead. The event was advertised as an open meeting to educate the Harvard community on why it should support Nixon and the war. Actually, it seems to have been an event designed by the Young Americans for Freedom (and perhaps the CIA as well) to entrap student demonstrators into disrupting an academic event, an infraction that could trigger serious disciplinary action at Harvard.

If so, it worked brilliantly. Among the people most eager to attend were the members of the campus chapter of the Students for a Democratic Society, better known as the SDS, which saw the teach-in as the perfect opportunity to rebuild their flagging chapter. Like many of my new classmates, I was opposed to the war in Vietnam and curious about SDS (which had stormed into the public consciousness two years earlier amid violent demonstrations at

Columbia). So when two of my new friends proposed to go protest the teach-in, I readily agreed. Together they made quite the pair. While one had shoulder-length blond hair that embodied the spirit of the times, the other wore a stunning pink shirt that was remarkable—even by preppie standards. Fortunately, I was dressed more conservatively, in a nondescript black ski jacket.

When we arrived, we found the theater jammed to capacity and a crowd of over a thousand screaming, "One, two, three, four, we got to end this fucking war" and "Ho, ho! Ho Chi Minh! The NLF is going to win!" That was before the crowd learned that in a few minutes the ultimate provocation—South Vietnam's Ambassador to the United States—would come out to offer an analysis of why President Nixon's policy of Vietnamization was working. Predictably, the crowd went crazy. One member of our trio was particularly creative in expressing his contempt for the pro-war argument, closing his eyes, putting his thumbs in his ears, wiggling his fingers, and screaming, "WHHoooo, you're a racist"—something he was wont to do whenever he was frustrated, which was frequently enough. Unfortunately for him, antics of this sort violated Harvard's code of conduct—and the Young Americans for Freedom had conveniently arranged elaborate video and still photographic services to record the event, a step that they would later say had been taken so that the speech could be widely disseminated. Naturally, his antics caught the attention of the photographers and television cameras. Within days, Harrison had unwittingly become one of the symbols of the demonstration (something he was very proud of).

When the Young Americans for Freedom subsequently reviewed tapes of the event, they couldn't miss his antics (or his hair, which made him easily identifiable), but they also couldn't miss the disruptions coming from another student in an alarmingly bright pink shirt. As a result, both of my friends were identified as students who had disturbed the proceedings and reported them to the administration. To make matters worse, one of them then missed the disciplinary hearing after being arrested at a protest against Medicaid cuts in Roxbury. (He had allegedly pulled a knife on a cop and demanded his gun.) It fell to me to go to Professor Donald G. M. Anderson, professor of mechanical engineering and head of the university disciplinary committee, to explain to him that my classmate had been "unavoidably detained" at a violent demonstration in Roxbury and was, "um, presently incarcerated." Professor Anderson, a mild-mannered pipe smoker, took this surprisingly well, allowing that under those circumstances it probably was quite unlikely that he would be back for his disciplinary hearing. We both agreed that it would be best to reschedule. I later testified at his criminal trial that he was a man of peaceful purposes with a reputation for decency and civility, which was generally true; I think he managed to get off without any jail time. Meanwhile, my other classmate was temporarily suspended, and to this day he says that I deserve to have been sanctioned as well—and would have been if I'd been a snappier dresser. "The only reason you didn't know about snazzy pink shirts and instead wore a fashionable black ski jacket," he'd occasionally tell me, is that "I went to St. George's and you went to Horace Mann."

In a bizarre footnote to this story, I later found out that during this period one of my friends dated a girl—a radical street person named Linda (whom he had met while she was living in Harvard Square)—that was in fact an agent with the FBI's infamous COINTELPRO program, an FBI program designed to "neutralize" political dissidents. Apparently, Linda's role was to press him and his somewhat more timorous classmates to undertake ever more violent actions. The trade-off of sexual favors for more radical action was something the FBI apparently saw as an irresistible lure to student radicals. Harvard Square, I learned, could, in its own way, be almost as risky as the streets of East Harlem or Mississippi.

As exciting as extracurricular activities at Harvard were, my real education ultimately came from a handful of innovative thinkers who were committed to working with highly engaged students. Bill Schneider was one. Another, in the government department, was Martin Kilson, the first African American professor to receive tenure at Harvard, a brilliant man—part intellectual historian, and part ward boss. I took a class with Kilson during my first semester on campus and, like hundreds of other students, I was captivated by his iconoclastic personality, his wicked sense of humor, and his amazingly varied skills. I remember one day walking into Kilson's office to talk to him about a paper I was working on to find him furiously arguing with a black student about who was behind a series of rapes in Medford. Kilson was screaming that the culprits were lower-class black hoods. The student was equally vehement in his insistence that whites were doing the raping. They argued back and forth until Kilson finally drew himself up and said, "Get your black ass out to Medford. I want to know what's happening. You're my scout, my courier. Don't just talk. Do some fieldwork!" He then turned politely to me and said, "Schoen, your paper." It was classic Kilson—irreverent, fieldwork focused, and incredibly versatile.

I had an odd but delightful relationship with Kilson. From the beginning, he made it clear that he would only have a limited amount of time for me.

"Schoen, there are plenty of professors who will work with you—a smart Jewish boy," Kilson would tell me. "I have to put my time into the black students."

He was also clear about his opinion of our respective intellectual abilities. There was only one person in our class with the intellect to be a Robert Merton (the eminent sociologist and theorist of media and criminality), he once said, and that was a young man named Cornel West. "All of you other people are going to have to go out into the field, talk to people, and really work for insights," he said. Getting out and doing research was an imperative for him, in part because he personally wanted to know what was happening in the black community, his community; sometimes he operated almost like a demanding ward leader and neighborhood elder with his students. He also felt it was important to go out, work, and do original research to be serious. My final paper for his class was just the sort of thing he liked: a report on how Charlie Rangel had defeated the legendary Adam Clayton Powell to win election to Congress from Harlem. In the course of writing it, I talked to virtually everyone—Rangel, community leaders, ward bosses, journalists. I even got mugged at knife point in central Harlem and had to beg my way onto a city bus to get home.

My experiences in Harlem while writing the paper fit perfectly with Kilson's Hobbesian view of urban politics. His worldview was summed up by pie-shaped diagram he had taped to his wall—a circle sliced into Irish, Italian, black, and Jewish cuts. "This is the city," he liked to say. "Everyone is fighting for power. What we're going to talk about is how you get your fair piece of the pie. That's what this is ultimately about. Never forget it." Over the course of my subsequent career in urban politics, I've never forgotten it.

To supplement this field learning, I decided to go to work for the student newspaper, the *Harvard Crimson*. Like my coursework (and my encounter with the Harvard administration), it was an amazing learning experience. During the 1972 presidential campaign, I got the opportunity to spend ten days on the back-up plane (known to the press corps as "the zoo plane") with Democratic nominee George McGovern. By then, McGovern was on the ropes. During those last weeks of the campaign, the one potentially hopeful thing McGovern talked about was something that was new to me—Watergate. On October 10, a front-page story in the *Washington Post* reported the following:

> FBI agents have established that the Watergate bugging incident stemmed from a massive campaign of political spying and sabotage conducted on behalf of President Nixon's re-election and directed by officials of the White House and the Committee for the Re-election of the President.

McGovern kept saying, "This brings it all into the White House, into the presidential campaign." But no one was listening or cared. After two weeks on the plane, I still couldn't figure out what McGovern's themes were. All I saw were the sullen crowds, a disorganized campaign, and a candidate who was strident and whiney.

McGovern lost to Nixon in a crushing landslide. For many, McGovern's loss was so devastating that it felt like the country had turned against our generation. Indeed, some of them were turning away from conventional politics altogether. However, I didn't see Nixon's triumph as simply a vindication of the reactionary status quo. The reality was far more nuanced. I saw two reasons for McGovern's defeat. First, he had run a terrible general election campaign—a stark contrast to his primary effort. During the primaries, McGovern had run a focused campaign with a clear message: he was against the Vietnam War and for a guaranteed income for every American. Unfortunately, as the campaign moved into the general election, that focus and discipline fell apart. The problem was not the electorate; it was McGovern.

The second reason for McGovern's defeat was more profound—and more troubling. Remarkably, it is a lesson that some in the Democratic Party continue to ignore even today: McGovern lost in a landslide because he had abandoned the center. While many Democrats (including some of my classmates) were content to dismiss McGovern's loss as "race hatred," I believed that the Democratic Party was ignoring a mortal threat: the growing dissatisfaction of white, ethnic voters, suburban voters, and Southern voters who had long been the base of the Democratic Party but now made up the core of Nixon's silent majority. It was a trend that I began to witness on my own in New York's outer boroughs; but I began to understand it in the living room of then-Professor Daniel Patrick Moynihan.

Moynihan had just returned to Harvard in the winter of 1972 after a stint in the Nixon administration as a special assistant to the president and head of the new Urban Affairs Council. He also had the kind of street-smart credibility that I appreciated. In addition to having been in the cabinet or sub-cabinet of the past three presidents, he too had canvassed door-to-door in New York election campaigns. Oddly enough, he had gone to high school in East Harlem just a block and a half from Fat Tony's clubhouse and the building where the junkie had terrorized Nardelli and me while campaigning.

Moynihan, along with Schneider and Kilson, had a determinative influence on my intellectual development, and he always made it clear that his intellectual edge had been honed on the streets of New York. The son of working-class Irish Americans, raised largely by his mother who ran a bar in Hell's Kitchen, Moynihan had shined shoes in Times Square while studying at Benjamin Franklin High School on Pleasant Avenue between 116th Street and 114th Street. After a stint in the U.S. Navy during World War II, he had gone on to college, gotten his PhD from Tufts, won a Fulbright to study in England, and then returned to New York City politics, serving proudly as an election district captain for the regular Democratic organization's club in the First Assembly District North in the late 1950s and early 1960s. Before shifting his focus to Washington, he had even briefly been a candidate for mayor and then city council president in 1965. I had ample reason to take him seriously. More wonderful yet was the fact that he took us students seriously, teaching an unusually personal seminar to undergraduate and graduate students in the living room of his house on Francis Avenue, a good walk from the Harvard Yard.

There was something wonderfully incongruous about Professor Moynihan, who often related stories about growing up in Hell's Kitchen while sipping sherry like a proper Bostonian. The class was more a monologue than anything else, with Moynihan using the news of the day as a springboard to talk to us about ethnicity, political change, and most of all, that *bete noir* of my classmates, Richard Nixon. Moynihan believed that we were foolish not to see the good Nixon had done. Nixon, he claimed, had accomplished a feat similar to what the British conservative leader Benjamin Disraeli had in 1867, when he flummoxed liberals with a reform bill that expanded the franchise in England to voters not normally seen as conservatives. Nixon, Moynihan explained, was likewise taking liberal issues away from the Democrats by expanding on many of the Great Society programs that voters largely identified with Lyndon Johnson and the Democrats. Moynihan dismissed Nixon's rhetoric as nothing more than political spin and urged us to understand that Nixon was changing America by governing from the center and isolating the Democrats on the Far Left. In a certain sense, I took a course in what Dick Morris came to call "triangulation" long before I even understood what the concept meant.

Moynihan wouldn't and didn't go too far in explaining what he had tried to do with Nixon, but it was very clear how proud he was of what he had done in the administration and how frustrated he was that the overly partisan (in his view) Democrats did not understand. He brushed off McGovern, instead arguing that we could all learn more from Nixon's initiatives on school integration, black capitalism, and general expansion of economic opportunity. At the time, I rather cynically took many of Moynihan's claims as the rationalizations of a Democrat who had compromised himself by working for a pro-war, Republican administration. It was only with experience (and being in President Clinton's private study plotting strategy) that I came to fully appreciate what Moynihan was trying to teach us. (Ironically, by the time I did, Moynihan had emerged as a critic of much of what President Clinton was doing; occasionally, I was even asked to sound him out to see how far he would go in opposing the president's move to the center. Moynihan would prove to be a particularly strident critic of President Clinton's welfare reform legislation, predicting "something approaching an Apocalypse" with hundreds of thousands of children sleeping on grates in the street because of it. The prediction proved not to be borne out, but it is safe to say that passing welfare reform was a crucial ingredient in Clinton's 1996 victory.)

The end result of Moynihan's semester-long seminar in his living room was that he provided a road map for how to construct a middle ground on issues, one which Morris and I would bring to President Clinton—a centrist political strategy that co-opted the other side's issues and built new coalitions. Other academic analyses of the evolving American electorate made similar points. In 1969, the Republican political analyst Kevin Phillips had written a book called *The Emerging Republican Majority*.[1] In it, Philips famously urged the Republican Party to abandon its liberal, northeastern Republicans—a dwindling pool of voters—in favor of pursuing conservative white Southern voters who had become estranged from the Democratic Party for a variety of reasons, not the least of which was President Lyndon Baines Johnson's civil rights legislation. It proved to be effective counsel. The alienation of the white ethnic voters from the Democratic Party, white Southerners' anti-Democratic anger, and suburbanites' growing affinity for the GOP all added up to a potential disaster for the Democratic Party.

Another book from that era that had an important impact on my thinking was Ben Wattenberg and Richard Scammon's *The Real Majority*,[2] which argued that the Democratic Party was losing touch with white, middle-class Americans. It urged Democrats against pandering to various ethnic groups and urged the party to refocus instead on "the unpoor, the unblack, and the unyoung"—in other words, your typical voter, whom the authors thought of as a forty-seven-year-old housewife from Dayton, Ohio. It was a prophetic book, now largely forgotten but relevant to this day.

From my experiences in New York, I could see it too: the Great Society programs of the 1960s were backfiring. Congress and municipal governments were focusing federal resources on only the poorest parts of the ghettoes, leaving lower-middle-class white neighborhoods in the cities and suburbs to fend for themselves. For many white neighborhoods, crime and encroaching ghettos were becoming the dominant political concerns. But instead of correcting the Great Society's problems and focusing on voters' concerns about social and cultural values, the Democratic Party was paralyzed, caught between competing interest groups. One of the things that seemed clear to me (as it certainly was to Bill Schneider, Ben Wattenberg, and Richard Scammon) was that there is a powerful tendency in American politics that seeks to resolve the excesses of the Left and the Right into a new centrist approach. Yet most Democrats seemed uninterested in claiming this vital middle ground. Nowhere was this clearer than in my hometown of New York City.

In New York, the growing animosity between outer-borough ethnic whites, minorities, and social-engineering liberals had broken into plain view a year earlier in the predominantly middle-class Jewish neighborhood of Forest Hills, Queens. In 1971, Mayor John Lindsay proposed building a massive public housing project—three high-rise towers with 840 apartments—on a vacant lot in the area. Residents of the neighborhood were horrified by the prospect of thousands of low-income and non-white New Yorkers teeming into a neighborhood they had struggled to build. The community turned out hundreds of protestors.[3] Manhattan liberals—many of them Jews—had been horrified by the sight of fellow Jews protesting against black and Puerto Rican public housing developments. One of the more prominent and unexpected figures to raise his voice on behalf of the protestors was a congressman from the Village named Ed Koch. While most Manhattan Democrats were quick to dismiss the Forest Hills incident as an embarrassment best forgotten, I saw it as something else: a sign that New York City Jews were quickly losing their appetite for 1960s-style integration and perhaps Great Society liberalism as well.[4] Certainly the growing popularity of Alabama Governor George Wallace and his surprising performance in the 1968 general election and then again in the 1972 Democratic primaries provided compelling evidence that the Democratic coalition was crumbling in ways that were hard to appreciate.

This shifting tide in New York's liberal population and the racial divisions that were altering the Democratic coalition at large soon came together for me in an unanticipated fashion. By early 1973, it was clear that incumbent Mayor John Lindsay, the charismatic liberal Republican, was on his way out. The two leading candidates to replace him were City Comptroller Abe Beame and Bronx Congressman Mario Biaggi. Beame—a diminutive and unprepossessing man—was the candidate of the Democratic Party regulars who'd rallied to Frank Rossetti's defense some two years before. Biaggi was the law-and-order candidate—a former cop who had been wounded ten times in the line of duty.

The crucial constituency in the election was the city's Jewish voters. In the early 1970s, Jews made up about 30 percent of the electorate in New York City's Democratic primaries. Jewish voters were particularly important because Italian and Irish voters, on the one hand, and Puerto Rican and black voters, on the other, tended to cancel out each other's votes. In contrast, the Jewish vote was very much in play. It was clear that Jewish voters were becoming increasingly concerned about crime; what was not clear was whether that concern would translate into support for Biaggi.

Two other candidates complicated the picture. One was Congressman Herman Badillo. His strength centered on Manhattan liberals, blacks, and Puerto Ricans. The other was state Assemblyman Al Blumenthal, one of Morris and Nadler's old foes. Blumenthal had little chance of winning: he had limited money, little name recognition, and a poor speaking style. However, he was expected to attract considerable support among Jewish voters.

It was an intriguing lineup. If Biaggi had gone on to victory, the Democratic Party might have awakened to the consequences of its dwindling support among white ethnic voters much earlier than it actually did. But it was not to be. That spring, Biaggi found himself embroiled in a scandal of his own making. Biaggi had been involved in a shooting fourteen years earlier and due to the uncovering of "new evidence" he was forced to testify before a New York grand jury, where he repeatedly took the Fifth Amendment. He angrily—and publicly—denied that he had taken the Fifth Amendment, furiously dismissing the allegation. But when the transcript of his testimony was released, it showed Biaggi hadn't just pleaded the Fifth—he had pleaded it repeatedly. Although Biaggi stayed in the race, his campaign never recovered.

The revelations about Biaggi put Beame and Badillo in the driver's seats (with Blumenthal in a very distant third place). Yet the fundamental question of the campaign remained unchanged: who would Jewish voters support, the progressive minority candidate (Badillo) or the more conservative comptroller Beame? That summer, Schneider, Dionne, and I set out to find the answer for an unusual client—*New York Magazine*.

In early 1973, Schneider had approached *New York Magazine* and proposed to field a different kind of poll that would not only reveal people's attitudes but also predict the outcome of the election itself and thereby accomplish two things: first, reveal how the political attitudes of the city's Jewish voters were changing, and second, to call the election itself. To make it work, Schneider turned to Dionne and me. One of our younger research assistants on the project was a fellow named Lloyd Blankfein—of whom Schneider predicted great things. Blankfein would later become a protégé of Goldman C.E.O. Jon Corzine, a future client, and in 2006 he became the CEO and Chairman of Goldman himself.

Our plan was an ingenious one. The key to fulfilling Schneider's promises was our ability to conduct a citywide exit poll of Jewish voters on the day of the primary. We would then use the results to develop a profile of Jewish voters in the city—results that would in turn offer telling insights into who was likely to win the runoff two weeks later. Moreover, we would do it almost instantly. We told *New York Magazine*'s editors that we would have results for them within twenty-four hours of the polls' closing. The journalist Richard Reeves would then take our numbers, turn them into a feature story, and predict the likely winner in November—all by the end of the week.

In the lead-up to the election, we selected eighty-six election districts in all five boroughs that reflected the demographic composition of the city as a whole. On June 4, primary day, our team of volunteers conducted 601 twenty-minute interviews with Jewish voters across the city. The next day, as we were cross-tabulating the data, the newspapers reported that Beame had edged Badillo by forty thousand votes—a victory, to be sure, but not a big enough win to avoid a runoff on June 29. No one in the city had a clear sense of who was going to win—except for us. Our exit polls showed that Jewish voters were sharply divided. While older ones strongly supported Beame, younger Jewish voters were overwhelmingly in favor of Badillo. Because Beame's support among white, largely Irish and Italian, Catholics effectively cancelled out Badillo's black and Puerto Rican support, the runoff election would be a contest between older Jews for Beame and younger Jews for Badillo. The question now became: how would the Jews who voted for Blumenthal and Biaggi cast their ballot in the runoff election?

Our polling showed that Blumenthal voters favored Badillo over Beame by 57 percent to 35 percent. However, according to our poll, Biaggi voters, were for Beame by an even larger 71 percent to Badillo's 19 percent. That left Beame going into the June 26 runoff with at least a ten-point lead among Jewish voters. All of the evidence pointed toward one conclusion: If Jews decide the runoff, which was very likely, the bottom line would read "Abe Beame is the Democratic candidate for the Mayor of New York" and, most likely, the next mayor of New York City.

In effect, we had done an in-depth, academic-style poll—but in twenty-four hours rather than six weeks. Our findings, which were published in *New York Magazine* a mere five days after the Democratic primary, set the city abuzz. The magazine was ecstatic about our work. Veteran political operatives were amazed. What we had done—conducting an exit poll on runoff day, tabulating the results immediately, and providing the information that would allow *New York Magazine* to predict the probable winner five days later—had never been done before.

Among our new admirers was Jack Rosenthal of the *New York Times*. Soon after the spread appeared, Rosenthal, who later became the *Times*'s editorial page editor, took Schneider, Dionne, and me to lunch to discuss our work. Over lunch, he offered Dionne and me jobs as reporters.

Both Dionne and I were editors at the *Harvard Crimson*. I was on track to become the executive editor next year. A starting job at the *New York Times* was a dream job for any aspiring journalist. Dionne accepted on the spot. It was a great hire. After an impressive career at the *Times,* Dionne was ultimately hired away by the *Washington Post.* Today, he is a nationally acclaimed columnist and author and a fellow at the Brookings Institution.

I said no, unequivocally. Although I had enjoyed my time at the *Crimson,* I was not interested in being a journalist. Politics was my passion. So instead of going to the *Times,* I decided to take our data to the Beame campaign. Our exit poll results gave Beame's advisers something they had never had before—concrete, statistically valid information on the themes that worked best for their candidate. Not surprisingly, they were thrilled with it, and quickly offered me a summer job targeting people at district level.

In college and the years immediately following graduation, I thought of myself as a political consultant first and a pollster only second. During the spring and summer of 1973, while E.J. Dionne and I were helping Bill Schneider put together the *New York Magazine* poll, I got my first job as an independent campaign strategist working for one of the most unusual politicians in New York City— Father Louis Gigante.

In a city in which extraordinary characters are commonplace, Father Gigante stood out. Though we worked together on the Nardelli campaign, it was only after moving back to the city that I really come to understand what an intriguing, unlikely figure he was. Gigante was the pastor of the St. Athanasius Church in Hunts Point, a neighborhood of tenement apartment buildings at the southern tip of the Bronx. For decades, Hunts Points had been the first stop for Italian, Irish, and Jewish immigrants who found work in the nearby factories and worked toward the moment when they could move out to better lives in Queens or, later, Long Island. During the 1950s and 1960s, however, that pattern changed. Residents from the slums destroyed by the construction of the Cross-Bronx Expressway flooded into the area along with a new generation of poorer, less skilled Puerto Rican immigrants. The South Bronx also became a dumping ground for the city's abundant welfare recipients. Already small apartments were subdivided even further, crime surged, and the South Bronx began its descent toward the burned-out, apocalyptic urban landscape.[5]

Hunts Point residents responded to this new influx by moving to Co-op City, Queens, or beyond. Those who were too poor to move out kept to themselves—and kept out of the crossfire—as much as possible. But "Father G"—as he was known in the neighborhood—was different. A former basketball star at Georgetown University, he mingled easily with the neighborhood's new residents on the neighborhood basketball courts. He founded the Southeast Bronx Community Organization, one of the country's first community development corporations, which ultimately built thousands of affordable housing units in blighted neighborhoods, and taught himself Spanish to go along with the English and Italian he already knew so that he could communicate with his new congregants. When the energetic priest decided to run for city council in 1973, he asked me to help the campaign.

I readily agreed and quickly recruited Dionne as well. Despite his unsuccessful run for Congress, Gigante was the epitome of the charismatic, activist priest and the kind of figure with whom I could truly build a winning campaign. However, I gradually became aware of the fact that Father Gigante had another connection that made him even more unusual: his brother, Vincent "The Chin" Gigante, was widely considered to be a top capo, perhaps even the boss of the Genovese crime family. Oddly enough, that meant that Father Gigante's brother could well have been the titular head of the crew who had opposed Nardelli and helped Frank Rossetti three years earlier. While it was never totally clear to me what their relationship was, I realized early on that the forces who supported Rossetti were now more than willing to support Father Gigante when *his* name was on the ballot.

The issue of Father Gigante's brother was not a topic of daily conversation, and Father Gigante maintained then, as he would for decades, that his brother was mentally ill, infirm, and living with his mother, Yolanda, on Sullivan Street in Greenwich Village. The only time I ever really heard him address the question of the Mafia was when I invited him to speak at Harvard in early 1974. There, Gigante made a big splash by defending "machine" politics and claiming that "the Mafia" did not in fact exist. Maybe, but Father Gigante did have some unlikely acquaintances for an activist-politician-priest.

The council seat that Father Gigante was running for, the Eighth District, was a triborough seat that spanned the South Bronx, East Harlem, and a tiny sliver of the Upper East Side and Queens. A more heterogeneous council district would have been hard to find. The district's Queens neighborhoods were conservative and still largely Italian and Irish. East Harlem and the South Bronx, which had once had similar demographic profiles, were now dominated by a large African American population, a fast-growing Puerto Rican community, and the dwindling Italian enclave around 116th Street. At Harvard, I had been interested in political coalitions. This race was the perfect laboratory for all that I had learned. It was my chance to see if it was really possible to build a diverse ethnic coalition.

Common sense and my early exposure to Dick Morris had taught me the importance of segmenting the electorate. However, Morris's techniques were labor intensive in the extreme and seemed better suited for the more compact and homogeneous Upper West Side. Success for Father Gigante would depend on our ability to capture votes from a very diverse electorate that was spread out across three boroughs. It was almost like running in Mississippi, Puerto Rico, and the Hamptons all at once.

To meet this challenge, Dionne and I decided to base Father Gigante's campaign on a relatively new technique—sophisticated, targeted direct mail. In our spare moments at the *Harvard Crimson,* we drafted Father Gigante's marketing campaign. For the conservative neighborhoods of Maspeth and Glendale, Queens, we produced red, white, and blue mailings emblazoned with the American flag that proclaimed the virtues of the "patriot priest" and positioned him as a defender of traditional neighborhoods and values. The mailings that went to apartments in East Harlem and the South Bronx praised him as a fighter and an activist. To voters on the Upper East Side, we described Father Gigante as a reformer and champion of—recycling.

Remarkably, it worked. Gigante narrowly eked out a victory in the Democratic primary. That November, Father Gigante swept into office, crushing his opponent by eight thousand votes. Four years earlier as a high school senior, I had been a dodging doorman on the Upper West Side. Now as a junior in college, I had elected my first politician. I was a political player in my own right, but even then I somehow knew that the world of New York City politics would not be a broad enough canvas for me to work on.

When I left Harvard in 1974, I knew that I wanted to be involved in politics, but I still didn't know precisely how I was going to do it. So when Bill Schneider encouraged me to apply for a scholarship to go to Oxford, I did. Based largely on my work for *New York Magazine* and my honors thesis, I got in. So in the fall of 1974, I went to Oxford University in England on a Knox scholarship. But practical politics was more in my blood than I had known.

After about a month in Oxford, I was bored stiff. Fortunately, Bill Schneider came to the rescue. He put me in touch with Bob Worcester, now Lord Worcester, who was at the time the chairman of Market Opinion Research International, a research firm with then-close ties to Britain's Labor Party. I called him about working at the firm, paid or unpaid. "You want to volunteer?" he responded, with a heavy emphasis on the last word. Once I made it clear that I was ready to work without pay, the job was mine. I even teamed up with my old friend, E.J. Dionne (who had won a Rhodes Scholarship and, thanks to Jack Rosenthal, spent the summer as a stringer in Paris) to work on a campaign to bring Great Britain into the European Common Market in the spring of 1975. Soon, Dionne and I were drafting questionnaires, tabulating the results on a daily basis, and even drafting the presentations that Worcester then delivered to the Labor Party cabinet that employed him. Indeed, so deeply were we involved in the campaign that even the usually parsimonious Worcester felt compelled to pay us at the end. Soon Dionne and I were receiving the princely sum of one pound sixty pence an hour for our efforts. When we won, Worcester was so delighted that he gave us each a twenty-pound bonus.

Our cash cow notwithstanding, I was eager to get back to New York. Fortunately, Patrick Cunningham, a colleague of Father

Gigante's, was now the chairman of the New York State Democratic Party and wanted me back too. My work with Mark Penn on the 1974 Carey campaign for governor had been a great success, and Cunningham, who had taken a liking to me, thought I could help the Democratic Party now that Governor Carey had appointed him state party chairman. Consequently, in the summer of 1975, Cunningham gave Mark and me a contract to do polling for the state party. Cunningham arranged for us to use the phone bank at the United Federation of Teachers on Park Avenue. It was up to us to do the rest. We hired people to make phone calls. We wrote the questionnaires, supervised the interviews, and did the data processing.

We quickly learned that the state leadership took our findings very seriously indeed. The first poll we did for the state party gauged the appeal of potential Democratic senatorial candidates for the 1976 election, a seat ultimately won by Daniel Patrick Moynihan. Our results showed great enthusiasm for Bess Myerson, Mayor Lindsay's popular commissioner of consumer affairs and, more important, the first Jewish Miss America, as well as growing support for a somewhat more unconventional candidate, my old professor, Daniel Patrick Moynihan. I knew I had arrived when, during the presentation, Cunningham sent the then-Councilman for Riverdale and future borough president Stanley Simon out to get me a turkey sandwich—which Simon dutifully did, asking solicitously whether I'd prefer mustard or mayonnaise.

Our second big poll in late 1975 was designed to measure the appeal of potential presidential candidates in New York State. Just before we were scheduled to field the poll, someone pointed out that we had forgotten about an obscure Southern governor named Jimmy Carter. When Mark asked me if we should redo the questionnaire, I told him not to worry about it: there was no way the little-known governor of Georgia would ever be nominated, much less elected president.

Our work for Cunningham and our polling for the state Democratic Party began to attract interest from other New York political players, the most important of whom was David Garth. Garth had begun his career as a sports producer, but by the early 1970s, he had become the most influential media consultant in the country—

the man who had beaten the Democratic machine in 1965 and put a charismatic Republican Congressman, John Lindsay, into Gracie Mansion.

Garth was impressed by our ability to do quick, cheap polling. That summer, he put me on his payroll at four hundred dollars a week to do polling for some of his clients, including Senator John Tunney of California. That was a pretty good sum of money in 1976, and in fact, Garth had originally planned to pass my cost on to his clients. However, Garth's plan ran into a problem: Tunney and a number of other Democratic candidates had already engaged another Harvard wunderkind, Pat Caddell, to do their polling. One year later, Caddell would become the country's first celebrity political consultant by propelling Jimmy Carter into the Oval Office.

Caddell's rising star (and success in tying up Democratic candidates) meant that Garth was stuck with my salary. Garth was not someone who liked to be on the bad end of a deal, and I suspect that he was eager to find a new client to pass our fees off on. That summer, just such an ambitious new client appeared—beginning a campaign that would reshape the face of New York politics for years to come.

CHAPTER THREE

THE INVENTION OF OVERNIGHT POLLING, ED KOCH, AND POLITICS IN THE BIG EASY

E d Koch was a former city council member and a junior congressman who represented New York City's "Silk Stocking" District, which stretched from the rough-and-tumble Lower East Side to the wealthy neighborhoods of the Upper East Side. It was not, at first glance, a district that one would expect to elect a working-class Jewish lawyer from the Village.[1] Republican John Lindsay had represented it in Congress before he was elected mayor in 1965. However, Republican support in the area had withered in recent years, and in 1968 Koch had handily upset the district's incumbent representative, Whitney North Seymour Jr.

Nevertheless Koch was not happy in Washington, and in 1973, he tried to run for mayor. His campaign was a flop. He raised only about a hundred thousand dollars for the race, and, as a reformer, the party bosses shunned him. Seven weeks after he entered the campaign, he was forced to drop out. But despite this botched campaign, Koch remained an intriguing figure. As a Jew with roots in the reform movement of the 1950s and 1960s (Koch had led the effort to oust Tammany Hall boss Carmine De Sapio in 1961), he could reasonably expect to run well among Jewish voters. Furthermore, Koch had a conservative streak than alienated him from the more committed liberals. In 1971, he had strongly supported the Forest Hills protestors. As someone who believed that even self-identified liberals were moving to the Right, I thought it was a smart move. While his tough talk on crime had attracted little attention during his abortive 1973 race, Mark Penn and I believed that Koch and his centrist message just might appeal to outer-borough voters.

Little did we know that Koch's campaign would launch our business and revolutionize the practice of politics by introducing something entirely new into campaigning—the instant, overnight poll.

Ed Koch certainly wasn't thinking about us during 1976. He was thinking about David Garth. Koch wanted Garth on board not just for his skills but also for the credibility he would gain from having such a professional in his corner. That's how good Garth was perceived to be: simply having him on your team made you a contender. Garth, however, had another candidate in mind—Mario Cuomo.

Koch's political appeal to outer-borough ethnic whites rested in part on his stance on the Forest Hills low-income housing project but also on his support for the death penalty. Ironically, Cuomo's career began in Forest Hills too. Cuomo had brokered the deal that settled the explosive standoff and secured his political reputation. Afterward, he had gone on to become secretary of state in the Carey administration. Now Cuomo was widely seen as an up-and-coming Democratic star. Governor Carey was eager for him to run, and so was Garth.

But Cuomo couldn't make up his mind. Even then, Cuomo was well on his way to becoming the Hamlet of American politics. He kept resisting entreaties from Garth and Governor Carey to get into the race. Garth, in turn, continued to stall Koch. But, perhaps with the cost of our salary on his mind, Garth did eventually direct us to field a benchmark poll to gauge Koch's appeal. This would allow Garth to keep a foot in the Koch campaign and earn enough cash to help offset my cost for the summer of 1976.

Late that summer, Mark and I fielded our first large-scale poll. We found that in a five- or six-way race with Manhattan Borough President Percy Sutton, Congressman Herman Badillo, Congresswoman Bella Abzug, Mayor Abe Beame, and Secretary of State Mario Cuomo, nobody had more than 20 percent support. That was the good news. The bad news was that at around 6 percent, Koch was dead last.

We presented our results one afternoon in early September 1976 in the apartment of Richard Menschel, who was a partner at Goldman Sachs. His wife, Ronay, was chief of staff to Congressman Koch. While no one was pleased with the horserace results, Mark and I emphasized the good news that our polling had uncovered. There was a path to victory. Voters were extremely angry with Abe Beame. They thought they had elected a technocrat who could fix the city's budget. Instead, Beame had proven to be incompetent. In addition, voters believed that he had bankrupted the city to give public sector unions and special interests a break. People were tired of clubhouse politics; Beame was vulnerable. The electorate wanted nonpolitical decision making and merit-based hiring. They wanted a nonpartisan, competent decision maker in office, somebody who would take on the public sector union, fix the city's finances, and do something to stop the tidal wave of crime that was engulfing the city.

Mark and I argued that despite his low standing in the polls, Koch possessed many of these characteristics. He understood the concerns of middle-class ethnic voters. He supported the death penalty. His roots in the reform movement allowed him to run as an outsider—as someone opposed to the clubhouse system. In short, we thought Koch had a shot if he positioned himself as an outer-borough liberal or, as Koch himself would later put it, "a liberal with sanity."

Koch and his coterie were heartened by our analysis. But the most important audience for our polling was David Garth. Our results convinced Garth that Koch was a viable candidate, one who might just be able to win. But before Garth signed on with the Koch campaign, he made one last effort to get a clear decision from Cuomo. Garth practically presented Cuomo with an ultimatum: get in the race or I'm going to work for Koch. Finally, Cuomo said he was out, and in late 1976, Garth agreed to sign on with Koch as his media adviser and strategist.

Although we had done Koch's benchmark poll, Penn and I fully expected to be replaced once the campaign got underway, since we'd never done polling for a campaign. When you got right down to it, we were just two recent college graduates with no experience polling for a huge mayoral campaign, no matter how good our other work had been. Also, in the fall of 1976, we were both first-year law school students. Mark at least was in the city (at Columbia); I was still in Cambridge at Harvard.

But for whatever reason, Garth never got around to replacing us. Perhaps we were all they could afford. Campaign treasurer, Bernie Rome, often referred to us behind our backs as the "nickel-and-dime researchers"—and sometimes not even behind our backs. Whatever the reason, it gradually became apparent that the job was ours. It was a turning point in our career. I would no longer be a volunteer, neighborhood strategist. Here was an opportunity to develop strategy for a major campaign that had a chance to win. We were now in the big leagues, and I scarcely knew where to begin. We had no office, no employees, and preciously little experience. Fortunately, I thought to call Theodore Kheel, the politically connected labor mediator who we had periodically rented a phone bank from, and ask him for office space. Kheel basically said, "I'm not going to let you pay for office space. I'm going to give it to you. But you've got to rent my phone bank." Which we were more than happy to do since we didn't have phone banks either. So in early 1977, we set up shop at Automation House, the East 68th Street townhouse that served as Kheel's base of operations at the time. Suddenly, we were two law school students with prime Upper East Side office space—not bad.

On March 4, 1977, Koch officially threw his hat into the ring. The media response was less than enthusiastic. The *New York Times* ran the story on the front page, but its second paragraph noted that Koch would have to "scotch skepticism over the seriousness of his candidacy." The *Daily News* reported the story in an even less flattering fashion. Clearly, the press wasn't going to put Ed Koch into Gracie Mansion. Fortunately, thanks to our polling, David Garth had other ideas about how to position Koch as an anti-special interest, anti-crime, fiscal conservative. He decided to start running Koch advertisements in late June, well before any other candidate went on the air.

Garth's strategy was highly unorthodox. At the time, there was considerable skepticism that television made a difference in political contests. Most campaigns saved their dollars until the final two or three weeks of the race. Twenty years later, when we used a similar early-media strategy for President Bill Clinton's reelection campaign, it was still considered unusual. Garth, however, felt that it was important to define Koch in favorable terms up front.

The first commercials aired in June. Photographs from Koch's tour of duty as an infantryman during World War II introduced him to voters as a strong leader. The next round of ads sought to frame Koch's positions on the issues in a way that reflected the findings of our benchmark poll. Many of these ads focused on the public sector unions. One ad attacked the Patrolman's Benevolent Association for rules that permitted only fifteen hundred police on the streets during any given shift. Another slammed the Board of Education and claimed that teacher salaries were exorbitant.[2] Drawing on our research, Garth developed a brilliant punch line for the commercials—and for the campaign: "After eight years of charisma, and four years of the clubhouse, why not try competence?"

At that time, a thirty-second television spot cost anywhere from five hundred dollars to a few thousand dollars, depending on the time of day it ran. The Koch campaign put a very large portion of the six hundred thousand dollars it raised in the run-up to the primary into Garth's ads. Ultimately, about 75 percent of the total funds raised in the campaign went to Garth for media and consulting fees.

Another reason for the early blitz was a disturbing new rumor that Mario Cuomo was planning to enter the race. One month later, the rumors proved to be true when Cuomo made it official, announcing that he was a candidate. Hamlet was back—and he was bringing some formidable competition with him. To compete with us, Cuomo brought in media consultant Jerry Rafshoon, the partner to the most famous political pollster in the country, Pat Caddell.

At first, we felt very much outgunned. Caddell had just masterminded one of the greatest upset victories in modern American politics—Jimmy Carter's win over President Gerald Ford. No pollster was more famous. However, it turned out that we had two critical advantages. First, we knew the city much better than Caddell and Rafshoon. Second, we had a technological edge—a polling innovation that would alter the course of American campaign history.

When Mark and I conducted our benchmark survey for Koch, it took us five nights simply to get enough responses to have a statistically significant sample. Once we had the results, we then had to input the data onto punch cards and tabulate the results overnight on a mainframe at Columbia University. That took a night—if you entered all of the data correctly. If you made a mistake with a punch card, you had to start all over again.

Garth was not sympathetic about computer mishaps of this sort. After several of these incidents, Mark and I decided that we needed to find a better way. So Mark said, "Let's buy a microcomputer for a thousand dollars." This was 1977—four years before the launch of IBM personal computers. Microcomputers were seen as little more than an expensive toy for hobbyists. That was the perception, and I confess that I shared it.

"Why would we want to spend a thousand dollars on a computer?" I responded. So Mark said, "Fine, I'll do it myself." And he did. Mark purchased a microcomputer "kit"—self-assembly required. Once he assembled the computer and got it running, Mark wrote a program that could compile the results almost immediately. At the end of the process, he had created something that no other political consultant in America had—a tabulation system that allowed a single person to input data directly into a microcomputer and get polling results almost immediately.

Mark's microcomputer was located in his two-bedroom apartment room near Columbia. That became our center of operations.

After the phone bank at Automation House completed its work, we'd bring the data back to Mark's apartment. We hired several yeshiva students to input the data for us overnight. Typically, they would come in at around 10:00 PM to start inputting the data and finish around 6:00 AM. Mark's program then cross-tabulated the results. Within an hour, we had results for Garth.

Mark's microcomputer didn't just speed the process up. It gave us the ability to do something absolutely new: poll for Koch on a daily basis. Instead of simply setting strategy at the beginning of a campaign, polls could now be used to evaluate tactics and determine the campaign's message and strategy on a day-by-day basis.

Garth put our polling breakthrough to good use. In particular, he began to use our polls to measure the effectiveness of his advertisements. Garth could now put an ad on the air during the day, have us poll that night, and have feedback by the next morning. Soon Garth insisted that Koch save one hour every morning to film new ads that responded to issues our polls detected. Campaign tactics that had once been developed over the course of weeks were now fine-tuned on a daily basis.

We also used polls to look for trouble spots. One potential issue was Koch's sexual orientation. Koch was single and a resident of the West Village. In his first run for city council, he had campaigned on a platform that called for an end to anti-sodomy laws, full abortion rights, and less restrictive divorce laws. (Opponents nicknamed his platform, "SAD.") These circumstances (and a nasty whispering campaign by the Cuomo campaign) gave rise to the rumor that Koch was gay. In some Cuomo strongholds, posters appeared proclaiming, "Vote for Cuomo, not the Homo."

Our polls showed that the public wasn't particularly interested in Ed Koch's sexual preferences. But Garth worried that it might become an issue later in the campaign. So he came up with a characteristically clever solution: he convinced the wildly popular Bess Myerson to become a highly visible part of Ed Koch's life. Our polls showed that Koch would benefit greatly from Myerson's support, and she soon became a fixture of the Koch campaign. Garth and Koch even encouraged speculation that the relationship might go further than mere political support. The Koch campaign was gathering steam.

That summer, the Koch campaign received significant boosts from two tragic events. On Wednesday, July 13 at about 9:00 PM, lightening strikes coupled with a Con Edison repair mistake plunged New York City into darkness, inciting riots, looting, and arson across the city. Although the police ultimately arrested more than three thousand people, much of the city experienced what *Time* magazine described as "a night of terror." At first, Mayor Beame's numbers soared as he "took control" of the emergency. But after the city lost power completely, Beame's situation changed for the worse. A few days later, we put the poll back in the field and Beame's numbers had plummeted.

Later that month, things got even worse for the incumbent mayor when police found the body of Stacy Moscowitz—the latest victim of a serial killer called "Son of Sam" who seemed to be taking much glee in terrorizing the city. These horrible and frightening events shifted the attention of the entire city to public safety, and only one candidate had been running from the get-go on a "tough on crime" platform—Ed Koch.

By early August, our polls showed that rising levels of fear and concern about crime were eroding support for the early frontrunners, incumbent Mayor Abe Beame and Congresswoman Bella Abzug. In contrast, Koch was beginning to move up, but he wasn't the only candidate showing signs of strength. Cuomo was also coming on strong, and he too had begun to emphasize law and order in his public appearances. Moreover, Cuomo had recently picked up an endorsement from the *New York Times*. It seemed that Cuomo had the momentum, and that it was only a matter of time before he would claim victory. That all changed in late August when Koch and David Garth struck a deal with a then-obscure Australian press baron named Rupert Murdoch—a deal that would change the course of the election and, perhaps more important, prove to be the first step toward the creation of an activist, conservative press in New York City.

In 1977, press baron Rupert Murdoch was still a newcomer to New York. While Murdoch had made his fortune by building a chain of tabloids in his native Australia, he wasn't a man to limit his ambitions to the land down under. In 1976, Murdoch made his move into the biggest media market in the world—New York—with a series of acquisitions that served notice to the city that a new player had arrived. That year Murdoch scooped up two of New York's most innovative magazines—*The Village Voice* and *New York Magazine*. He also purchased a prestigious but declining afternoon newspaper, the *New York Post*. While Murdoch's acquisitions of the two magazines that had done so much to define the "new journalism" of the 1960s and 1970s received the most attention, it was the *Post* that held the key to Murdoch's agenda.

Although it may be difficult for present-day readers of the *Post* to believe, until Murdoch's coming, the *New York Post* had been a generally liberal paper, known for its excellent arts coverage and the polished columns of the journalist Murray Kempton and the sensible liberalism of James Wechsler. With Murdoch in charge, things quickly changed. First, the *Post* was transformed into a morning paper—a move that clearly announced his intention to complete with the larger and more successful *Daily News*. Then Murdoch brought in a troupe of Aussie transplants who specialized in the attention-grabbing tabloid style. The effect was electric. It was as if William Randolph Hearst had suddenly returned to town.

But Murdoch wasn't just looking for business success. More than anything, Murdoch wanted to be a player. That was the true appeal of owning a New York newspaper like the *Post,* and now that he had the power of a daily paper, he intended to wield it. Murdoch was particularly determined to take down New York City's powerful unions, which he and many others blamed for the city's fiscal crisis. He was looking for a tough, anti-union, mayoral candidate, and Ed Koch seemed the candidate who fit that description best.

By August, Murdoch was reportedly inclined to support Koch. However, one of the things he reportedly wanted was a commitment from Koch to appoint Edward Costikyan, a lawyer and long-time player in the reform wing of the Democratic Party who had briefly entered the race in his own right earlier in the year, as deputy mayor. Costikyan had endorsed Koch after his withdrawal, so relations between the two men were good. An agreement in principle, brokered by Garth, was supposedly struck. Koch agreed to make the appointment;

Costikyan agreed to serve as deputy mayor; and the *Post*'s endorsement was his.[3]

Although I was not privy to the behind-the-scenes negotiations, I vividly remember the scene when the deal was sealed. Koch and Garth were waiting in the media consultant's office for the Australian press baron to arrive. When Murdoch did show up, he looked exhausted. His jacket was over his shoulder; his tie was askew. The men quickly went into an unair-conditioned conference room—it was a brutally hot August day—with a few aides. About half-an-hour later, they all emerged smiling and backslapping, despite the heat. The deal had been done. Murdoch soon delivered on his portion of the deal with typical pizzazz.

On Friday, August 19, the *Post* splashed a three-column endorsement of Koch across its front page. Our polling showed that Koch's support went up about five or six points almost immediately. But even more important than the *Post*'s endorsement was the coverage it provided during the remainder of the campaign. In the remaining weeks of the campaign, the *Post* became a reliable champion of Koch, juxtaposing flattering articles and attractive pictures of candidate Koch with vitriolic attacks on his opponents, particularly incumbent mayor Abe Beame and Bella Abzug. The *Post*'s enthusiasm for Koch went so far that more than fifty *Post* staffers actually sent Murdoch a letter protesting the one-sided coverage. Murdoch responded by inviting the protestors to quit. No wonder Koch, Garth, and the other participants in the epochal meeting with Murdoch that had sealed the deal had emerged from Garth's office with looks of elation.

The deal between Koch and Murdoch marked a turning point in the campaign. Murdoch's support more than offset the bump Cuomo received from the *New York Times*'s endorsement. A few days later, Koch received another unexpected bonus—an endorsement from the *Daily News,* a paper that had actually urged Cuomo to enter the race in April. The *Post*'s impact also confirmed Murdoch's position as a potent player in city politics. The only person who didn't benefit from the deal was Edward Costikyan. Soon after the deal was struck, Koch held a press conference announcing the need to reorganize the city government and appoint a powerful deputy mayor to supervise his operations. To Murdoch, the subtext was presumably clear: I am ready to make good on my part of the deal. But it never happened. While Costikyan did receive a high-profile position in the transition, he was pushed aside after the election.

Mark and I had little to do with the deal that Garth brokered with Rupert Murdoch during this period. We were too busy with the campaign itself. While we were spending much of our time writing questionnaires and putting polls in the field, we were also developing a model of what made New York voters tick.

One of the most common misperceptions about polling is that success depends primarily on asking the right questions. Though you do need fair, unbiased questions, the key to meaningful results is having the proper analytical framework with which to interpret the numbers. Understanding the context of poll results is the most important job of any political consultant. My experiences from high school onward in East Harlem, in Mississippi, and at Harvard had helped me develop a keen understanding of how racial conflicts were disrupting long-established patterns of politics. I knew outer-borough white ethnic and Jewish voters—their hopes and their fears—in a way that an outsider like Jerry Rafshoon never could. I knew that they were the crucial swing voters in New York City elections. Most important, I knew what could change their votes. Ultimately, polling isn't about predicting what will happen. It's about trying to determine what you need to say in your communications to get voters to react the way you want them to.

It's also important to make sure that you are targeting the right people—the people who are actually planning to vote, not just the people who will pick up their phone and answer your questions. To ensure that we were polling the right people, Mark and I also developed a precinct model to help us understand who would vote. We picked sample precincts—every fiftieth or hundredth precinct—and developed turnout models based on previous elections to help us understand who the likely voters would be. Our polls made sure that Koch's messages were reaching these key voters; our model allowed us to predict who would go to the polls, how they would vote, and what the outcome would be.

Primary election day was Thursday, September 8. Going into the home stretch, polls conducted by the *New York Times* and CBS continued to show incumbent Mayor Abe Beame and Congresswoman Abzug tied for first place, with Cuomo and Koch locked in the battle for second place. We saw things quite differently. By doing polling in conjunction with our turnout model, we were able to predict that Koch would win the primary, with Cuomo coming in a close second. We were right on target. At the end of the day, Koch received 180,914 votes versus Cuomo's 171,100 votes, with Beame, Abzug, Sutton, and Badillo bringing up the rear.

It was a heady experience, to be sure. Using our model, Mark and I had told Koch and Garth within half-an-hour of the polls closing that Koch would finish first, with Cuomo a close second. I remember the look of disbelief on their faces as we calmly projected the results. Although Mark and I were confident in our methodology, we were nonetheless greatly relieved when the final results were tabulated an hour or so later and our prediction was confirmed.

K och's first place finish was a triumph of perseverance and of poll-driven positioning and skillful advertising. Garth had positioned Koch, once a West Village liberal, as a tough-on-crime, outer-borough Democrat. In doing so, he had topped the one candidate who really was an outer-borough Democrat—Mario Cuomo. But we still hadn't won the nomination. Since Koch failed to win more than 40 percent of the primary vote, he and Cuomo were forced into a runoff election on Monday, September 17.

The key to the election would be who got the support of Mayor Beame and the county bosses. Despite having run as an outsider, the Koch campaign mounted an aggressive effort to win the backing of Beame and Brooklyn county boss Meade Esposito as well as the city's black establishment. Cuomo conducted a characteristically ambivalent—and unsuccessful—campaign for their support. Garth also persuaded the popular but mercurial Bess Myerson, who had largely vanished from the campaign in August, to rejoin Koch for the final two-week push. When Democratic voters returned to the polls on September 17, Koch defeated Cuomo again, this time by a vote of 432,000 to 355,000.

Despite the victory, Cuomo wouldn't go away. Though he had lost the Democratic nomination, Cuomo decided to stay in the race and run as the Liberal Party's mayoral candidate. Koch and Cuomo would meet for a third time on the ballot that November.

In preparation for this final showdown, Mark and I stepped up our polling operation. We were now fielding polls every night. Garth used our results to shape advertisements and decide which endorsements should be used where. In areas where our polls showed Harlem kingmaker and former Manhattan borough president Percy Sutton's endorsement would help, we trumpeted his support. In areas where South Bronx Congressman Herman Badillo's support helped more, Garth went with Badillo instead. In effect, we were using polls to segment the electorate in much the same fashion that we had done on the Upper West Side eight years earlier.

Rafshoon and Cuomo had no inkling of how sophisticated our operation had become. To most people, the idea of doing an instant poll was simply inconceivable, but we were giving Garth a daily snapshot of what the electorate was thinking and how it was reacting to our campaign effort. It was a view that most other campaigns wouldn't have for another ten to fifteen years.

Ironically, Koch himself was quite distant from the process. He listened to the polls, but his interest in them was sometimes minimal: he tended to trust his instincts and, occasionally, delight in controversies that most politicians would have preferred to avoid. Koch had enormous confidence in Garth and left the details of the campaign to him. Indeed, it wasn't at all clear that Koch understood exactly what we were doing. Yet our polls gave his campaign an unprecedented tactical advantage. Every night we took the pulse of the city. Every morning, Garth would plot the campaign accordingly, filming and airing commercials to suit the public's mood. It was a system that was quickly proving our recipe for success.

However, in the final weeks of the campaign, a problem arose that resisted this formula. Our polls had consistently shown that a large segment of the electorate was looking for an outsider. In the primaries, Koch had been seen as the man of change because he wasn't the establishment candidate. But when the establishment began to rally around him during the runoff, that dynamic changed. Our polls showed that now Mario Cuomo was seen as the outsider. The final week before the election was a nerve-wracking time. Cuomo was closing steadily. Our final poll showed that Cuomo was only eight points behind Koch. If our election modeling and polling were correct, Koch would win—but narrowly.

At the very end of the offensive, Cuomo and Rafshoon went on the attack, launching an all-out blitz of negative ads that attempted to link Koch with former Mayor John Lindsay. It didn't work. Garth's media campaign had inoculated Koch so effectively that this desperate effort to link him to the failed policies of the past never had a chance. Koch edged Cuomo one final time, by a vote of 712,551 votes to 587,196 votes.

The Friday after the election, Mark and I wrote an op-ed for the *New York Times* entitled, "Koch's Narrow Win." The piece discussed how the growing identification of Koch with the establishment during the runoff had almost cost him the election. Three years later, another "outsider" politician—Ronald Reagan—would ride this same sentiment into the White House.

Garth failed to appreciate the importance of the point we were making. When our piece appeared in the *Times*, Garth went nuts. He threatened to take Mark's mini-computer and throw it out the window on to Fifth Avenue. Fortunately, the breach was not that serious. The polling style we had developed was just too good.

In many ways, Koch's post-runoff effort was the first truly modern political campaign. But as potent as these tools were, as we would soon discover, they were not infallible.

During the Koch campaign, Mark and I pursued two careers. Polling for Garth occupied much of our days and many of our nights. In the time that remained, we were law school students, Mark at Columbia, me at Harvard. With the Koch campaign, we had enjoyed a stunning moment of success, but it was still not at all clear that we could replicate our successes in other parts of the country or make a living as pollsters.

In 1979, both Mark and I finished law school—and reached a moment of truth. We had to decide whether we would be lawyers or political consultants. Our hearts were in politics and polling; however, we wanted to make a decent living as well. If we could make about as much as young attorneys fresh out of law school, we were willing to give political consulting a try. But at the time, that seemed like a pretty big "if." Garth kept us busy and made sure we were paid well, but it was not clear that there was real money in political consulting until 1979, when I was called down to Louisiana to meet Edgar B. "Sonny" Mouton.

It was not my first trip to New Orleans. Four years earlier, I had been introduced to politics, New Orleans-style, by one of my more unusual classmates at Harvard, Mike Early. A former Catholic priest, Early was a midcareer fellow at the new Kennedy Institute of Government when I was an undergraduate. In 1976, I had helped elect Early to the New Orleans city council with a Dick Morris-style door-to-door campaign. It had been an eye-opening experience. Soon after I arrived, Early had taken me over to meet then-Lieutenant Governor James Fitzmorris, known to Louisianans simply as "Jimmy Fitz." We walked into the lieutenant governor's New Orleans office; Early introduced me; and after a few pleasantries, Jimmy Fitz started in on the topic that he was clearly most interested in—the price of pardons. I was stunned. For all Jimmy Fitz knew, I could be a young lawyer at the Justice Department, yet he was talking about felonies that he was clearly planning to commit as if it were a perfectly normal conversation. For a first-year law school student, it was an eye-opening experience. I had no great desire to work for Fitzmorris when he announced that he would be running for governor, but when the opportunity arose to work with Mouton, I was curious to see if we could replicate our New York success in Louisiana.

"Sonny" Mouton was from Lafayette, the capital of Cajun country. He was a small, bald, and frankly not very attractive man with a Cajun accent so deep that at first I thought I was going to need a translator to understand him. These characteristics had not, however, impeded his career in Louisiana politics. In 1979, Mouton was one of the deans of the Louisiana state senate, but he had bigger dreams: he wanted to be governor. My job was to figure out a strategy that would help Sonny break out from his political base, Lafayette, and break away from the perception that he was just another slick Cajun politician.

So that spring, I went down to Louisiana once again. After arriving in New Orleans, I rented a car and set out for Lafayette. It was a destination that meant nothing to me. Just a midsized Louisiana town, I supposed. Had I paid a bit more attention to the business pages, however, I might have realized that Lafayette was the center of Louisiana's booming oil industry, and as such was a "veritable gold rush" city. During the 1970s, no state had more money sloshing around than Louisiana, and no state spent as much of its money on politics.

Despite the immense oil wealth pouring through the area, Lafayette at first appeared to be a ramshackle, run-down Southern town. I met Mouton, his campaign manager Eustice ("Useless") Corrigan, his treasurer, and a cadaverous oilman named J.Y. Foreman who had some definite but never entirely clear connection to Mouton. I ordered a steak, not expecting much, and tried to keep from staring at the astonishing bouffant toupee perched precipitously atop J.Y.'s emaciated body. I quickly learned that he was dying of cancer. Indeed, Mouton spent much of the meal saying, "Don't die on me, J.Y. don't die." The company was strange, but the food was good. What arrived at our table was one of the best meals of my life—a steak that Brooklyn's Peter Lugar steakhouse could have been proud of. At that moment, something clicked in my mind: I realized that Lafayette was not what it appeared.

I decided to pitch these guys the full package—a strategy-setting poll for $21,500. They agreed without blinking an eye—and suggested that I come back to New Orleans in late May to present the results. So I did. This time I was whisked to an enormous suite at the new Hyatt Regency Hotel. Before my presentation, we met for dinner in the French Quarter. All of Mouton's cronies were in evidence except for J.Y. Foreman. It turned out J.Y. had in fact died.

Despite the campaign treasurer's death, the money still seemed to be flowing freely. I remember that at the end of the meal one of Mouton's sidekicks—a man the waiter called "Monsieur Butch"—

requested an extra five sticks of bread. I must have looked surprised because he turned to me and said, in his deep Cajun drawl, "It's so much
easier to buy your bread at restaurants than to have to suffer the indig-
nities of a grocery store." When it came time to pay for the two polls I
had done, "Monsieur Butch" just rounded the total up to twenty-two
thousand dollars. "It's easier to keep count of that way," he told me.

Clearly, the good times were rolling. What I didn't realize at the
time, however, was that J.Y.'s death had indeed been a catastrophe
for Mouton. Without his moneyman, Mouton had to rely even more
on a figure that even Louisianans viewed with suspicion, Governor
Edwin Edwards's chief of administration, Charles Roemer. Although I didn't fully understand the connection, Louisianans did.
The poll results I had collected showed that Mouton was suffering
from his close association in the mind of the public with Edwards's
corrupt chief of administration—he was a terrible drag on Mouton's
campaign. In order to win, Mouton had to redefine himself as a
fresh face with new ideas. That meant that he had to disassociate
himself from Roemer. This Mouton was unwilling to do.

"I just can't do that," he told me.

"Why not?" I asked.

"I just can't."

"You have to," I told him.

"You'll have to take this up with Charles's son, Buddy," Mouton
told me. So I called Buddy. I explained the situation; Buddy listened
in silence. Finally, I stopped. "You'll have to talk to my Father,"
Buddy told me, "thanks very much." And click. So I called Charles.

I knew that Charles Roemer fancied the idea that he himself was
a skillful pollster and political operative. I also suspected that he
wouldn't be too happy to hear that he was not just a liability to
Mouton but an albatross. I was right. Buddy and his father listened
while I presented my results. Neither man said a word or gave even
a hint about what they were thinking. Then, silence. Finally, Roemer
père spoke up.

"Listen," he said. "I wouldn't piss across the street for a poll done
by people like you who associate with David Garth in New York."
That ended my involvement with the Roemers. A decade later,
Charles Roemer was convicted of bribery to obtain state insurance

contracts along with reputed New Orleans Mob boss Carlos Marcello. (The convictions were later overturned.) Ironically, Buddy Roemer would be elected governor in 1987 after running as just the kind of "outsider" that I had attempted to position Mouton as a decade earlier.

Still, the pay was good. I definitely wasn't the only person benefiting from the well-financed Mouton campaign. Once during the campaign, I asked about a payment owed me and was told I should pick it up Friday morning at a small hotel in the French Quarter that served as Mouton's headquarters. I arrived early for breakfast. By 10:30 AM, more than thirty other people had showed up as well. Among the crowd were two of Louisiana's most controversial black leaders, Don Hubbard and Sherman Coupland. Everyone seemed to be just hanging around. Finally, I asked Jean Nathan, a New York media personality whom David Garth had inserted into the Mouton campaign, what in the world are all these guys doing here. She told me nonchalantly that they were waiting for their handouts and "walking-around" money.

Mouton ultimately finished fifth in the primary. That fall, David Treen became Louisiana's next governor, the first Republican governor since Reconstruction. But despite its ups and downs, Louisiana convinced me that there was money to be made in political consulting. In 1979, Mark and I decided to work full-time as partners and pollsters. Penn & Schoen was now our business.

Still, it wasn't always easy. It was sometimes difficult to distinguish between opportunities and pitfalls—a lesson I learned well that October, when I got what looked like a great break, thanks to the inept comments of Massachusetts Senator Edward Kennedy.

In the summer of 1979, Ted Kennedy was the undisputed star of the Democratic Party, the heir to Camelot and an enticing alternative to the embattled Jimmy Carter. Rumors were swirling that Kennedy would challenge President Jimmy Carter for the Democratic presidential nomination. Taking on an incumbent president is always a dicey situation, but Carter's high unfavorability rating made him vulnerable. House Speaker Tip O'Neill even announced that the nomination was Kennedy's for the taking, but Kennedy was having a hard time taking it. The event that brought us into contact with Kennedy's campaign was the candidate's disastrous October 12 interview with CBS correspondent Roger Mudd—an encounter that has gone down as one of the greatest blunders in campaign history.

Earlier that fall, Kennedy had agreed to do sit down with for two interviews with Mudd, whom he saw more as a friendly social acquaintance than a ruthless journalist. The first interview took place on September 29. Mudd flustered Kennedy with questions about Mary Jo Kopechne's death at Chappaquiddick ten years earlier and about the state of Kennedy's marriage. It was clear that Kennedy had been completely unprepared. What was remarkable was that when he sat down with Mudd for his second interview two weeks later, he did even worse.

Disaster struck when Mudd asked Kennedy the easiest of all questions: "Why do you want to be president?" Kennedy's answer was what may well be one of the most incoherent statements in American political history: "Well," he began, "were I to make the announcement, and to run, the reasons that I would run is because I have a great belief in this country, that it is, there's more natural resources than any nation of the world; there's the greatest educated population in the world; greatest technology of any country in the world, and the greatest political system in the world. And yet I see at the current time that most of the industrial nations of the world are exceeding us in terms of productivity, are doing better than us in terms of meeting the problem of inflation that they're dealing with their problems of energy and their problems of unemployment."

Kennedy's breathtaking inability to answer this simple question in a coherent fashion scandalized the nation—and delighted the Carter campaign. To deal with the fallout from this fiasco, Kennedy's campaign manager, Steve Smith, called Garth to ask for advice. Garth told Smith that the campaign needed to do a poll to get a read on the situation. Remarkably, however, Kennedy didn't really have a pollster.[4] Garth recommended that he call us. Within twenty-four hours, we had poll numbers that showed just how much the comments had hurt Kennedy. Steve Smith professed himself delighted with our speedy work.

It didn't take long for Kennedy to stumble again. The same night that Kennedy was flubbing his second interview with Roger Mudd, Iranian students were overrunning the U.S. Embassy in Tehran. As the nation rallied around President Carter, Kennedy criticized the Shah of Iran. Rupert Murdoch's *New York Post* splashed Kennedy's picture across its front page under the devastating headline, "The Toast of Tehran." This time I was summoned down to Washington to present my results in person.

I was excited about the possibility of sitting down and strategizing with my first serious American presidential candidate. So the next day, fresh poll numbers in hand, I flew down to Washington, and caught a cab out to the Kennedy homestead in Mclean, Virginia. When I entered the house, I found Ted and Joan in the kitchen. They were not speaking, and Kennedy made no motion to rise. Instead, he glumly pointed me to the back room. There I found Smith and the rest of Kennedy's inner circle—Paul Tully, Carl Wagner, and a few others. I was disappointed that Kennedy himself wasn't participating, but I went ahead and gave my presentation.

My points were simple: my poll showed that Kennedy's gaffes were killing his campaign. I saw Kennedy's gaffes as disasters, not mere mishaps, and argued that the campaign needed to treat them as such. Kennedy needed to apologize, now, before he went off to campaign in Iowa for the upcoming caucuses. Kennedy's advisers listened politely but seemed not the least bit concerned. At the end of the presentation, Steve Smith came out to thank me for more excellent work. Then came the kicker. "Can I get you a cab back to the airport?"

"No," I replied. "I don't want to go to the airport."

"Oh," he said, looking puzzled. "Where would you like to go?"

"To your headquarters."

"And why would you want to go there?" he asked.

"Because I want to get paid!"

"Get paid?" he asked, as if such things were beyond him.

"Yes," I replied. "We need to get paid for the first poll that we did on the Roger Mudd interview and for this poll."

A big smile washed over Smith's face. "Oh, so you know our reputation?" He seemed almost pleased by the fact that Kennedy had gained a reputation as a cheapskate.

"I do, but the bottom line is that we've done two polls; each poll is ten thousand dollars; and we need our money," I said firmly. So instead of going to National Airport, I drove in to Kennedy headquarters, an old car dealership in Rockville. There I waited for several hours. Finally, a check was produced. Suffice it to say that was the end of my work for Ted Kennedy.

Fortunately, other campaigns soon beckoned. One was for a brilliant young politician in Washington, D.C., a veteran of the civil rights movement named Marion Barry. I'd never met a smarter politician; Barry struck me as an absolutely brilliant man: charismatic, thoughtful, up-to-speed on polling, strategic, a man at ease with people and issues in much the same way that Bill Clinton was. Unfortunately, he was also surrounded by corruption. After handling the polls for Barry's successful campaign, Ivanhoe Donaldson, Barry's campaign manager and chief of staff, commissioned us to do an exhaustive forty-five thousand dollar poll on the issues facing the city. Imagine my shock when two years later the U.S. Attorney for Washington, Joe DiGenova, subpoenaed all of our records—every letter, records for every phone call, notes on every conversation, everything—as part of a criminal investigation. After a long and costly legal struggle (which ended with no charges being filed against us), I had a chance to ask DiGenova why in the world he'd come after us. Apologetically, he explained that Donaldson had paid us our forty-five thousand dollars and then billed the city the same amount for *his* work on the poll—and that wasn't the only example of expense padding that DiGenova added. His prosecutors had found massive numbers of bills for fictitious or nonexistent projects. Indeed, there were so many fake projects, he told me, that at first he hadn't believed that ours was real at all.

Another more propitious assignment was the long shot campaign of a New Jersey multimillionaire, Frank Lautenberg, who aspired to become a member of the U.S. Senate. Garth thought that Lautenberg, as a Jew, didn't have a shot against the well-respected, liberal Republican Congresswomen Millicent Fenwick. We thought Lautenberg did—if he went negative. Over Garth's objections, we teamed up with media consultant Robert Squier to craft a series of ads that stopped Fenwick cold and led to the biggest upset of the 1982 election year.

However, New York City remained our bread and butter. In 1982, we vainly counseled Ed Koch against a run for governor. Despite our best efforts, he lost the Democratic primary to the ultimate winner of the election—Mario Cuomo. Ultimately, he was done in by his unwillingness to go negative early enough and a series of gaffes about upstate New York, the most infamous of which was a *Playboy* interview quoting Koch describing the suburbs as "sterile," "nothing," and a way of "wasting your life." To make matters worse, Koch's relationship with African American New Yorkers, who made up nearly 30 percent of the city's Democratic electorate, had deteriorated badly. Many observers believed that the Reverend Jesse Jackson's strong showing in the 1984 New York Democratic presidential primaries heralded the fall of Ed Koch a year later.

There was only one way Koch could win the city again in 1985: he would have to shore up support among his voters and turn them out in huge numbers. It was the kind of challenge that the old Tammany machine had once excelled at. However, by 1985 the clubhouse system had decayed to the point where it was no longer much of a factor on election day. If Koch wanted to have a machine behind him, he would have to build it himself. So in the spring of 1984, that's just what he decided to do. Instead of crafting polls, we would organize the city.

Koch entrusted the campaign to an old aide, Jim Capalino, who had been with Koch since his days in Congress and had organized Koch's street campaign during the 1977 mayoral race. Capalino would focus on putting together a volunteer organization of Koch loyalists in every precinct in New York. Mark and I would do just about everything else. Our offices soon became the campaign's de facto headquarters, and Mark and I threw ourselves into the task of developing direct mail for the campaign, setting up the phone banks, and raising money. We even came up with a novel way to capitalize on Koch's celebrity, spread his message, and raise money, all at the same time: when Koch published his first book *Politics,* a mixture of memoirs and reflections, we convinced him to barnstorm the city and sell autographed copies for a hundred dollars each.

These events proved to be so popular that they gave rise to a new campaign ritual—the politician's preannouncement book tour.

Our nontraditional campaign for Koch was proving to be a remarkable success. One year earlier, many intelligent and experienced observers of the political scene had viewed Koch as damaged goods. By the spring of 1985, it was clear that our opponents, City Council President Carol Bellamy and Manhattan Assemblyman Herman D. Farrell, didn't have a chance. When Koch formally filed for reelection in July, our machine and what was left of the city's regular Democratic organizations came up with 159,694 nominating petitions—five times the number of petitions filed by his opponents. Mark and I had assembled a field organization with thousands of volunteers and raised hundreds of thousands of dollars for a man who one year earlier had been widely viewed as unelectable. That September, Koch crushed Bellamy and Farrell to win the Democratic primary showing, as the *New York Times* put it, "overwhelming strength in every section of the city." A month later, Koch drubbed Bellamy (who like Cuomo reappeared on the ballot as the nominee of the Liberal Party) again to win a third term.

Our triumph in 1985 marked an important moment in my partnership with Mark: all of the political skills that we had refined for the past decade and a half really came together. We had built a structure and organized a campaign better than anything New York City had ever seen. We had taken the techniques that Dick Morris developed on the Upper West Side and used them to build what Morris had only imagined—a citywide organization of canvassers and volunteers. We had employed the direct mail strategy that I had first tried in 1973 for Father Gigante, using it to shore up support for wavering Koch voters across the city. We had developed the ability to do daily polling ten years before the rest of the field and had the edge that an extra decade of familiarity conveyed. I was now confident that we were in perfect command of a powerful set of tools. We could play this game anywhere, not just as pollsters, but also as strategic consultants. We were ready to move to a bigger stage.

PART TWO

GOING GLOBAL

CHAPTER FOUR

POLLING IN THE PROMISED LAND

By early 1981, I had worked on campaigns in New York, Louisiana, Mississippi, West Virginia, and the United Kingdom. As a result, I considered myself to be a fairly experienced strategist, one who could adjust well to different political environments. This belief would meet its ultimate test in the fractured, heated politics of Israel, and the reelection campaign of Menachem Begin.

It didn't take me long to realize that Israeli politics was unlike anything Mark Penn and I had ever dealt with before. There's an old joke in Israel that the country has one president and four million prime ministers. Israelis take their politics extraordinarily seriously, exhibiting a passion and knowledge like nothing we'd ever seen. Having never known a day of peace in its history, Israelis viewed politics as a life-and-death issue. What's more, the diversity of political and cultural perspectives was truly daunting. There was everything from Communists to borderline fascists like the Kahane movement, blocs ranging from Israeli Arabs to the Orthodox religious parties who were far more concerned about the religiosity of the country than they were with geopolitical issues. Among secular

Jews, the divide was just as great: the chasm between European (Ashkenazi) and Arab and other non-European (Sephardic) Jews sometimes seemed as large as the distance between New Yorkers and Mississippians. In 1981, it was a new and strange world to me.

The experience of working in Israel would open up extraordinary new opportunities for not only our firm but also for political consulting at large. We were venturing in uncharted waters, exporting the American model of political campaigns around the world. As we would soon discover in Israel and elsewhere, political consultants were capable of playing a critical role not only in exporting political ideas but also in safeguarding and improving democracy itself.

Though I was raised in a Jewish home, I had very little attachment to the state of Israel. Growing up, it was never a focus of conversation at the dinner table. I wasn't terribly familiar with the nation's politics, and I had only two relatives in the country (one of whom ended up working on the campaign). I had visited the country only once, during the summer of 1979 after I received my law degree from Harvard. All in all, Israel wasn't really on the radar screen—until I received that call from David Garth asking me to help him rescue the political career of Israel's most controversial founding father, Menachem Begin.

Menachem Begin was more than simply a political leader: he was a political lightning rod. For many Israelis, he was—and always would be—a terrorist. As the leader of the Jewish guerrilla group, Irgun, he was held responsible for the bombing of the King David Hotel in July 1946, which killed not only British officers but also numerous innocent Jewish and Arab civilians. In spite of that, in 1977 he had pulled off the most stunning upset in Israeli political history—unseating Israel's Labor Party, which had ruled the nation without interruption for the first three decades of the country's existence. It's difficult to overstate the political earthquake that Begin's victory represented. Labor's enshrinement as the country's political elite was practically sacrosanct in Israel. Never before had it been seriously threatened politically. But in the wake of the disastrous 1973 Yom Kippur War, which caught Israeli political and military leaders unaware, the party's vulnerability, particularly among the disadvantaged Sephardic population, was exposed. The world held its breath to see what the conservative Begin would do. It was a sign of the man's complex nature that one of his first actions was to make peace with Egypt.

By 1981, however, Begin's political fortunes had dramatically shifted. Israel's economy had collapsed; inflation was running as high as 133 percent, and Begin and his Likud Party were wildly unpopular. For two years, poll after poll showed the Labor Party and its leader Shimon Peres with a double-digit lead. When Begin announced in January 1981 that elections would be held in June, polls gave Labor a lead of fifty-eight seats to Likud's twenty. No serious political commentator gave him a chance of winning. A *New York Times* article in March said of Begin that he was "probably in his final months as Prime Minister." The piece even speculated that Begin could potentially make a deal on Palestinian autonomy in the West Bank and Gaza in order to confirm his legacy as a peacemaker.[1]

We were facing an uphill battle—but not an unwinnable one. First, Peres was so far ahead that it was practically inevitable that his support would erode. His strong poll numbers notwithstanding, there wasn't much love for Peres among ordinary Israelis: even those who supported him often disliked him. Peres had a very cerebral personality, which ran counter to Begin's image as a fighter, warrior, and, ironically, peacemaker. Begin had brought Israelis their first taste of peace with an Arab neighbor when he signed the Camp David Accords with Egyptian President Anwar Sadat in 1979. For many Israelis, this was proof that Begin could be more than just a warrior. Our early polling showed that for Begin to have any shot at winning, he would need to play on both those evident strengths.

Having seen such great success with daily polling in the 1977 Koch race, we figured it would work in Israel as well. Not surprisingly, in 1981, such methods were unheard of in Israel. We saw this as an opportunity to bring to Begin something that Israeli politicians could only have dreamed about—a daily snapshot of the mind of the Israeli electorate. We arrived in Israel and soon discovered the challenge of polling in Israel brought a set of new obstacles.

When we worked overseas, we made it a priority to work with a local firm that knew the ins and outs of local polling. We called an

Israeli company named Dahaf. When we met with Director General Mina Zemach, we explained what we wanted to do.

She looked at me for a good, long time and finally said, "CIA or PLO?"

I was thrown for a bit of a loop by that one. "What do you mean?" I asked.

She said, "Look, you're not telling me who your client is and the only people with the money to do this are the CIA and the PLO. I'm going to have to check you out."

So somewhere in the files of the Mossad is a background check on myself and Mark. Luckily, we passed muster. But getting past Mina Zemach was only half the battle: we also had to find a way to get our surveys out and bring back their results from around the country.

The solution we devised was ingenious. We decided to use an old-fashioned taxi service to gather information from around the country. This wasn't exactly the Battle of the Marne, but at the time there was really no other way to get the information we needed. We had interviewers for Dahaf taking polls and via the taxis bringing the forms to us in Tel Aviv. We then had the data input into Mark's microcomputer—the same one we bought in 1977. One of my most vivid memories of that campaign was poring over the numbers with Mark in our hotel room in Tel Aviv's then-shabby Sheraton hotel. This was in the days before pay-per-view movies, and there were only one or two movies in the system. One of them was "The Sting," and I swear we must have watched that movie about twenty or thirty times. I think I could recite the lines verbatim, even today.

Our entree into Israeli politics came in large measure from one of Begin's closest advisers and former Irgun compatriot, Yaacov Meridor. Along with another young adviser, Ehud Olmert (who would go on to become prime minister in 2006), Meridor quickly saw the importance of modern political tactics; it was he who was the driving force behind the decision to bring Garth, Mark, and me onboard. Meridor had been introduced to Garth by a mutual friend named Zev Furst, the former head of the Anti-Defamation League office in Jerusalem, now a partner in Gath's firm. Furst had originally brought Garth to Israel in order to brief Begin about the U.S.

political scene and the 1980 election in particular. By 1981, however, Begin needed more than information; he desperately needed Garth's help.

On the surface, Begin didn't seem like the type of politician who would listen to political consultants or read polls. At first, we couldn't help but feel that he was just humoring Meridor and Olmert with their "American consultants." In reality, however, Furst told us that whenever he came into the prime minister's office with the latest poll numbers, Begin's face would light up "like a little child." As if that wasn't enough, Furst reported that nothing seemed to make Begin happier than seeing himself on television.

In 1981, outside consultants were still a new phenomenon for Israeli campaigns (although Labor had also brought on board the famed U.S. political consultant David Sawyer for the 1981 race). Elections then were more of a family affair—or so it seemed. I remember meeting Yitzhak Shamir, who was Begin's campaign manager, and thinking what a nice, grandfatherly man he was. Only later did I discover he had been a terrorist, responsible—or so it has been alleged—for the murder of UN mediator, Count Folke Bernadotte, in 1948.

Not surprisingly, most of those around Begin were not inclined to follow our advice. It was a constant struggle—as it often is in foreign campaigns—to get his advisers to accept our suggestions. Those around Begin were resentful and dismissive of our work; they simply couldn't accept that their own political judgment could be trumped by the read-outs of a microcomputer. However, whether it was on purpose or by accident, Begin and those around him couldn't have followed our advice any better than they did.

First, his finance minister, Yoram Aridor helped stem the financial crisis by cutting taxes and allowing subsidies on luxury goods. Almost immediately, prices on everything from color televisions and stereos to milk and beer began to fall. For the first time, Israelis began to see a light at the end of the economic tunnel.[2] At a time of mushrooming deficits, it was one of the more egregious examples of economic sleight-of-hand that I've ever seen, but it worked brilliantly. By early May, Begin was already being hailed as the new front-runner.

Second, Begin attacked Labor, and in particular Peres, as being soft on security and in particular in dealing with the Palestinians—a line of attack that still resonates in Israeli politics today. Begin also embraced the idea of a so-called Jordanian Option that would involve an agreement with Jordan's King Hussein to take back the West Bank and half of its population in an attempt to promote Likud as the only bulwark to a Labor-initiated Palestinian state led by Yasir Arafat. A new slogan soon emerged—"Save Israel from Labor."[3]

But it wasn't just Labor that Begin was saving the country from, it was also Shimon Peres. Likud and Begin, in particular, attacked Peres in highly personal terms, calling him a liar and hypocrite and saying that the country couldn't trust the Labor leader on security.[4] Fair or unfair, Peres's image of untrustworthiness, slickness, and indecisiveness allowed the attacks to reverberate. Begin co-opted the peace image for himself, running on a platform of "Peace and Security." Considering that it was Begin who had brought Israel its first peace agreement with the Camp David Accords, voters responded quite positively to the approach. In the end, the election became one of public images—the hawkish, tough-on-security Begin versus the peacenik, dove Peres. As it turned out, a personality-driven race played directly into Begin's hands.

While playing up his peace bona fides, Begin never shirked from accentuating his strongest asset—the perception of him among Israelis as an outsider, a fighter, and an iconoclast. For many Israelis, Begin's appeal was that he refused to shy away from international opprobrium and would take whatever measures he deemed necessary to defend Israel. In May, with unprecedented bluntness, he attacked the presidents of Germany and France when they made sympathetic statements regarding the Palestinians and in particular the Palestine Liberation Organization (PLO). Begin chided German Chancellor Helmut Schmidt, who had served in the German Army during World War II, for never breaking "his oath of allegiance" to Hitler until the bitter end of the war. He then went after French President Valery Giscard d'Estaing, saying of the world leader that he had "no principles whatsoever."[5]

It was amazing stuff for a world leader to say, but Israelis responded positively. Later in the month, Begin solidified his position on security issues by threatening military action against Syria if it refused to remove surface-to-air missiles along the Lebanese border. He told the Syrians, in no uncertain terms, that if they didn't remove the batteries, the Israeli Air Force would do it for them.

Yet for all of Begin's inflammatory rhetoric, he was also a man of action. As he showed repeatedly, he was not afraid to take whatever measures he deemed necessary in order to safeguard Israel's security. For some time, Begin had been concerned about a French-built nuclear reactor that was slowly taking shape in the Iraqi desert. While French advisers and Iraqi officials asserted that the facility had peaceful purposes, those who were familiar with Saddam's pattern of cruelty and megalomania knew differently. Why would a country with some of the largest known oil reserves in the world need a nuclear power plant? There was clearly no good reason, and as Begin would later explain to us, he could not allow it to happen. Begin understood that if the reactor became fully operational it would only be a matter of time before Saddam Hussein would be able to produce a functional nuclear device—and threaten the State of Israel with destruction. On June 6, 1981, Begin made sure that would never happen.

In a daring raid, Israeli Air Force jets, flying low over entirely hostile territory, wiped out the reactor in Osirak, Iraq. In a mere two minutes, the Israeli Air Force had removed the danger of Saddam Hussein having nuclear weapons. Predictably, the international response was heated: even the United States criticized Israel, joining their UN Security Council colleagues in condemning the attack. To Begin's credit, he understood, even then, the extraordinary threat that Saddam Hussein represented, not only to Israel but also to the region as a whole. It was a danger that would become quite evident a mere decade later.

There were those in Israel who labeled the attack a political ploy to ensure Begin's reelection, but in my view, it had nothing to do with politics. I remember going into Begin's office a day or two after the attack. He looked exhausted and somber—as if the weight of the world was on his shoulders. I told him that his lead in the polls had jumped from three or four points to nine points, and I'll never forget his response. "Look," he said, "I didn't do this for political reasons. I know your government is going to condemn me and the world is going to condemn me. But someday you will thank me for doing what I've done. I've kept Saddam Hussein from acquiring nuclear weapons for at least five to seven years. It was the right thing to do; it will preserve peace. Why would I do it three weeks before an election? If it was political, I wouldn't have done it in this way. But understand, it will take the world a long time to appreciate why I did what I did."

In this case, history is certainly on Menachem Begin's side. For all those who condemned his actions in 1981, what we know today about Saddam Hussein provides ample evidence that to have allowed him to control a nuclear arsenal would have risked incalculable damage not only to the region but also to international stability as a whole.

In many ways, Begin's election presented some of the toughest political conditions I had ever worked in—tougher even than East Harlem. The tone of the campaign was vicious. Commentators at the time said it was the dirtiest in Israeli history. There were stories of Peres being jeered and pelted with eggs and tomatoes by Likud supporters, and Labor activists being threatened and even attacked. Then there were the constant chants at Likud rallies of "Begin, Melech Israel" or "Begin, King of Israel," which wasn't exactly the most democratic sentiment. Likud even ran ads featuring nothing but this omnipresent chanting.

The *New York Times* went so far as to publish an article in late June ruminating over the "Growth of Fascism" in Israel.[6] At the time, Begin was perceived as a warmonger in the United States, and I couldn't help but confront my own fear that I was working for someone who was really inimical to the interests of peace. Looking back, however, what he did with Iraq and the Camp David Accords balances the record pretty well. In the end, I remain convinced to this day that, fundamentally, Begin believed in peace—but only in a peace that truly protected Israel's security. In his mind, this meant maintaining control of the territories, Judea and Samaria as he always referred to them. For Begin, control of the territories and maintaining Israeli security were both part of the same equation. It was a view that would soon become dominant in Israeli political discourse.

Many people have argued that the attack against Iraq's nuclear reactor won the election for Begin, but Likud's numbers had been solidly trending upward since April. There was certainly a bump from the bombing, but within a week it had evaporated. In the end, bombing the Osirak reactor became a metaphor for the type of man that Begin was and the steps he was willing to take in order to safeguard Israel's security.

The raid on Osirak did reinforce Begin's base of support among more hawkish Sephardic voters who had long been marginalized by the Ashkenazi-dominated Labor Party. Starting in 1977, Sephardic voters formed the bedrock of Begin's support, and our polling indicated that they were essential to his hopes for reelection. The irony was that Begin was not only of Polish descent, but he was the quintessential Ashkenazi Jew. Never a day went by when Begin didn't stride into the prime minister's office wearing a coat and tie, which in Israel is practically unheard of. His formality stood in sharp contrast to the voters who provided his base of support.

Nonetheless, Begin understood that Sephardim voters wanted to be treated with respect but at the same time yearned for a fatherlike figure. While Peres, who was also of Ashkenazi descent, came across as aloof and arrogant, Begin connected with Sephardic voters. As a politician who for nearly three decades had been laughed at by Labor and disrespected, he—to paraphrase one of our later clients—"felt their pain." Moreover, Begin also understood the power of demagoguery: it was no accident that those chanting "Begin, King of Israel" at Likud rallies were primarily Sephardic voters.

As the race tightened toward the end, we urged Likud to make an even greater push for Sephardic votes, but first, we got a little help from Labor. At a rally for Peres only a few days before the election, a prominent Ashkenazi comedian made an ethnic slur against Sephardic voters. At the same time, Begin was campaigning even more heavily in Sephardic neighborhoods and surrounding himself at rallies with the few Sephardic members of the Likud party, including David Levy, who would later serve as foreign minister. In the end, more than 60 percent of these voters ended up casting their ballot for Begin.[7]

Begin's effectiveness in appealing to Sephardic voters was matched only by his efforts in attracting religious voters. In private, Begin was not a religious man, but in public, he played the role to the hilt. He refused to schedule meetings or drive on the Sabbath, and he always kept Kosher in public. Moreover, Begin consistently spoke in the language of Jewish tradition and values. He appealed to Jewish nationalism and pride, which were key concerns of many religious voters. Begin was quite effective in using this message of Jewish strength to co-opt the supporters of one of the key religious parties, the National Religious Party (NRP)—a right-of-center religious, Zionist party. Fearful of a Labor victory and clearly enthralled by Begin, NRP voters began defecting to Begin and Likud. For Likud, it wasn't a hard sell because religious voters were generally suspicious of Labor and its less-than-positive attitude toward religious voters. But even more so, the emerging religious/settler movement, of which Begin was a strong supporter, began to view the fate of the West Bank and Gaza Strip as defining issues for Israeli security. Since Peres and Begin took opposing positions on the subject, religious voters began moving more and more to the Right and into the arms of Likud. This affiliation among religious voters with Likud's muscular views on security and the future of the territories would become a dominant theme in Israeli politics over the next twenty years. More and more, religion and security became conflated into a defining political ideology for the Right.

But the critical factor in putting Begin over the top may very well have been a key question that we inserted into one of our later polls, "Would you be more likely to vote for Begin if he appointed Ariel Sharon, defense minister?" The response was a resounding yes, and right before the election Begin announced that he would give Sharon the defense portfolio if he won the election. What makes the story most interesting was that the man responsible for that question being included in our questionnaire was none other than Ariel Sharon. He was one of the few Likud advisers who came to understand the importance of polls and the degree to which they could be effectively utilized for largely political reasons. Meeting in a barren office, Sharon harangued Zev and me about the results. Over and over he said to us, "Tell him, tell him the results. Tell Begin."

It was also clear to us that Begin and Sharon enjoyed an unusually close relationship. Begin didn't always strike us as the most sentimental man, but when it came to Sharon we would often hear him express concern over Sharon's weight and diet, worrying that he would have a heart attack or stroke. While Sharon's prominence certainly helped Begin win the election, the result of the decision to elevate Sharon would have negative consequences—the Lebanon War, the Sabra and Shatilla massacres, and Begin's resignation from office in 1983.

On the eve of the election, every Israeli commentator agreed that the race was simply too close to call, with a slight advantage for Labor. We saw it differently. The morning of the election, we confidently strode into a tense prime minister's office and informed Begin that he was going to be reelected—by a one seat parliamentary margin. Begin was skeptical; everything else he was hearing told him that Peres was going to win. We brought out the results and began poring over them at Begin's desk. In those days, printouts from computers were a bit different than today—the paper came out in small scrolls of contact paper. With their strange appearance, we dubbed them the "Dead Sea Polls." Combing through these tiny scrolls of paper, we tried to convince the prime minister of Israel that he had eked out a victory.

Begin's advisers were also unconvinced, so much so that Begin himself urged us to go and share our numbers and tell everyone else what we were seeing. We did, but no one was listening. Some of them even laughed at the numbers and told us that we were undercounting the smaller parties. Things went from bad to worse for us soon after the polls closed. Israeli television called the race for Peres and Labor. Joyous crowds at Labor headquarters began celebrating and preparing for a return to power. We could feel the long knives being unsheathed, but we weren't quite ready yet to run for cover. "Be patient," we counseled. Israelis, however, are not known for their abundant patience, and Mark and I began to consider an exit strategy. Luckily, it wasn't necessary. An hour later Israeli television announced that there had been a computer error; Begin had in fact been reelected. His margin of victory was 49 percent to 48 percent—just as we predicted. As we suspected, the minor parties that everyone was so concerned about had been squeezed by the tightness of the race. Israelis were concerned that in such a close election a vote for a smaller party would be wasted. As a result, they pushed their support to the two biggest parties.

With Begin's successful reelection in 1981, our work in Israel was done. In the end, I think it would be a stretch to argue that we were responsible for Begin's victory, if only because Begin's advisers remained habitually suspicious of the tactics that we were extolling. Nonetheless, the decision to appoint Ariel Sharon Defense Minister before the election, a decision shaped in part by our polling, had undoubtedly played an important role in Begin's victory. Our efforts to coordinate the campaign's messaging and paid media, as the events that were organized based on our polling results, also contributed to Begin's narrow win. As one observer of the campaign noted, "In the Likud campaign, the degree of coordination between press and TV slogans and themes, and the techniques utilized in the television ads betrayed the American influence."

In the end, our work helped to provide one of the first examples of how American consultants could aid foreign campaigns. We had shown that while our methods had been developed in the U.S. politics, they could work far beyond America's shores. (Understanding voters and what makes them tick is useful information anywhere politicians are trying to influence the electorate.) The key was to create the right model on which to base public opinion research. Our methods had helped create a template for how it could be done anywhere in the world. This process would gather true momentum more than a decade later.

For most of the next decade, Mark and I made overseas campaigns a key aspect of our business, and we continued to follow Israeli politics with particular interest, while staying in touch with many of our former colleagues. In 1984, we were hired by Ezer Weizmann, the legendary father of the Israeli Air Force, who was trying to put together a centrist party, which has long been a goal of Israeli politicians seeking to bridge the Labor/Likud divide. We took a poll for Weizmann, which showed support for the idea, but also that such an effort faced enormous challenges. In the next election, the new party only got about three to four seats in Israel's parliament, the Knesset. It never ended up being a big player in the Israeli political scene.

In the years after 1981, Israeli politics found itself in complete deadlock. In 1984 and 1988, Peres faced off with Likud leader Yitzhak Shamir. (Begin had resigned from office in 1983 in the wake of the disastrous Lebanon War.) In those contests, neither side was able to cobble together a working majority. Instead, the two parties formed governments of national unity—in effect dividing up the government, with each leader taking turns as prime minister for two years. By 1992, however, Israelis were looking for a fresh face. They found it in Yitzhak Rabin.

To be sure, Yitzhak Rabin was not really a fresh face. He had briefly and disastrously served as prime minister from 1974 to 1977, when he had been forced to resign over an illegal bank account in the United States in his wife's name. However, he was a former army chief of staff, who promised progress on the peace front with ironclad guarantees for Israeli security. After defeating Peres in the 1992 Labor Party primary, Rabin promised Israelis true "Peace with Security." With his strong military reputation and his resonant message of change, he handily defeated the grandfatherly Shamir.

Rabin quickly delivered on his promises of peace. In September 1993, he signed the Oslo Accords with the PLO to begin work toward a Palestinian state, swallowing his enormous pride and shaking hands with his sworn enemy, Yasir Arafat, on the White House lawn. For the first time a region soaked in conflict saw its first glimmer of a peaceful future.

But there were those on both sides who would stop at nothing to halt the peace process. As the slow progress of negotiations began, Palestinian terrorist groups like the Islamic fundamentalist Hamas and Islamic Jihad stepped up their attacks against Israelis, killing dozens in horrific suicide bus bombing attacks. The outrages were not limited to the Palestinian side. In 1994, a right-wing Jewish settler named Baruch Goldstein massacred more than two-dozen Palestinians in a mosque in Hebron. The tenor of Israeli political debates also began to take an ominous tone. Chants of "Death to Rabin," and images of the prime minister as a Nazi became *de rigueur* at anti-peace rallies. After one such gathering, crowds marched on the Israeli parliament, Knesset, attacking Labor ministers in the parking lot.

By 1995—with elections less than a year away—it was unclear if Rabin would be able to convince Israelis to give him the four more years he needed to move the peace process forward. However, by then, the wheels of reelection were already in motion. In the fall of 1995, I got a call from our old friend Zev Furst asking us to do polling for the reelection campaign. As a strong supporter of the Oslo process and Rabin's political efforts, I was excited by the opportunity.

Certainly, everyone understood Rabin was facing a tough battle against Likud leader Benjamin Netanyahu. I vividly remember putting the finishing touches on a benchmark poll for the election. The date was Friday, November 3. We had gotten Rabin and all of his key advisers to sign off on it, and we were preparing to send it out the next Monday. Unfortunately our plans abruptly halted on November 4, when tragedy struck. After appearing at a pro-peace rally in Tel Aviv's Zion Square, Rabin was returning to his car when a young, right-wing Israeli stepped out of the shadows with a gun. Rabin was struck down by two bullets and a few hours later died at a local Tel Aviv hospital. On both a professional and personal level, I was devastated. But tragedy, notwithstanding, there was still important work to be done in continuing Rabin's legacy.

With Rabin gone, the mantle of leadership had been passed to his foreign minister, long-standing rival and our former nemesis, Shimon Peres. Though we had worked to defeat Peres in 1981, he asked us to stay on board for his race against Netanyahu. There were no hard feelings over our efforts fifteen years earlier and if anything Peres couldn't have been nicer.

I recall meeting with Peres in January 1996 when I made a lengthy presentation to him in his spartan prime minister's office explaining the issues and my concerns that it was going to be a very close race. I made suggestions about outreach to specific groups, such as a women's campaign. His response was, "Doug, we don't really do that here." As I quickly discovered, this blasé reaction would be the norm, not the exception, during the prime minister's race.

Peres was quite relaxed and even charming. He was up seventeen or eighteen points at the time, and he simply couldn't fathom the notion that he might lose—particularly to someone as radical and polarizing as Benjamin Netanyahu. When I suggested that he could realistically be beaten, Peres' response did not bode well for the campaign, "I can't lose to somebody like that," he dismissively responded. In Peres's view, "Bibi" as Netanyahu was known, was a lightweight, a second-rate politician and a charlatan.

There's an old adage in politics that one side rarely makes any mistakes and the other makes nothing but. Never was that more true than in the 1996 campaign for prime minister. After the Rabin assassination, support for Peres was at high levels, not simply because he was seen as the natural heir to Rabin, but also because many Israelis, particularly on the left, blamed Netanyahu for the poisonous political debates that preceded Rabin's assassination.

Unfortunately, Peres never quite figured out how to effectively utilize Rabin's legacy. He told me that January evening that he had wanted to avoid calling early elections in order to prevent the race from becoming a referendum on Rabin. He wanted his own mandate for peace. Had he called an early plebiscite, it seems almost certain that he would have won. At the same time, the campaign could never decide how to use Rabin's legacy and the Right's role in creating an environment that made that assassination possible. This issue sparked extraordinary debates within the campaign. The wounds of Rabin's death were still raw, and there was enormous fear of exploiting Rabin's death. The campaign even had video of Netanyahu speaking at anti-Oslo rallies doing nothing to stop protesters who were knifing pictures of Rabin and holding "Death to Rabin" posters. It was incendiary material, and in the end, Peres decided not to use it, preferring instead to win the race on his own merits, not on Rabin's tragic coattails. As time passed, however, the old doubts that had long surrounded Peres returned. Quite simply, a large segment of the population didn't trust him on the issue of security. They felt that he didn't have the stomach to stand up to terrorists. The irony was extraordinary. Peres was responsible, in part, for creating Israel's nuclear deterrent capability; he had negotiated a significant and critical arms transfer from France in the 1950s; and in the 1980s as finance minister, he single-handedly righted the Israeli economic ship. For all his accomplishments, as the *New York Times* put in March of 1996, Peres "could never shake the image of an indecisive dreamer and a shifty politician."[8] Very quickly that notion was tragically reaffirmed.

On January 5, Israeli forces, using a booby-trapped cell phone, assassinated Yahya Ayyash, a leading member of the Palestinian group Hamas. Ayyash, who was known as "the Engineer," was responsible for a number of grisly terror attacks against Israel. Hamas wasted little time in responding. On February 25, two bus attacks in Jerusalem and Ashkelon were launched. Twenty-five people were killed and scores were injured. A week later, the bombers struck again. First, a bus bombing in Jerusalem that killed nineteen people, followed the next day by an attack in the heart of Tel Aviv. In all, more than sixty Israelis were killed, hundreds were injured and the country's fragile confidence in the peace process was shattered.

Peres's poll numbers quickly plummeted, and a race that few people thought Netanyahu could win was now neck and neck. Our advice was clear and unambiguous: Peres had to find a way to burnish his hawkish profile. One of our first suggestions was for Peres to announce that Ehud Barak, the former Israeli Army chief of staff and the most highly decorated soldier in the history of Israel, would serve as defense minister in a new Peres government. Just as we had bolstered the Begin campaign in 1981 with Ariel Sharon's potential ascendancy to the Defense Ministry, we believed that Barak could have the same positive effect for Peres.

I had met Barak several years earlier when, of all things, he was working as a consultant for Slim-Fast foods. I had been hired by a friend named Danny Abraham to do a presentation for the company, and I went to Florida to meet with Barak, Abraham, and Ed Meyer, who was then chairman of Grey Advertising. I was immediately impressed. For a former military man, he had an extraordinarily instinctive understanding of how to sell diet products; he also understood what it would take for Peres to win the 1996 race.

Of all Peres's advisers, Barak was by the far most perspicacious. Not only was he one of the smartest people I've ever met, but he also had a very systematic take on politics, which dovetailed nicely with our approach and led him to embrace our efforts. However, Barak was a Rabin protégé, and Peres was suspicious and afraid of his machinations. It didn't help things that the man running Peres's campaign was Haim Ramon, who was locked in a death struggle with Barak to become Labor's heir apparent.

In April when Hezbollah guerrillas in Lebanon started shelling Northern Israel, Peres reacted with atypical fury—and did what he thought was necessary to demonstrate his security bona fides. For eleven days, Israeli forces pounded Hezbollah positions. The result was a diplomatic and political disaster. Arab leaders as well as some European and American politicians attacked Israel and accused Peres of overreacting. Even worse, Peres's tactics were ineffectual. The violence didn't stop, and Hezbollah continued to fire rockets into Northern Israel. There was criticism of the military tactics both inside and outside of government, and popular calls for even stronger responses grew. Things soon went from bad to worse. On April 18, more than a hundred innocent civilians were killed at a UN refugee center in the Lebanese town of Qana, leading to the end of the offensive. Peres's support dropped further.

We went back to Peres and told him in early May 1996, in no uncertain terms, that he would lose if he didn't do a more effective job of looking more hawkish and burnishing an image of toughness. We told him that Lebanon was not the top concern for Israelis. They wanted Peres to go after terrorists directly and force Arafat to do more in clamping down on the suicide bombers in his own midst. These were difficult moves for Peres, who was committed to a peaceful vision for the Middle East that he had laid out in print— one of interdependent economic relationships between friendly states. Nonetheless, we believed that this vision could only be achieved if Peres was elected and we laid out a series of steps that he should take including cracking down on terrorists and deemphasizing the notion of a Palestinian state.

I'll never forget Peres's reaction.

"We should do everything Schoen is saying," He said. Then he turned to Ramon and said, "But I know you're not going to do it."

Ramon shook his head and said, "That's right." This astonishing comment laid bare Peres's tragic and irredeemable flaw: for all his amazing gifts, he simply lacked the drive to do what it took to win. Peres's response was indicative of the aura of foreboding and malaise that surrounded the campaign from day one. During the 1981 race, I had the sense that Begin would do everything necessary to claw his way to victory. The 1996 Peres campaign was the complete opposite and was one of the most dispiriting I've ever worked on. Peres never actively engaged in the campaign because he never truly thought that he could lose. To make matters worse, his key advisers misled Peres about the danger that he was in, telling him that he was ahead and refusing to carry out his instructions when he did bestir himself to action. Decisions would be made by Peres and then Ramon and Teumin would do the opposite, whatever they thought was best or, quite frequently, nothing at all. In the end, Peres never took the difficult, but essential steps to impose discipline and get a handle on the campaign.

Instead of taking our advice, Peres started developing his own ideas on how to turn things around. It involved one of our other high-profile clients, President Bill Clinton. In the spring of 1996, Clinton was the most popular figure in Israeli politics. In every poll we took, he consistently bested both Peres and Netanyahu. Peres thought the key to reassuring Israelis about security was to get the United States to agree to a joint defense treaty with Israel. Peres adviser Moshe Teumin called me in April and asked us to go back channel with Clinton (we were already heatedly involved in Clinton's reelection effort) and work an agreement out. Clinton and Sandy Berger both thought it was a terrible idea and refused to go along. In their view, it wasn't good for Israel, and it would do little to help Peres politically.

Clinton, however, did desperately want Peres to win. With his extraordinary and instinctive understanding of Israeli politics, he used me to channel political advice to Peres and his campaign staff. In fact, just several days before the election, Clinton put out a statement that was basically an endorsement of Peres. The wording and tone were based almost solely on our polling and messaging advice.

Needless to say, all of these steps were taken very discretely. There was already an enormous sensitivity in Israel to the use of American political consultants. Netanyahu had brought on board Arthur Finkelstein, a Republican political operative who had worked on campaigns for right-wing candidates like Al D'Amato and Jesse Helms. The presence of Finkelstein led many commentators to complain that Israeli politics were imitating American elections—an accusation that was very close to the mark.

The 1996 election was the first ever in Israel in which the prime minister was directly elected. As a result, the race was, in many ways, more personality driven then any other in Israeli political history. Image took on a new and important role—and it played directly to the strengths of Netanyahu, who had made his career defending Israel to U.S. television audiences. He understood how to speak in sound bites and work the modern media apparatus. For Peres, television was still a bit of a mystery.

Netanyahu, however, knew more than just how to be effective in front of a television camera—he understood the effectiveness of images and the role of television in winning campaigns. His political ads were provocative, incendiary, and highly effective. One showed Arafat leading Peres by the hand. The other displayed the gory aftermath of one of the bus bombings in Israel. They were unlike any ads that had ever been seen in Israel, and they played directly on Israeli fears of terrorism. In addition, with Finkelstein's backing, Netanyahu stayed on message with a single, consistent message: "I'll never compromise on security." Moreover, he continued to assert that Peres would divide Jerusalem and give the eastern half to the Palestinians as the capitol of a future state. With Peres running a campaign that made clear he would take risks for peace, the race became that much more daunting.

Netanyahu not only hammered away on the security issue but also played up his position as the political outsider. Of 1996's many firsts, few were more important than the newfound prominence of Russian voters. In the late 1980s and early 1990s, millions of Russian Jews had immigrated from the Soviet Union to Israel. By early 1996, the Russians—led by the charismatic, former refusenik Natan

Sharansky—had become a potent political force making up 14 percent of the population. For the first time, politicians on both sides of the political fence were forced to cater their messages to them directly. Netanyahu had by far the more effective message. Just as Begin used an anti-elitist, anti-Ashkenazi, anti-Labor argument to appeal to North African voters, Netanyahu delivered the same message of alienation to Russian voters, blatantly exploiting the Russians' fears and hatred of Communism. Referring to Peres as "Comrade Peres" and "red Peres,"[9] Netanyahu argued that Labor had neglected the Russian community and that Likud would do better if put into office. The message resonated. When Sharansky announced his support for Netanyahu, it was clear which way the Russians were going to vote. In the end, more than 65 percent cast their ballots for Bibi.

Instead of cultivating the Russian vote, Peres focused in part on Israel's Arab voters who traditionally supported Labor candidates, spending countless hours and precious resources to energize them. In a direct election, their votes could make all the difference. However, the Lebanon excursion in May, as well as decades of neglect by indifferent Labor leaders, led many of them to tune Peres out. In the end, tens of thousands of Israeli Arabs, who voted in the parliamentary election, left blank their ballots for prime minister. Had most of them bothered to cast a ballot (most likely for Peres), we would have been toasting victory.

Despite Peres's efforts to appeal to the Israeli Arab population, terror and security remained the two most important issues of the campaign for the Jewish electorate, and Peres had no effective answer for them. By May 1996, the images of burned-out buses and dead Israelis—not those of a slain prime minister—were the most prevalent in Israeli minds. In the weeks and months after Rabin's assassination, a bumper sticker began appearing on Israeli cars. It said simply, "Shalom Chaver" or "Goodbye, Friend," a reference to a line uttered by President Clinton on the day that Rabin died in November 1995. As more and more Israelis died in horrific attacks, a new bumper sticker appeared—"Shalom Chaverim," or "Goodbye Friends." In a nutshell, that small spelling change best summed up the political transformation in Israeli society.

Nonetheless, for all of Peres's missteps, the race stayed close until the final days. Two days before Election Day, the two candidates engaged in their only face-to-face debate. Peres barely prepared for it and appeared tired in front of the camera. Nevertheless, he still could have won the debate if he had done one simple thing: talk about security. We begged him to dwell on this subject at length, but he didn't do it. Two days later he lost his fifth election in Israeli history—by a mere 29,457 votes.

History has never been kind to Shimon Peres. Once again on election night, Israeli media outlets called the race for Peres, just as they had in 1981, only to reverse themselves hours later. That night in 1996, I was in the White House with Clinton and our colleague Dick Morris. After the initial call for Peres, Morris tried frantically to arrange a call between the president and the apparent victor. I called Furst, but Peres demurred, "I can still lose this," he said. Peres and Clinton never did talk that night.

The topsy-turvy nature of election night was indicative of the extraordinary frustration that defined the 1996 race. We had a good strategy and smart ideas for motivating the electorate, but we could never get focused. In the end, Peres was powerless in the face of events that seemed beyond his control. Later, Furst recounted an evocative postscript to the campaign. The day after the election, he went to Peres's office in the Defense Ministry. He was surrounded by his closest advisers, but Furst demanded to talk to him alone. There was a small bedroom behind the minister's office and sitting at the end of the bed, Furst told Peres, in no uncertain terms, that his advisers had consistently lied to him and fed him misinformation. With tears streaming down his face, Peres just shook his head in frustration. It was a sad coda to an even sadder campaign.

Everyone who worked on the 1996 campaign was dispirited by the outcome, but our efforts in Israeli politics continued. In early 1997, we received a surprising call from none other than Ariel Sharon. He was interested in taking on Netanyahu and wanted our advice on the best way to proceed. With Zev Furst taking the lead, we helped conduct a poll on his behalf. Based on the research, a clear course of action was evident: sit back and do nothing. The poll results showed that Sharon's negatives were high because of his actions in Lebanon and his hawkish reputation. With Netanyahu increasingly unpopular in Israel, Sharon was better off letting Bibi destroy himself and waiting for political events to come to him. It was going to be events and circumstances, not his own behavior that could potentially propel him to power. Sharon got the message and actually played the role of loyal soldier for Netanyahu during his tempestuous years at the helm of Israel. Four years later, Sharon found himself in the prime minister's office—in a course of action that was pretty much exactly what our poll numbers predicted.

Our dalliance with Sharon, notwithstanding, we maintained a close relationship with Ehud Barak. From the moment Peres lost, Barak was already putting the pieces in place for a prime ministerial run of his own. In May 1997, he commissioned an exhaustive and comprehensive poll of the Israeli electorate. Our results spelled out a clear direction for the campaign: Barak needed to develop a strong domestic agenda, with specific policies on education and security. In addition, we told Barak that he needed to create his own, more prominent profile in Israeli politics. Everyone seemed very pleased with the results—but that was the last we heard from the campaign for more than a year and a half.

Then in the fall of 1998, word leaked out that Barak had hired, Stanley Greenberg, Bob Shrum, and James Carville to run his campaign. Over the years, Mark and I had developed a somewhat antagonistic relationship with Stanley Greenberg. Part of the antipathy is professional: we often compete for the same clients. Part of it reflected our different visions for the Democratic Party. We tend to be more centrist; he tends to be more populist and liberal. When we heard they got the Barak job, we were clearly disappointed.

It didn't take long, however, for us to discover that Barak hadn't quite cut us out of the picture. One of Barak's key advisers was a man named Isaac Herzog, the son of the former president Chaim Herzog, and he quickly passed word to us that "Ehud wants you." And so began one of our most fascinating back-channel campaigns ever: Shrum, Greenberg, and Carville were the public face of the campaign, and Zev Furst and I were working behind the scenes, polling continuously, and funneling our own advice and that of President Bill Clinton.

It seemed an odd arrangement, but Barak used to tell us that as a military man, he liked to have "lots of information from different sources."

"Your role is important," he reassured us. "You're just playing a different kind of role than usual." Barak liked the idea of having prominent American consultants. In his view, it gave him credibility among Israelis—a far cry from 1981—and one couldn't get more prominent than having James Carville on your team. But at the same time, Barak wanted to make sure he was doing the right thing so he asked us to conduct about five or six polls for him that basically served as a check on what he was being told by Greenberg and company.

We later found that he had hired a French consulting team as well.

On the whole, we did a pretty good job of keeping our role quiet. There was even a front page article in the *New York Times Magazine* about the role of political consultants in the 1999 campaign that made no mention of either me or Mark. But it's never easy keeping a secret and about a week before the election, word leaked out. Greenberg and Shrum heartily denied it, but after they were given our polls, they quickly saw what we were doing. Suffice to say, not many fences were mended between our two camps on that campaign.

One thing was for certain: the leak didn't come from Barak's camp. What most impressed us about Barak was that he never let on to what the other groups were doing. He was incredibly effective at internalizing the advice he was given, keeping his own counsel, and then unveiling a strategy that brilliantly combined the disparate pieces of advice he was receiving. Barak was a top-notch strategist, which is often unusual for a political leader. He could remember every number we put in front of him. When I would brief him on the latest numbers, it was like defending a PhD dissertation: he peppered me with questions and demanded a fresh perspective on the data we were showing him. At one point, during one of my "lectures" he turned to one of his advisers and asked, "Why isn't Greenberg giving us this information?"

Unlike Peres, Barak learned a valuable lesson about how to deal with advisers. While he surrounded himself with excellent people (in many cases, former army buddies), there was never any question about who was the boss. He ran a tight ship and the internal discord of 1996 was a faint memory by 1999. In fact, Barak's preoccupation with victory later came back to haunt him.

In 1999, there were two elections rather than just one for a party list as there had been before 1996: one for prime minister, the other for the party makeup of the Knesset. While running on the Labor ticket, Barak was clearly more focused on his own campaign then the fortunes of the party. In fact, Barak submerged Labor under an umbrella organization called One Israel that combined several parties under a left-of-center mandate. In the end, Barak's attention to his own race, above all else, ended up costing Labor seats in the Knesset and imperiled his later governing position.

Barak's greatest limitation was his own belief that he was the smartest guy in the room. To be sure, there was some justification for this view—Barak was uniquely brilliant, with extraordinary political instincts. But his accompanying self-confidence would, in time, take its toll.

However, as the 1999 campaign began, that was the least of our concerns. In fact, Barak was almost a secondary figure in the race. This race for prime minister was in many ways far more clear-cut than any other we worked on in Israel: it was quite simply a referendum on Benjamin Netanyahu. Since becoming prime minister, Bibi had ricocheted from one crisis to another, both personal and political. Relations with the Palestinians and even the United States had reached new lows of mistrust and recrimination. Social tensions in Israel—between Russians and non-Russians, secular and religious, Sephardic and Ashkenazi—had grown more acute. The economy was in tatters, with growth stagnant and unemployment on the rise. According to one of our first polls, by a margin of 66 percent to 27 percent, Israelis thought the country was on the wrong track.

Netanyahu faced his greatest challenge on the personal level: his insufferable nature and personal arrogance had alienated numerous Israelis. An article in the *New York Times,* on the eve of the election, summed up well people's views of Bibi. In explaining his grudging support for Barak, a potential voter noted, "I want to be able to go out in the street and say I have a Prime Minister who is not a paranoid, a cheater and a liar."[10]

The dislike of Netanyahu wasn't just restricted to Israelis—it could also be found at 1600 Pennsylvania Avenue. To put it mildly, Bill Clinton viewed Benjamin Netanyahu with great suspicion. Part of it was personal. Clinton had long admired and venerated Yitzhak Rabin, and he saw Bibi as having usurped the throne after Rabin's assassination. Most of all, however, their abundant differences were political. Clinton blamed Bibi for the stagnation in the peace process and believed that, with Barak at Israel's helm, peace would be that much easier to achieve. Even in the midst of the Kosovo war, Clinton played an active behind-the-scenes role in the campaign, consistently finding the time to pore over our poll numbers and offer advice to Barak.

Our strategic polls indicated that Clinton's view of Netanyahu was quite similar to that of most Israelis. Quite simply they didn't trust Bibi, and they believed that as long as Netanyahu remained in the prime minister's office, security would be weakened. Our recommendations and those of the president were clear: run against Bibi and on peace.

This was actually a tricky argument, since terrorist attacks had actually decreased during Netanyahu's years in power. However, our message was that if you're constantly antagonistic to the Palestinians (which Bibi was) and if you only reluctantly make compromises for peace, Israel was never going to have true security. Israel needed someone like Barak, a former soldier, who would make true concessions for peace but never would weaken Israel's security. Keep in mind, this was at a time when most Israelis saw Arafat as a peace partner and were open to the peace process. At that time, 69 percent of Israelis considered Palestinian statehood to be inevitable and 82 percent wanted to move forward with the peace process. (Today, quite clearly, times have changed.)

More than just security, Barak understood how to appeal to individual constituencies in ways that Peres never grasped. Early on he realized that the Sephardic bloc was key, and he began reaching out to these voters, who had long been ignored by Labor. In fact, he offered an unprecedented apology to Sephardic voters for Labor's past behavior, demonstrating a degree of solicitousness that was unusual for Israeli politicians.

Our polling numbers also showed a sliver of opportunity among Russian voters, due to years of unfulfilled promises from Netanyahu. The problem Labor faced was that Russian voters were fairly hawkish and without any sort of initiative by Barak they threatened to break three to one for Netanyahu. When you consider that Russian voters were approximately 14 percent of the electorate, a break of three to one meant approximately 10 percent for Bibi and 4 percent for Barak—a six-point advantage for Netanyahu and in a close election, a potentially huge margin.

In 1996, Bibi had effectively co-opted the Russian voters by highlighting Labor inattention and promising cabinet positions and a Ministry of Absorption. Barak took a similar tact, going on the offensive, making promises of more money for Russian immigrants, and criticizing Bibi for doing little to address their concerns. He spent significant time on the ground, talking to voters and making his presence known. Russian newspapers were generally hostile to Barak, but his campaign found ways around that problem by placing glossy pro-Barak pamphlets in the papers and even publishing biographies of Barak in Russian. In addition, Barak's focus on the economy and social issues played well among Russians, who were largely nonideological. While traditionally they tended toward the Right on security, Barak's military background defused many of their concerns on that issue.

Near the end of the campaign, Barak openly promised the Ministry of Interior to Sharansky. Not only was it a bold move to attract Russian voters, it represented a stunning rebuke to religious parties, who had long controlled the Interior Ministry. With tensions already high after a prominent religious leader launched a tirade against "Russian criminals and prostitutes," Barak's gambit produced quick results.[11] In a poll we commissioned in January, Barak received only 17 percent of the Russian vote in a four-way contest. Even in a head-to-head match-up with Netanyahu he was down twenty-three points. By the time of our last major polls in May, Barak led by four points among Russians. In the end, he won 57 percent of their votes.

Barak's embrace of the Russians was in sharp contrast to his attitude toward religious voters. As in 1996, religious voters were more inclined to support Netanyahu, but unlike 1996, there were gains to be made by attacking the religious parties. A new party named Shinui had formed, which ran on a strongly secular platform that aimed to weaken the influence of Orthodox Jewish parties in Israeli life. Tensions between religious and secular groups had been growing for years, and Barak was not above playing on these concerns. At first, he reached out to religious voters, but sensing the way the political winds were blowing, he began, later in the campaign, to talk about forcing religious students to serve in the military (they previously did not) and removing some of the large subsidies for religious schools. In a polarized political environment, it was an effective message for disaffected secular voters, particularly Russians, who resented the power and prominence of religious parties in the political system.

While the voter outreach was effective in burnishing Barak's credibility with swing votes, it paled next to Barak's most effective strategy—attacking Benyamin Netanyahu. Our research consistently showed that the more negative the campaign was toward Bibi, the more Barak's numbers improved. Israelis blamed Netanyahu for the lousy economy; they blamed him for the failure of the peace process to move forward, and they had little confidence that he could bring Israelis together and fix the country's problems.

To be sure, a candidate's job is much easier when the opponent's unfavorability numbers are worse then the favorability ratings. In fact, we even discovered that positive information about Barak made little difference in the minds of voters. This was an election about Netanyahu, and it was the electorate's views of him that would decide the winner. So taking our advice, Barak continued to hammer away at Netanyahu and point toward a brighter future under his leadership.

I remember specifically being grilled by Barak about our conclusions at a meeting late in the campaign at the Hilton Hotel in Tel Aviv. The data clearly showed that the electorate was not persuaded by any of the arguments that we made about the benefits that Barak would bring to the country. Instead, it was only when Netanyahu was subjected to the most intense criticism, labeling him dishonest and untrustworthy, that the electorate moved in Barak's direction. I was unequivocal in my advice: Barak needed to go negative at once. Barak asked Furst to review the data for him twice and ultimately was persuaded. He then told his campaign staff, who had joined the meeting, to follow our directions.

There was, however, one final risk that threatened to undermine Barak's candidacy. The 1999 race, like 1996, featured a direct election for prime minister. That year, instead of the normal two combatants, there were five: two fringe candidates and Yitzhak Mordechai, a centrist politician who had served as defense minister under Netanyahu. Mordechai had entered the race under the mantle of the new Center party in order to nominally provide Israelis with the long-debated "centrist" alternative. But the reality was quite opposite. Mordechai couldn't stand Netanyahu and had entered the race for pretty much one reason only—to defeat him at the polls. In an April debate with the prime minister, Mordechai had unleashed a stirring set of attacks—uncharacteristic for even the rough and tumble of Israeli politics—that left the normally loquacious prime minister tongue-tied. It was actually a major turning point of the campaign, severely undermining Bibi's credibility and puncturing the image of Netanyahu as a supreme political debater.

Despite Mordechai's effective attacks, by the eve of the election in May, it was clear that support for him was evaporating. His numbers had fallen to the single digits, dropping ten points in the month before the election. Nearly all those voters were moving to Barak. Nonetheless, Mordechai was still draining away enough votes to undercut Barak's chances for reaching 50 percent, which he needed to avoid a runoff with Netanyahu. There was widespread concern in the Barak camp that not only Bibi would fare much better in a one-on-one matchup, but that he would be able to galvanize voters fearful of a center-left victory. However, if Mordechai were to drop out of the race, it was clear that Barak would reach the 50 percent marker.

On the eve of the election, it became imperative to find some way to get Mordechai out of the race. Armed with our numbers,

Furst went to see him. While we certainly had Barak's blessing, he didn't want his fingerprints on Mordechai's departure, so Zev went to the former general as a nominally "concerned pollster." In no uncertain terms, Zev told the former general that if he stayed in the race, Bibi could potentially win if there was a second round of voting. At first, Mordechai demurred, telling Zev he would think it over. Certainly, no politician likes to be asked to get out of a race, but considering Mordechai's enormous antipathy toward Netanyahu, it wasn't a hard sell. By the next day, Barak had received word that Mordechai was dropping out.

Mordechai's defection only intensified the tone of a campaign that had become increasingly vicious during its final weeks. In an attempt to argue that Barak would be weak on security, Netanyahu brought back ads from his 1996 campaign that showed the fiery aftermath of bus bombings. He began lashing out at the press, accusing them of ganging up on him. (Netanyahu was probably right about that since journalists in Israel generally couldn't stand him.) Finally, he sharpened his attacks against Barak, at one point even hinting that he was anti-Zionist. In the end, however, these attacks had no impact on Barak. The campaign ran numerous ads showing Barak on the wing of a Sabena jet in 1973 moments before he stormed the plane and freed Israeli hostages from the clutches of terrorists. This was a man who had dressed in drag during a raid in Lebanon that took out a key Palestinian terrorist leader. Intimating that Barak would be weak on security simply wasn't going to stick. Moreover, Netanyahu's credibility was so weak at that point that few took his attacks seriously.

By that point, it seemed clear that Barak was in the driver's seat, but with a still large number of undecided voters and the polls fairly tight, things could still go badly. In the end, however, Barak won by a 56 percent to 42 percent margin. It was one of the largest electoral victories in the nation's history. By modern Israeli standards, which tended to feature tight races, it was a landslide victory.

Ultimately, the deciding factor was the undecided voters who broke almost completely for the challenger—as they often do in tight races. Israelis had tired of Netanyahu, and after years of political instability and a lack of progress on the peace process, people wanted a change. Barak had run unabashedly on a platform of peace, and voters clearly responded. For many on the left, Barak's election represented a piece of unfinished business in Israeli society. On election night, tens of thousands of Barak's supporters spontaneously gathered in Tel Aviv's Rabin Square, where the former prime minister had been killed. Many shouted into the air, "Rabin, we have won."[12] Barak joined the crowd for a second victory speech and dedicated his victory to the slain Rabin, who for a brief moment, it seemed, had not died in vain. It was a stirring moment, and one couldn't help but believe that the end of the Arab-Israeli conflict was around the corner. Of course, as we would soon discover, a lasting peace was more elusive then any of us had imagined.

Looking back on my work in Israel, it is extraordinary to think of how far we had come. In 1981, consultants and daily polling were not only unheard of, but they were viewed as much a hindrance as a help. By 1999, things had radically changed. Both campaigns featured high-powered American consultants who were not only polling on a daily basis but also integrating those results into campaign strategy, messaging, and media. For Russian and Sephardic voters, whose needs had been ignored for so long in Israeli society, strategic polling helped ensure that their concerns would be raised and that they would be actively courted by both sides.

Today in Israel, polling is part of governing. The result is that elected officials have a far better understanding of voter attitudes toward major issues. What's more, Israeli elected officials have become far more responsive to the concerns of ordinary voters. Considering the life-and-death nature of Israeli politics, it is simply essential for politicians to understand where the voters are, before making major concessions that could affect the future of the State. It was a lesson that Ehud Barak would learn two years later—when he lost his chance at reelection.

Interestingly, in recent years, polling has actually ended up driving political change in Israel. As an interesting addendum to our efforts in Israel, in 2004 we received a call from Shimon Peres. He was interested in doing a poll that looked at the efficacy of creating a national unity government between Labor and the governing Likud

Party. This was quite a different situation then we had encountered years earlier in our polling for Ezer Weizmann. There was strong, majority support for the option. In fact, our results indicated that a marriage between Labor leader Shimon Peres and Likud leader Ariel Sharon was the smartest political option for both politicians.

While the Labor Party was deeply mistrusted in running the country, their policy positions, such as disengagement from Gaza, the building of a separation fence, and the resumption of direct negotiations between the Israeli government and the Palestinian Authority, enjoyed broad support. Amazingly, Peres had become the most popular politician in the country, followed by Sharon. In addition, voters were increasingly disenchanted with all political parties in Israel, creating, in my view, an opportunity for a centrist political alternative that would enjoy strong support. Our polling made clear that if Peres and Sharon joined forces, it would be the most effective means of capitalizing on Israel's new political realities.

As a result, we were not at all surprised when Sharon stunned the political world by announcing that he would leave the Likud Party and form his own centrist slate, called Kadima. After all, as a December 2005 article in the Israeli paper Haaretz made clear, Peres had shown Sharon our poll and encouraged him to make the move. Tragically, Sharon would not see the results of his bold gambit. In January 2006, a massive stroke would fell the prime minister and put him in an irreversible coma.

The reigns of leadership quickly passed to Ehud Olmert, who became the new head of Kadima. Not long after his ascendancy, Olmert called Zev Furst seeking advice on the best political strategy for this new party. Our relationship with Olmert went way back as he had been one of the top Begin advisers. In fact, he had been one of the few Begin advisers who had supported bringing us aboard that campaign twenty-five years earlier. When we met him in mid-February, he fondly recalled our work in 1981 and joked about the difficulty of getting Begin to take political direction in the waning days of the campaign.

At the behest of some of Olmert's advisers, we took another benchmark poll and came to a number of surprising conclusions. We found that Olmert had several critical strengths: he was seen as the candidate best suited to keep Israel secure, to unify the country and to provide strong leadership. Despite these positive attributes, voters lacked specific information about what he planned to do as prime minster, so we counseled Olmert to run on the key themes of leadership, security, and unity and to outline a clear plan for Israel's future. We suggested that his vote would be squeezed from the Left and Right if he did not offer a specific plan for Israel's future. Olmert was in a difficult position, given that Sharon was still alive, and he did not want to offend his supporters. Nevertheless, he heeded our advice and later in the campaign outlined his own plan for a unilateral disengagement from the West Bank.

Olmert operated in an environment of great uncertainty. Once when I was with him, he turned to me and wearily said something that really brought home the challenges that the prime minister of Israel faces.

"You know Doug," he said in a composed but tired voice. "Every day an aide comes into my office with a scrap of paper seeking a decision on life or death matters. The decision could go unnoticed—or it could provoke an international incident. There is no way of knowing. Every day. That's something that only someone who has sat in this seat could understand."

Olmert wasn't boasting or revealing a secret, but he was showing more of himself than political leaders usually do to outside advisers. Yet I imagine this combination of heavy responsibility and great, sometimes impenetrable, uncertainty is something other leaders share as well. Did Ariel Sharon know what would come of his decision to invade Beirut in 1982? Surely not. Did Hezbollah leader Sheik Nasrallah understand that kidnapping Israel soldiers last summer would trigger massive Israeli retaliation? Based on his own comments, he did not. Acting as a political consultant, my job is to understand and explain voters' thinking. Public opinion research gives me the tools to do that. It's a far cry from a glass ball, but it's also a lot better than operating in the dark as so many politicians typically do.

While Olmert enjoyed a healthy advantage on issues of security, economic concerns were an emerging liability. We were shocked to find that the economy and jobs as well as poverty were right behind terrorism and security as top concerns among Israeli voters. People thought the country was headed in the wrong direction economically; they were concerned about growing poverty, and they were looking for a leader who spoke to those issues. While voters showed little patience for the Labor Party due to their views on security issues, the party leader, Amir Peretz was seen as the candidate best suited to creating jobs, spurring the economy, and bridging the gap between rich and poor while Olmert ran a poor third on these questions. It seemed clear to us that Israelis were prepared to address their rising concerns over economic policies that seemed to exacerbate the nation's growing inequality.

We tried to impress on Olmert the necessity in talking about pocketbook issues and offering a detailed economic plan. To our dismay, no such plan was forthcoming. Zev and I sent Olmert an urgent note as the campaign was winding down urging specificity on both security and the economy. Olmert, we argued, had to be his own man if he was to maximize his vote. To be sure, Olmert was walking a tightrope, but ultimately he missed an opportunity to address the key hidden issue of the campaign.

Unfortunately for Olmert, our assumptions were accurate. Instead of the big political win he was expecting, Kadima received only twenty-nine seats in the Knesset—nine more than the Labor Party. The Likud Party, whose leader Benjamin Netanyahu had been associated with the failed economic policies of the past, was decimated with a mere twelve seats. Most shocking of all was the performance of the Pensioner's Party, which was focused on increasing social benefits for the country's senior citizens. It received a stunning seven seats.

In the end, Israelis had voted their pocketbooks, and while Olmert remained in power, he had won fewer seats than he had hoped. As a result, he was forced to join what ultimately became a broad based, sometimes-unwieldy coalition. Had Olmert been better able to refine and articulate his message during the 2006 campaign, his governing coalition might well have been stronger and provided him with a firmer base from which to operate. It was a cautionary tale of what happens when politicians are not fully responsive to public opinion.

CHAPTER FIVE

PUBLIC OPINION RESEARCH AND THE FALL OF SLOBODAN MILOŠEVIĆ

W hen the Berlin Wall collapsed in 1989, Yugoslavia was the wealthiest and most open country in the Communist world. Liberated from the Nazis by guerrilla warfare rather than by the Red Army, its citizens enjoyed a level of freedom and prosperity unknown behind the rest of the Iron Curtain. No Communist country in Europe seemed better prepared for a capitalist future.

A decade later, Yugoslavia was a country in ruins. In eight short years, its citizens fought four wars—first in Slovenia, then in Croatia and Bosnia-Herzegovina, and finally in Kosovo. Over a ten-year period, more than two hundred thousand people were killed, and more than three million people were displaced from their homes. The world stood by while concentration camps returned to Europe. In a five-day period in July 1995, Bosnian Serb irregulars systematically slaughtered more than seven thousand

Bosnian Muslim men and boys in Srebrenica—civilians who were supposedly residing in a United Nation's "safe area." More than three million land mines were laid that continue to claim victims to this day. Residents of the former Yugoslavia now earn less than half of what they made in 1989.[1]

The Balkan Wars were the final tragedy of the tragic twentieth century. Many view what happened there as a kind of natural disaster—an "eruption" of ancient ethnic animosities. Historical grievances certainly came into play, but the killing and ethnic cleansing that swept Croatia, Bosnia-Herzegovina, Serbia, and Kosovo were part of a deliberate political strategy. Despite these deeply rooted antagonisms, what is truly difficult to acknowledge about events in Yugoslavia is this: it didn't have to happen.

In the fall of 1992, just six months after Bosnia-Herzegovina declared independence from greater Yugoslavia—a declaration that ignited a war that led to tens of thousands of deaths—I created a highly unorthodox campaign that came within a hairsbreadth of stopping it. My client was one of the most unlikely politicians of the post-Cold War era, a Serbian American entrepreneur from California who had unexpectedly become the prime minister of Serbia that summer, Milan Panic. Our opponent was the man responsible for the greatest bloodletting in Europe since the time of Hitler and Stalin—Serbian President Slobodan Milošević.

The story of the campaign to elect Milan Panic in 1992, the outright electoral theft that stole the presidency from him, and the unaccountable failure of will by the international community that left Milošević in power has never really been told before.

Similarly, Milošević's ultimate removal from power has also been long misunderstood. When Milošević began his fourth war of the decade in early 1998, by butchering ethnic Albanians in the province of Kosovo, the United States finally responded. On March 24, 1999, NATO went to war for the first time in the organization's history. Seventy-eight days and thirty-eight thousand air sorties later, Serb forces withdrew from Kosovo as NATO warplanes pummeled the civilian infrastructure of Belgrade. By October 2000, Milošević was out of power, and U.S. military strategies had found a new mantra: air power works.

Milošević's ouster was one of the most important achievements of the post-Cold War era. Success in Serbia emboldened policymakers in Washington to add "regime change" to the list of American foreign policy goals—and to rely on air power to achieve it, paving the way for the air war in Afghanistan and the war with Iraq.

Unfortunately, these foreign policy lessons from Serbia have been twisted and misunderstood. Contrary to what military hawks would like to believe, air power did not lead to strategic victory in Serbia. While the bombing did force Milošević to pull out of Kosovo, it did not force him out of office. On the contrary, at least in the short-run, NATO's decision to target the civilian infrastructure of Belgrade actually appeared to strengthen Milošević's standing with the Serbian public. That's why just over a year after the last bomb fell, Milošević called a surprise general election. In his mind, the air campaign that the allies were engaged in had *strengthened* his political position. Whereas just a year earlier he had looked vulnerable, now that the bombs were falling, he thought he could win. In short, bombing may have been justifiable as a way to save the lives of Kosovars, but it did not bring about regime change in Serbia.

What ultimately ousted Milošević in 2000 was the daring and unprecedented application of public opinion research by a courageous group of politicians and protesters. The story of the overthrow of Slobodan Milošević is the story of how a Serbian American millionaire, Milan Panic, a small group of students, opposition politicians, and, ultimately, the U.S. government put together an effort to defeat the Serbian strongman at the polls and then thwart his effort to hang on to power. This chapter tells the story of how public opinion polls, not cluster bombs, ultimately ended one of the last great tragedies of the twentieth century. Our work in Serbia illustrates how public opinion research techniques can play a critical role in U.S. foreign policy. The techniques we pioneered in Serbia are, in their own way, every bit as powerful as the new generation of "smart bombs" recently rained down on Iraq. What *should* emerge from our work in Serbia is a new understanding of how public opinion research and strategic messaging can be utilized to achieve vital foreign policy goals without the collateral damage of war.

These techniques don't always work. Our efforts to force strong-man Robert Mugabe to relinquish his stranglehold on Zimbabwe in the winter of 2001 ended in failure. Some regimes govern with no regard for public opinion. Others, such as Hugo Chavez's Venezuela, have developed ingenious ways to steal elections and appropriate electoral legitimacy. Nevertheless, at a time when "regime change" has become an American foreign policy goal across the world, it is critically important to understand that the battle for public opinion is often the most important battle. This chapter describes how to fight such battles—and what it takes to win them.

How could it happen? Fifty years after the Holocaust, how could a country in the middle of Europe engage in ethnic cleansing while the countries of the West stood by and watched? The answer is in the analysis. When the fighting began, some historians and journalists who ignored more than a half-century of peaceful coexistence attributed the conflict to ancient, smoldering enmities between Orthodox Serbs, Catholic Croats, and Muslim Bosnians. The policy that politicians took away from books such as Robert Kaplan's 1993 *Balkan Ghosts* was a convenient one: these people hate each other so much, there's nothing we can do.[2]

If the international community had understood Serbian public opinion better, it would have realized that this simply was not true. In the late fall of 1992, Serbia and the world got a chance to end the violence and save hundreds of thousands of lives thanks to the appearance in Serbia of one of the most unusual politicians of the post-Cold War era.

Milan Panic was born in Belgrade, Serbia's capital city, in 1929. As a young man, Panic made a name for himself as a champion cyclist. Then, in 1956, his family won permission to immigrate to the United States, and Panic left Serbia behind. They settled in California. Three years later, Panic established a pharmaceutical company named ICN Pharmaceuticals. He soon became a millionaire many times over.

In person, Panic was the picture of the CEO—tanned, trim, vigorous, and fit, with just a touch of the charming rogue. He spoke English with a thick Eastern European accent and Serbian, I was told, with a thick American accent. He was an impulsive man, a gambler, and a person fond of the big gesture—or at least that was his act. It would be hard to imagine a more unlikely politician, either in the United States or in Serbia. Yet I soon learned that a politician—or rather, a statesman—was exactly what Panic aspired to be.

My introduction to Panic came through former Indiana Senator Birch Bayh, father of current Indiana Senator Evan Bayh, whom I would help elect to the U.S. Senate in 1998. The Bayhs were a midwestern political dynasty, and they were also old friends of Panic's. Evan's father, Birch, had been in the U.S. Senate from 1963 to 1981. In 1976, Birch had entered the Democratic presidential primaries as the darling of the Left; however, his campaign had stumbled in the face of an unexpected challenge from Georgia Governor Jimmy Carter. One of the few good things that did come of his campaign was a close friendship with Milan Panic. Panic eventually put Bayh on the board of ICN, and the two men often discussed politics and their shared desire to lead their countries. In the summer of 1992, Panic got his chance in just about the most unexpected way possible: he was invited by Slobodan Milošević to return to Yugoslavia and become prime minister.

Why would Milošević turn to a Serbian American millionaire from California to run the country? In a word, desperation. That spring, the United Nations had voted to impose economic sanctions on Yugoslavia. Milošević had headed Serbia's state-run bank in the early 1980s; he knew that sanctions would devastate the Serbian economy and weaken his hold on power. He believed that putting an American face on the Serbian regime could help him lift sanctions and reverse Serbia's growing international isolation. Indeed, as far as Milošević was concerned, lobbying the West to lift sanctions would be Milan Panic's only task as prime minister; Milošević had no intention of actually turning power over to a pharmaceutical kingpin from California.

Though it seemed odd that Panic, a hugely successful entrepreneur who left Serbia as a child, would take on this hopeless job in the service of a brutal regime, in truth, Panic was a deeply patriotic man. He loved Serbia but had little use for the rhetoric that characterized politics in the region, such as the incessant discussion of

Prince Lazar's defeat at the battle of Kosovo in 1389, a traumatic event burned into the Serbia national consciousness. Indeed, during one address to the Serbian parliament, Panic would describe Lazar—Serbia's national hero—as a leader who picked "a stupid fight" and then warned parliament that Serbia wasn't just fighting the Turks, it was fighting the entire world—a stream of comments that drew gasps and boos, and then laughs. Such candor would be a feature of Panic's time as prime minister.[3]

In many ways, Panic had the classic, successful, first-generation immigrant's outlook on his homeland; namely, he viewed it as a rather rinky-dink place, a Mickey Mouse country that was alienating the world and destroying its economy by supporting a brutal war in Bosnia-Herzegovina that was pure folly. Panic hoped to sweep into Yugoslavia, reason with Milošević if he could, and if he couldn't, take control of the situation. Although the country was under sanctions intended to isolate it from the outside world, Panic received permission from the U.S State and Treasury Departments to take office in Belgrade. That would be the only assistance he'd receive from the U.S. government.

Panic and his team, a small group composed primarily of non-Serbian business executives from ICN Pharmaceuticals in Southern California, had parachuted into Belgrade five months earlier, in early July. As tact was not Panic's strong point, the honeymoon with Milošević did not last long. Upon arriving in the capital, the new prime minister announced his intention to take control of the key power ministries—the Defense Ministry, the Interior Ministry, and the Ministry of Foreign Affairs—a move he described as part of his plan to marginalize Milošević. If that didn't give Milošević second thoughts about his choice, Panic's first address to the Yugoslav parliament after he was confirmed as prime minister surely did. His speech called for an immediate cease-fire in the fighting in Bosnia and the withdrawal of the Yugoslav federal army's heavy weaponry from the area. He also promised to recognize the independence not just of Slovenia, Croatia, and Macedonia but also of Kosovo, the historic birthplace of the Serbian nation and the site of increasingly serious guerrilla warfare between the growing Muslim-Albanian population and Serbs. Applause was scant. During the speech, Milošević sat, stone-faced, in the first row.

In the beginning, such remarks by Panic had little weight. While the Yugoslav constitution did technically give the prime minister control of the power ministries, his ability to actually exercise power depended on his ability to command a majority in parliament. That Panic did not have. The ruling Socialist Party, which had approved him as prime minister, was firmly under Milošević's control. Milošević was able to defuse Panic's threat to oust the heads of key ministries and frustrate most of the new prime minister's other initiatives. As time went on, he also increasingly threatened Panic directly, on one occasion sending troops into the prime minister's office, on another, detaining Panic for hours at a military checkpoint.

But Milošević's situation was not all-powerful. For one thing, Yugoslavia's complicated federal system meant that in the early 1990s there were still other bastions of power inside the government. Milošević was the president of Serbia and the leader of the ruling Socialist Party, however, he was not the president of Yugoslavia. That post was held by Dobrica Cosic, a respected Serbian writer who was probably the most popular politician in Serbia at the time. Cosic had begun his career as a friend and ally of Milošević, but once he became president, Cosic quickly grew disillusioned with the Serbian president.

Another even more serious threat to Milošević's hold on power was the growing popular dissatisfaction with his rule. Students, priests, and city-dwellers were fed up with Milošević—and were increasingly willing to defy the regime to make their displeasure known. Indeed, on the day of Panic's arrival in Belgrade, more than a hundred thousand Serbs marched through the streets of the capital to demand that Milošević step down. To effectively challenge Milošević, Panic would have to gain the trust and support of the growing numbers of these dissatisfied citizens. At first, however, the opposition didn't quite know what to make of their new prime minister. It was clear that he was not a stooge of Milošević, but his ability to challenge Milošević remained a question.

The West didn't know what to make of Panic either. From the beginning, Panic made a major effort to line up international support behind his regime. Despite his overtures to the United States and Western governments and his claims that reforming the regime was possible, his efforts were met mostly with polite skepticism. Western governments were more intent on striking a deal with Milošević that would bring the fighting in Bosnia to an end than in challenging him. Ultimately, the supine response of the West would emerge as a major obstacle to our effort to bring about regime change.

Unlike the Western governments, Milošević lost no time trying to figure Panic out; he quickly identified the threat that his new prime minister represented and sought to check his power. Milošević blocked the government from recognizing Croatia as an independent country and vetoed Panic's effort to replace Foreign Minister Vladislav Jovanovic, who was closely associated with the war in Bosnia. Western governments reacted to these actions not by seeking to shore up the prime minister but by writing him off. When British Prime Minister John Major convened a peace conference in London in late August to find a solution to the war in Bosnia, the organizers did not invite Panic, despite the fact that conference protocol clearly called for the prime minister to head the delegation. Only strong lobbying won Panic his rightful seat at the table.

In London, the animosity between Panic and Milošević broke out into the open. The night before the conference, Panic's retinue snuck into the conference room and rearranged the seating assignment so that Milošević was isolated at the end of the table like the defendant at a trial.[4] When the conference got underway the next morning, and Milošević rose to rebut a point made by British Prime Minister John Major, Panic brusquely ordered him to sit down, saying, "You don't speak for this delegation." Milošević stared back at Panic—and then sat down.

Back in Yugoslavia, the incident made headlines. Milošević was reportedly enraged and humiliated. He lost no time in striking back. In early September, the ruling Socialist Party unexpectedly called for a vote of no confidence against Panic, charging that the prime minister had betrayed Serbia by unofficially agreeing to the outlines of a peace plan for the region. The proposed no confidence vote, which would force Panic to step down if it were to pass, threw parliament into an uproar. After hours of negotiation—and a forceful last minute intervention by Yugoslav President Dobrica Cosic, a one-time Milošević friend who had soured on the Serbian president—the motion was withdrawn. Panic was still alive, and he was determined to strike back before Milošević could hit him again, but to do that, he would need help. Though Panic was an accomplished businessman, he had no experience in politics. Nor did he understand how politics in a machine system worked. To have any chance of enacting his agenda, Panic would need a skilled adviser at his side. For that, Panic's old friend Birch Bayh came to me.

B irch Bayh's initial request seemed like a modest one. In November 1992, he asked me if I would go to Belgrade and conduct a two- or three-day assessment of Panic's operations and his political standing. Although I had no special insights into the situation in Yugoslavia, I agreed. It had been a grueling election season. While Evan Bayh had been reelected Governor of Indiana, our effort to oust New York Republican Senator Al D'Amato had fallen two percentage points short. It had been a very frustrating defeat. Our candidate, New York Attorney General Bob Abrams, had made several costly mistakes during the final stages of the campaign that probably cost him the election. I was upset at the loss and wanted a break. So instead of going on vacation, I agreed to go to Serbia. There I found a challenge like nothing that I had ever encountered before.

My first challenge was simply to get there. Earlier that summer, the UN Security Council had imposed a ban on air travel into Serbia, so I couldn't just fly to Belgrade. First, I had to receive a license from the Treasury Department allowing me to go to Serbia at all, then I had to fly to Budapest, the closest major airport. Panic's foreign policy adviser, Jack Scanlon, former U.S. ambassador to Yugoslavia, kindly offered to pick me up there so that he could fill me in on the situation during the six-hour drive to Belgrade.

True to his word, when I cleared customs, Scanlon was there to meet me. As we climbed into his BMW, Scanlon introduced me to my first key contact—his chauffeur, a powerfully built forty-year-old with a big smile named "Branko." Branko had been Scanlon's driver since his time as ambassador to Serbia during the Reagan administration. As we roared off toward the Serbian border, Scanlon dove right in to the particulars of life in Serbia. Branko should be my go-to guy, Scanlon told me. He could get me whatever—and he meant whatever—I needed. I should also use Branko to change money, since such operations were best done on the black market,[5] and I should also treat him as a valuable source of information. The moneychangers Branko dealt with were exquisitely attuned to shifts in the public mood and were wonderful sources of information. I was beginning to understand that work in Serbia would be very different from my more recent campaigns.

When we arrived in Belgrade later that day, the outskirts of the city looked like some blighted section of Queens—a mishmash of crumbling concrete apartment blocks and stained and discolored buildings. The center of town was only slightly more attractive. Scanlon, however, lived in a luxury apartment complex downtown across the street from the Intercontinental and the ultramodern Hyatt hotel in the part of the city known as New Belgrade where I would be staying. Panic had an apartment in the complex too. Scanlon invited me to stay for dinner, and I gladly accepted. As we sat down to a dinner prepared by his housekeeper, he casually observed that we should watch what we say because his apartment and indeed the entire building were undoubtedly bugged.

After a very nice dinner and a somewhat cautious conversation, Branko drove me over to the Hyatt. I walked into a lobby abuzz with activity—beefy Serbian men smoking cigarettes, drinking rakija (a kind of plum brandy) or beer, and noshing on pieces of sickly looking boiled meat while their sullen blond girlfriends sat beside them looking bored. Beside the check-in desk, I saw a sign, written politely in both Serbian and English, that made the following request of guests and visitors:

Please Check All Weapons at the Front Desk

Evidently, there had recently been a gangland slaying in room 389. I was in room 391. It was a moment that brought back memories of my old adversaries in the Italian social clubs that lined Pleasant Avenue. Of course, I had wisely avoided entering those clubs. Here I would be entering the heart of Serbia's Mafia scene.

The next morning, I walked over to Government House, a crumbling Soviet-style office building that housed the prime minister's offices. I introduced myself at the security desk and was promptly taken upstairs to a dingy little office, where I found a stack of polls conducted by local media outlets. I didn't have much time to look at them, however. Soon after I sat down, one of Panic's aides appeared and informed me that it was time for the cabinet meeting. I was a bit confused by this statement, but I left with the aide. Sure enough, I was soon ushered into a large conference room, where I met Milan Panic for the first time—and the entire Serbian cabinet, their aides, and a scattering of Americans for good measure. It was like they were all waiting for me. Panic introduced me to the group and gestured for me to sit down and jump right in (when the conversation shifted to English) if I had anything to add.

It was clear that Panic had only a vague idea of what a political consultant was supposed to do. Here I was in a foreign country whose culture and politics were still new to me and I was being invited to sit down and start governing. To Panic, I was just another member of his team. Needless to say, I saw my role rather differently. I suggested that perhaps I should not actually take part in cabinet meetings. Panic genially acceded to this suggestion, and I was shown out of the meeting, ending my career as a Serbian cabinet member.

What I really wanted to do was assess Panic's political standing. The results I had seen from a handful of media outlets suggested that Milošević was a deeply unpopular figure, with only a 20 percent to 25 percent approval rating. In the United States, a politician with that kind of approval rating was a politician who needed to start looking for another line of work. Think Nixon during the height of the Watergate scandal. Yet almost everyone I spoke to, even Panic's Serbian aides, seemed to believe that Milošević would never lose an election. Were the polls wrong? I decided to conduct a poll of my own.

To assist me with this task, I turned to Panic's press secretary, an American named David Calef. Calef suggested that we ask the University of Belgrade to do the poll and offered to go visit its director with me. I agreed, curious to meet my local partner.

The University of Belgrade was located in downtown Belgrade. Its campus was a forlorn jumble of concrete buildings that would not have looked out of place in the South Bronx during the 1970s. We arrived and found our way to a crumbling building, with no lights in the hallways and few signs of activity. At first, I thought that the building was abandoned and that we had been given the wrong address.

The head of University of Belgrade's polling group was a beaten-down woman named Liliana Bacovic. Bacovic's attitude reflected her surroundings: she was a very gloomy woman with the energy of a 12-watt light bulb. She could indeed conduct a poll for us, but it would be a difficult and time-consuming process. "How long a process are we talking about?" I inquired, with considerable anxiety, for my desire to get away from New York was rapidly yielding to a desire to get out of Belgrade.

Lilliana said it would take about ten days just to field the poll and get the data back. Then perhaps a few more days to collate the data, so in total about two weeks. While I gave David a look of despair, David just smiled and pulled a large roll of Swiss francs out of his pocket. It seemed Calef always carried a thick roll of bills in his pocket for just this type of contingency. A few thousand francs exchanged hands; an agreement was reached that we would have tabulated results in a week. Calef's roll of francs—"Swissies," he called them—would become a familiar friend during my time in Serbia, and as Panic's efforts to oust Milošević intensified, the roll of "Swissies" would soon give way to a plastic shopping bag.

Even an abundance of Western currency could not, however, get me out of Belgrade quickly. After all, there was no shuttle from Belgrade to the United States—or to anyplace else, for that matter. I would have to wait for the poll results, which meant that I would be spending Thanksgiving in Yugoslavia. My drop-in assessment of the situation in Belgrade was turning into a more extended adventure.

While I waited for my poll results, I spent my time boning up on Serbia politics. At the time, the only polls I had to look at were polls done by the few local media outlets not under the sway of the ruling Socialist Party. Although their methods were uncertain, the picture they painted was hopeful. Milošević was definitely vulnerable, and at least one politician could clearly beat him—Dobrica Cosic, the president of Yugoslavia.

Cosic was an interesting figure. As a writer, Cosic had played a major role in flaming Serbian nationalism and resentments during the 1980s. However, by 1992, Cosic had soured on his old comrade. Cosic welcomed Panic's arrival and made it clear that he was eager to see Milošević ousted. There was just one problem. As much as he disliked Milošević, Cosic wouldn't enter the race. His excuses were manifold: at the age of seventy, he was too old to mount a vigorous campaign; it was time to pass the baton to the next generation, and so on. In truth, he simply didn't have the hunger or the fearlessness necessary to take on Milošević, leaving only one viable candidate—Milan Panic himself.

I believed strongly that Panic should get into the race. The polls all seemed encouraging. They showed very clearly that Milošević had lost the confidence of the Serbian people, that the electorate wanted change, and that despite the fact that he had returned to Yugoslavia only five months earlier, Panic was considerably more popular than the incumbent. Indeed, Panic was probably even more popular than the polls showed, given the fact that the citizens of quasi-authoritarian states typically overstate their levels of support for the ruling regime.

A few days before Thanksgiving, I seized the chance to talk with Panic—and escape from Belgrade. Panic, who spent much of his time as prime minister traveling to Western capitals seeking to persuade European governments to support him as an alternative to Milošević, was about to go to Berlin and then to Geneva for talks with the UN High Commissioner for Refugees. I prevailed on Calef to let me join them. En route to the meeting, I told Calef that I thought Panic should get into the race. Calef agreed, and we decided to put the case for running to Panic at the earliest possible opportunity.

Our opportunity came on the flight back to Belgrade from Geneva. Panic was in good spirits. It was Thanksgiving, and Panic had somehow managed to procure and prepare a turkey. Brushing past a throng of Serbian and foreign reporters, we sat down Panic in the private cabin at the front of Panic's 737. I dove right in. "Look, Milan," I told him, "you're at 70 percent, he's at 25 percent. If you get into this race, I know you are going to win."

Panic looked skeptical.

"You don't understand the situation," he replied. "I'm a Serb; I've been here; and it's not that easy."

Panic's Serbian advisers were telling him that Milošević would never lose—or admit to losing—an election. He would play dirty, rig the results, and generally do whatever it took to stay in power. Running against him would be an act of futility. The campaign would invariably fail; Panic's political career would be destroyed—or worse, he might be killed. The pessimists argued that even entering the race would cause Panic to lose the credibility that he had slowly gained as a result of simply enduring in the face of opposition from Milošević. In short, a race would be an act of folly that would lead to disaster.

I didn't buy it. While it was true that I was new to Serbia, by the early 1990s, I had played a central role in campaigns in the United States, Israel, and more than a dozen other countries. I knew what a 20 percent to 25 percent level of support for a political figure meant—he was finished. Moreover, while Serbia was an authoritarian regime, it was not a total dictatorship. There would be elections. The world would be watching. With only bloodshed and economic collapse on the horizon, there was no way Milošević could rebound.

"All the polls show Serbs have had it with Milošević," I told him. "The economy is in shambles; they don't want war; your image is great."

"Well, that's because the state media has been promoting me," Panic replied.

If the official press turned negative, it would undoubtedly hurt, but I figured Panic had the money to defend himself. "Moreover," I continued, "Look, you're well ahead. You need to run. And if you run, you can win."

Panic asked what his chances of winning were. I predicted that if he got into the race, he had better than a one-in-two chance of winning. "You're at 70 percent favorability," I told him again. "He's at 25 percent. I know you are going to win."

It was clear that Panic wasn't taking our idea all that seriously, but I could also see that he was at least thinking about it. "Look," he said. "I don't have an organization. I don't have a campaign. I don't have a campaign manager." Even getting on the ballot would be a challenge: Serbian law required candidates for the presidency to collect ten thousand nominating signatures, and the registration deadline was only four days away.

Calef jumped in. "If we got you on the ballot, would you do it?" he asked. Panic responded casually. "Yes, sure." Then he got up to go carve a turkey in the main cabin. Although Panic himself didn't know it, that offhand reply meant the race was on.

We touched down in Belgrade late Thursday night. The next morning, I got up and went down to the hotel lobby for breakfast. When the elevator doors to the lobby opened, I stepped out not into the normal gathering of mobsters and mob-wannabes but into a sea of young volunteers. At the center of the action was Calef (and his wad of Swiss francs). The campaign to defeat Slobodan Milošević had begun.

Calef may not have known much about Serbia, but as a former organizer for the Democratic Party in California, he certainly knew how to run a campaign—and how to put money to good use. He had already commandeered two large conference rooms and turned the business center into our campaign headquarters. Flip charts adorned the walls; Calef was on the phone, calling activists and opposition leaders and organizing a cadre of students to serve as petition coordinators. Later that morning, Liliana Bacovic delivered our first poll, which confirmed what we had told Panic the night before: Milošević was extremely vulnerable. That evening, hundreds of volunteers, mainly students, packed into the hotel ballroom. Calef and his ubiquitous bankroll were everywhere. Calef's rag-tag army was ready to roll. By Monday morning, volunteers had collected nearly fifty thousand signatures in support of Panic's candidacy. I was ecstatic. Come December 20, I was convinced that Europe's last Communist government was going to fall.

Panic himself seemed a bit surprised by this turn of events. I don't think that he had ever really made a conscious decision to get into the race. I doubt he had anticipated how easily Calef and his cadre of volunteers would surmount the ten thousand-signature ballot hurdle. Nevertheless, Panic was true to his word. Now that he was on the ballot, he was ready to throw himself into the race. What was needed now was a campaign strategy. That was my brief.

To win the election, I developed a strategy that combined the best aspects of Ronald Reagan's 1980 campaign and Bill Clinton's 1992 campaign—both campaigns were against incumbent presidents with troubled economies. Our message to the voters was simple: are you better off now than you were four years ago? The answer to this question could only be no, and our polls showed that most voters blamed Milošević for this situation. I believed Panic had an excellent shot at winning the election. By 1992, Mark and I had worked campaigns in eighteen countries. We understood the pitfalls of transplanting American assumptions to foreign cultures, the tricks politicians can play, the shortfalls of local polls, and the unexpected challenges that can arise working in different cultures. In short, I thought I was ready for Slobodan Milošević. Sadly, I was wrong.

The first challenge came in parliament in the form of another motion of "no confidence" in the government. This time the measure passed easily in the lower chamber. In the upper chamber, it failed by a single vote, eighteen to seventeen.

Milošević's next attack was more devious. That fall, the Socialist Party slipped legislative language into a bill that set residency requirements for the presidency that Panic did not meet. The legislation passed, its "poison pill" unnoticed, until the Socialists brought a motion before the Serbian electoral commission that asked the commission to declare Panic ineligible, which it obligingly did. This was a potential catastrophe, but fortunately Milošević's control over the apparatus of government was not complete. We appealed the ruling to the Serbian Supreme Court, and on December 5, it handed down a ruling in Panic's favor. The legal obstacles to the race had been cleared. Unfortunately, the political obstacles were not so easily overcome.

The first big problem was the media. Milošević had complete control of the television airwaves, and every night, Radio Television Belgrade attacked Panic with scathing remarks that questioned his loyalty and portrayed him as a tool of the West, a mole sent in to infiltrate and destroy Serbia. The only way we could counter this message was through paid advertising. However, Milošević-controlled outlets refused to even air our commercials until Panic was officially certified as a candidate for the presidency. That meant that our first commercial did not air until December 10. With the active campaign scheduled to end on December 17 and the election itself scheduled for December 20, we had just seven days to make our case to the voters—and overcome the nightly barrage of pro-Milošević propaganda.

Still, I thought it was enough—if we could get the right message on the air. To help with that, Calef turned to a friend and former chairman of the California Democratic Party-turned-media-consultant named Bill Press. (Press would later represent "the Left" on CNN's *Crossfire*.)

Press arrived at campaign headquarters one morning in mid-December after a harrowing overnight bus ride from Bucharest. He looked a bit stunned and had more than a few chicken feathers on his clothes (he had evidently been one of the few people on the bus from Bucharest to Belgrade who wasn't in the poultry business). Nevertheless, he was ready to work and there was no time to lose. Calef had located a studio for us to work in at the end of a dank alleyway on the outskirts of the city; we set off for it at once. Years of canvassing in East Harlem and Hunts Point had taught me that this was not the kind of alley one wanted to walk down, but I did anyway, finding a ramshackle building that contained a single dilapidated soundstage. The soundstage was run by a five-foot-tall Serb named Alek who seemed capable of breathing only through the end of a Camel cigarette. It was clear we made Alek very nervous, at least at first. Later, he somehow became my Branko, always ready to drive me around town in his twenty-year-old VW bug.

Despite the facility's deplorable condition, Alek's uncertain help, and Bill's state of mind, within a few days we had come up with a brilliant ad as good as any American attack ad. The spot began with a picture of Milošević and asked where he had really improved the quality of life and the economy. The commercial then panned to Serbs with looks of quiet desperation queuing up in bread and gas lines. Finally, the ad cut to footage of a vigorous and confident Panic promising voters a healthy economy, peace, and a more democratic society. It was a bravura production given the time and the conditions under which we had labored. I saw it as our knockout punch.

It never ran. Radio Television Belgrade banned it as too inflammatory. I was first surprised, then furious. I told our ad representatives to do whatever they had to do to get it on the air, even if it meant taking out some of the references of Milošević. They ended up taking out every reference to Milošević, which left only pictures of the bread lines. Again, the state-controlled media turned it down.

One official was even candid enough to tell us that if they ran it, Milošević would lose. What was really remarkable was that some of Panic's own advisers were also relieved by our inability to get negative ads on the air. "You have to understand," our (Serbian) finance chairman finally explained to me, "if we lose, you will not be here on December 21 to face the consequences, but we will." Ultimately, the only ad we were able to put on air was a simple biographical ad.

Other signs of foul play abounded. One evening during dinner at a posh Belgrade restaurant, a journalist with Radio Television Belgrade came by and complimented us on one of our new radio spots—a spot that we had come up with just that morning in our headquarters at the Hyatt business center and had not even told Panic about yet. It was a clear sign that our offices were bugged. Or perhaps the raft of suspiciously well-qualified campaign "volunteers" who had recently appeared throughout the campaign—volunteers who spoke English, had extensive professional degrees, and seemed eager to spend as much time as they could working on the campaign—had something to do with the revelations.

Whatever the source of the leak, it was clear that the Panic campaign would keep no secrets. Given this fact, I decided to handle poll results in an unusual fashion. Instead of distributing numbers to only a handful of senior campaign staff, I publicized them as widely as possible. My reasons for doing this were twofold. First, I wanted to underscore that we could win. Every poll showed that a plurality of Serbs preferred Panic to Milošević. Yet when our pollsters asked people who they thought would be the next president of Serbia, a four-to-one majority thought the incumbent would prevail. It was critically important that our supporters understand that they could win and that it was worth voting.

The second reason I publicized our poll results was to prevent— or at least reduce—election fraud. None of us expected Milošević to play fair. We knew he would cheat. As a result, we didn't expect to win in the first round: we knew the Socialists would most likely steal enough voters to prevent that. The key was to prevent him from stealing the election outright and to force him into a runoff. Making public the polls that showed Milošević was trailing was one way to keep him honest—or at least, not too dishonest. If he announced a big victory, the world would know it was a fraud.

What Panic really needed was international support. A majority of the Serbian public was clearly dismayed at where Milošević was leading the country, believing that war would be economically ruinous and lead to Serbia's complete isolation from Europe. Panic believed that if the United States or Europe presented a clear "out" to the Serbian public—elect Panic, carry out his promises, and we will welcome you back into the fold of civilized countries, he would gain substantially in the eyes of the voters. To win, Panic needed election observers in large number and constant international attention and support.

Yet all our attempts to secure international support were met with polite indifference. I remember calling Cyrus Vance, the former secretary of state and current UN mediator who had been given the difficult mission of coming up with a peace plan acceptable to all the parties to the conflict. Panic can win, I told him; the conflict would be over. But he can only win the election and claim the presidency with the support of the international security. Vance was polite, even sympathetic. He told me he thought it was great that I was involved and wished me luck—but it was clear that he would not be able to provide any assistance. Neither would the outgoing Bush administration, despite the fact that then-Secretary of State Lawrence Eagleburger was a Serbia expert and himself a former Ambassador to Yugoslavia. Policymakers seemed to view Milošević as a necessary evil, the only person who could make peace in the region, rather than as the instigator of violence that he actually was.

At first, the U.S. failure to offer support puzzled Panic. As the threats against him escalated, however, puzzlement turned to anger. "Where is the United States?" Panic would rage, pacing the floor of his office in the leafy, suburban villa he had made his home. "I can't win without the United States, and the United States is absent! Milošević is a killer, and he's going to get away with it. They will regret the day they didn't help me."

Though incoming President Clinton had yet to take office, he gave no signal in the months leading up to his inauguration that he would shift his policy from that of George Bush. "I have given money to Clinton; he's very nice to me; but where is the help?" Panic would fume. "We can't win it with out them. Even if we run a great campaign, Milošević will steal it." To be sure, Clinton would not even take office until January. Nonetheless, as president-elect, he had the ability to help focus the world's attention on the Serbian election, and Panic was profoundly frustrated that he chose not to.

Fortunately, the United States and the United Nations weren't our only defenses against fraud. Our other safeguard was a then-little known international organization called the Conference for Security and Cooperation in Europe (CSCE), an organization that would later take on a prominent role in the region as the Organization for Security and Cooperation in Europe (OSCE). The CSCE had been formed to monitor and enforce the Helsinki Final Act, which provided for the territorial integrity of all of the countries in Europe as well as human rights within those countries. To the Soviet Union, the Helsinki Final Act brought international recognition to Cold War states such as East Germany. To the United States, the treaty obligated the Soviet bloc to adhere to Western standards of human rights. Of course, in practice it did no such thing. For the first decade and a half of its existence, it was a fairly toothless organization—more a forum for meetings than a player in international affairs. Then in 1991 the Soviet Union collapsed, and CSCE had the opportunity to actually do something. One of the tasks it took on was election monitoring.

We were particularly concerned about what would happen after the ballots were collected. Once the votes were cast, they would be shipped to the central election bureau where they would be tabulated. Trouble was the election commission was firmly under Milošević's control. To ensure that no acts of fraud were committed, Serb officials would have to be watched closely. All that would stand before us and outright theft would be CSCE election monitors.

A few weeks before the election we learned that the CSCE would be sending us just eighty observers to watch thousands of polling stations. It was an absurdly inadequate number of observers. The campaign protested, both publicly and privately, this seemingly inexplicable decision—to no avail.

Despite setbacks of this sort and Milošević's ongoing skullduggery, we justifiably hoped for victory. We had the momentum, and during the final weeks of the campaign, Panic toured the industrial heart of the country, receiving a rock star's welcome everywhere he went. One such visit was to Nis, a tough industrial city in southern Serbia that had long been a center of pro-Milošević support. It was a trip I had been concerned about for quite some time, since we didn't have organizers on the ground there, and we hadn't been able to send an advance team in. To make matters worse, the local officials were all in Milošević's pocket. It was just the sort of place where a group of Milošević's thugs might embarrass Panic—or worse.

Our local supporters assured us that we could expect a few thousand people to turn up for the speech. When we pulled up at the Ambassador Majestic Hotel at about 3:00 PM, we found a crowd of more than twenty thousand people there to greet us. It was clear that the crowd was giddy with delight at Panic's presence. When it came time to leave, they rocked our cars and cheered wildly. It took nearly a half-hour to get out of the central square.

In contrast, Milošević scarcely campaigned at all. He ran no advertisements, appearing instead on the news. Despite flattering news reports, Milošević looked like the out-of-touch thug that he was. Even indoors, he often wore a bulky raincoat over what was obviously a bulletproof vest. In the last days leading up to the election, Milošević did hold a few, Soviet-style rallies, complete with rows of grimly applauding *apparatchiks,* but for the most part, there was no sense that he was running a serious campaign. To the extent that the Milošević campaign had a theme, it was that Panic was a spy and a traitor. However, polls showed that most of the public simply didn't believe this allegation.

We, however, were headed toward a picture-perfect finish. All the polls showed that the race had tightened to the point where it was virtually neck and neck. Our polls showed Panic with a slight edge; others had the race tied. I made sure that these results were widely circulated both domestically and internationally. It was clear the race was headed to a runoff, and I wanted to make certain that Milošević didn't attempt to claim a victory. Once we got into a runoff, I felt confident that Panic would pick up the margin of victory he would need.

Panic was less certain. Whenever I would bring him my poll results, he would often say to me, "But you don't see what they say about me on television every night. You don't understand." But properly conducted polls don't lie: both campaign and media polls clearly predicted at best a win and at worse a tie, which would lead to a runoff.

Sunday, December 20, was Election Day. A heavy fog blanketed much of the country. In Belgrade, we awoke to find that Panic posters had been defaced across the capital with graffiti saying, "U.S.A." Turnout was very heavy in most parts of the country. The voting had its glitches—in many places young Panic supporters arrived at the polls to find that their names had been stricken from the voting lists—but there were no reports of obvious fraud or intimidation at the polling stations themselves.

To keep the process honest, we also arranged exit polls at sixty polling stations across the country to monitor the results. We scheduled two waves of exit polls, one for the morning, one for the mid-afternoon. The morning wave of exit polls, which came in mainly from the provinces, reflected the sentiments we had seen in all of the polls leading up to the election: it showed a virtual dead heat. The afternoon wave showed something very different, particularly in Belgrade: virtually every vote cast was for Panic. It was clear Panic was experiencing a huge surge. Given his popularity with younger voters, who tended to cast their ballots at the end of the day, I knew that the prime minister's lead could only widen in the late afternoon. Using the most conservative possible assumptions that gave every statistical break to Milošević, we had a tie. More likely, Panic would win, though probably not outright, which meant that there would be a runoff election in two weeks time.

Of course, the state media told a different story. Radio Television Belgrade selected precincts where Milošević had done well and reported them one after the other to create the impression of a strong showing for the president. It was an obvious attempt to conceal the overwhelming truth of Panic's intimidating performance. To counter this misinformation, at the end of the day, we released our results at a news conference and predicted a runoff.

The next day the Milošević campaign announced that it had won a smashing victory. Later in the day, election commission officials announced that after counting 19 percent of the votes, Milošević had nearly 57 percent of the vote to Panic's 33 percent. That is when we realized that Milošević was going to attempt to steal the election outright. If Milošević had his way, there would be no runoff. Panic hastily released a statement that pronounced these results were invalid "because of fraud, theft, and cheating in the counting of ballots."

As best I could tell, the votes were probably being stolen at either the regional counting center or perhaps even at the national vote-counting center in Belgrade. Either way, it was critically important that no more votes be stolen. At first, the CSCE seemed ready to recognize the problem. The day after the election, CSCE observers issued a statement describing the election as "riddled with flaws and irregularities." They also noted at least 5 percent of eligible voters, mostly young people, had been barred from voting. Such statements were helpful, but inadequate. What we really needed was for the CSCE to step in and supervise the vote-counting operations. Instead, something absolutely remarkable—something so reckless as to defy comprehension—happened. On December 23, most of the European observers in the CSCE mission went home for Christmas. The result: no one was watching the election workers who were actually counting the ballots—and stealing the election.

When the central election commission announced that Milošević had carried every precinct in Belgrade, I knew the results were completely rigged. On Thursday, December 24, while the CSCE observers were at home enjoying Christmas with their families, the central election commission pronounced Slobodan Milošević the winner outright. There would be no runoff election. In retrospect, I should have forcefully urged Panic and his advisers to lead his supporters into the street to protest the fraud. However, when it came to direct action, I did not have the necessary experience, and honestly, I was also a bit scared. Panic himself was disinclined to do it; none of his advisers would have advocated this course of action either. So I backed down. Before leaving Belgrade, I appeared on both CNN and *Good Morning America* to talk about the theft I believed was occurring, and back in the United States I continued those efforts. But no one was interested. Western policymakers still saw Milošević as someone they could do business with. Even Panic's advisers were strangely quiescent in reacting to the theft of the election. Panic's childhood friend and former Yugoslav Politburo member Tosha Olic warned me, in a fairly typical reaction, not to be "too strident" in protesting the election results. It was clear to me that back in Belgrade, Panic was surrounded by men giving him bad advice. If he had rallied his supporters and taken to the streets, I believe he could have taken control of the capital. Instead, at the advice of his Serbian advisers, Panic was silent. On December 29, Communists and ultranationalists teamed up and voted Panic out of his position as prime minister.

By creating a sense that his reelection was inevitable and by playing on the impression that Panic was merely an eccentric American businessman, Milošević had generated an environment in which the world saw his continuation in power as a given. He then gambled that if he stole the election in a sophisticated way, neither the European Union nor the United States would have the guts to take him on. He bet right. The threat to Milošević's power was over. Another seven years of carnage was about to begin.

After Panic's ouster, I tried to persuade him to stay involved with the process as the public face of the opposition—with little success. Panic's stint as prime minister of Serbia had soured some important stockholders at ICN; retaining control of his company was now his most pressing challenge. However, Panic did agree to sponsor a series of ongoing polls in Serbia. He also worked behind the scenes to create a unified opposition to Milošević.

This was not glamorous work. All of the leading opposition leaders had serious flaws. Over the course of the next couple of years, I would get a chance to know many of them personally. Vuk Draskovic, the head of the ultranationalist Serbian Renewal Movement and novelist, was a moody and messianic figure who seemed to talk only of his hatred for Milošević. Draskovic had been imprisoned and tortured for opposing Milošević, so I could understand his feelings. What was harder to figure out was why he joined the Milošević government a year later.

The most promising of the lot was the head of the Democratic Party, Zoran Djindjic. Djindjic's background was unusual. He had been trained in Germany as an academic philosopher by the eminent neo-Marxist thinker Jurgen Habermas. When he returned to

Serbia, Djindjic became an active opponent of the regime, which eventually resulted in his arrest. After that, Djindjic spent much of the 1980s in Europe, participating in pan-European socialist politics and, more surprisingly, building a thriving business importing machine tools from East Germany, a business that many believe led to close ties with Yugoslavia's ubiquitous criminal syndicates. He was clearly very bright and, with his steel gray buzz cut and steely eyes, projected a tough, analytical clarity. His critics in Serbia, of whom there were many, accused him of arrogance and close ties to the Mob. I saw him as a can-do guy. I much preferred him to Draskovic's ranting and moodiness.

Djindjic had also learned from our mistakes in 1992, and he understood the need to build a truly unified opposition. With funding from Panic, Djindjic had put together a fledgling coalition of democratic parties called the Alliance for Democratic Change. During the December 1996 municipal elections and in the months that followed, he had turned out tens of thousands of Serbs to protest against the regime, at one point staging eighty consecutive days of rallies against Milošević in Belgrade. Despite these efforts, Djindjic lacked the support to topple Milošević on his own. While he could be a force if he connected with the other opposition leaders, unfortunately, effecting such an alliance was no easy matter. Draskovic and Djindjic disliked each other at least as much as they disliked Milošević. In July 1997, Milošević was elected president of Yugoslavia after serving the maximum two terms as president of Serbia.

The Serbian opposition wasn't my only connection to Serbia. During this period, I also enjoyed the occasional opportunity to speak to President Clinton directly about the situation in Serbia. Mark and I had started polling for President Clinton in 1995 at virtually the same time Richard Holbrooke, the president's special envoy to the Balkans, was attempting to negotiate what came to be known as the Dayton Peace Accords with Milošević and other Balkan leaders. I told Clinton that Milošević was not popular with the Serbia electorate and that a move against him would probably not inflame nationalist sentiment. I also told him that I believed Milošević himself would respond only when his own personal safety was threatened. Clinton never really engaged me in conversation on the subject—I was his political consultant, not a foreign policy adviser—but when NATO war planes finally targeted Milošević's personal compound four years later, I was gratified by the possibility that my insights into Milošević's motivations might have played a small part in President Clinton's decision.

For the most part, however, my involvement in Serbia during those years was limited to keeping my finger on the mood of the Serbian public via the occasional poll. That too offered little hope. During 1997, our polling in Serbia showed *rising* nationalist sentiment. I was so alarmed by this trend that I arranged to have breakfast with Richard Holbrooke to show him the results. In 1995, Holbrooke had negotiated the Dayton Accords that brought the fighting in Bosnia-Herzegovina to an end. Now Holbrooke feared that rising nationalism would unleash another tide of violence.

These fears proved to be well based. In early 1998, clashes between ethnic Serbs and Albanians in Kosovo began to intensify. The Kosovo Liberation Army (KLA), a separatist guerrilla movement dedicated to liberating Kosovo from Serbian control, was stepping up its operations, and Serbian security forces were responding brutally. The death toll on both sides was rising. Serbian grievances about mistreatment at the hands of ethnic Albanians in Kosovo had fueled Slobodan Milošević's rise to power in the late 1980s. It was clear to me that the situation in Kosovo was a powder keg waiting to explode.

Throughout 1998, Serbian security forces fought a low-intensity war against the KLA. Serbia atrocities threatened to draw NATO into the fray. That fall, NATO air strikes were averted at the last minute when Richard Holbrooke negotiated a deal that sent two thousand unarmed observers into the province. Unfortunately, that did not stop the fighting. In January 1999, the bodies of almost forty ethnic Albanians were found at a scene of recent fighting in southern Kosovo. They appeared to have been executed en masse. Once again, NATO threatened Milošević with air strikes if the killing did not stop.

Western leaders were confident that Milošević would back down, as he always had when faced with the prospect of military attack. Instead, he dug in his heels. On March 24, NATO launched a series of attacks against military and industrial targets in Serbia and

Kosovo. But instead of folding, the Serbian Army swung into action, driving tens of thousands of ethnic Albanians out of their homes and villages, summarily executing some and displacing many others.

The military campaign began tentatively. News reports said that many officials in Washington believed that as soon as the first bomb fell, Milošević would pull back. I thought differently. After one of my meetings with President Clinton, I repeated my earlier advice: Slobodan Milošević, I told him, will give way only when he personally is at risk. If he moves, he will move the minimum amount necessary to remove the pain. I remember making the same point once to Sandy Berger, Clinton's national security adviser. "Sandy, I know this man," I told Berger. "I've been working against him since the early 1990s, and I know that the *only* thing that is going to make him react the way we want him to react is when he personally feels unsafe."

Berger gave me a long look. Finally, he said, "Doug, I'm not going to talk about military strategy, but let me ask you one thing: have you ever tried to decide what targets to hit and how to conduct a war with sixteen other nations?"

Point taken. Nevertheless, I was delighted when two weeks later U.S. warplanes took out first Radio Television Belgrade, the prop of the Milošević regime, and then Milošević's family house. NATO was finally making it clear that Milošević himself was at risk. I was not surprised when Milošević offered to withdraw Serbian forces from Kosovo soon thereafter.

Serb atrocities in Kosovo achieved what the war in Bosnia had not. In 1995, many Western policymakers had seen Milošević as a man they could do business with—odious, to be sure, but a figure whose cooperation was essential to peace in the region. When the West was forced to go to war against Milošević to stop atrocities in Kosovo, Western policymakers finally accepted the fact that they were dealing with a psychopath and a monster. After nearly a decade of slaughter, Western politicians figured out at last that removing Milošević from power was the key to peace in the Balkans. The question now was how to get him out.

A few weeks after the Serb withdrawal from Kosovo, I got a call from Milan Panic. "It's time to finish the job," he told me. "What do you mean?" I asked. "Well," he responded, "I went to the National Democratic Institute in Washington and told them to hire you to oust Milošević."

It turned out that the U.S. Agency for International Development (USAID) had been entrusted with over $40 million to support nongovernmental groups trying to oust Milošević. It was just a pittance compared to the roughly $3 billion that was spent in the air war in Kosovo; however, it was this $40 million-plus, not the smart bombs rained down on Belgrade from above, that ultimately brought down Milošević. One-tenth that amount eight years earlier could well have saved tens of thousands of lives.

USAID divided the money into three big pieces. A significant portion of funds went to Otpor ("Resistance"), a student opposition group that had originally been formed to protest a purge of faculty members at the University of Belgrade in 1998 but had soon become the fastest growing opposition group in the country. The National Republican Institute received money to train Otpor activists and others in, of all things, the principles of effective nonviolent resistance. The National Democratic Institute was in charge of public opinion polling. That work fell primarily to me.

In late September 1999, Penn, Schoen & Berland fielded what was in effect the benchmark poll in the campaign to oust Slobodan Milošević. It was an important poll. The two leading opposition politicians, Vuk Draskovic and Zoran Djindjic, had very different ideas of how to oust Milošević. Djindjic wanted a street strategy. He believed that Milošević would always find a way to steal an election and that only a well-organized, grassroots movement could topple him. Draskovic insisted on an electoral campaign. He believed that the Serbian people would never support a putsch, even against a leader as disliked as Milošević. Our results would play a critical role in determining which path the opposition opted for.

Our numbers backed Draskovic's argument. Serbian voters were extremely angry with Milošević. (His unfavorability ratings were at 70 percent.) They blamed him for the country's failing economy and two-thirds wanted him to resign before the end of his term. However, if Milošević refused to step down, as he certainly would, Serbs did not want the opposition to attempt a mass uprising of the sort that had toppled Communist governments at the end of the Cold

War. They wanted an election. But here too our poll numbers contained an unexpected surprise: only 20 percent of Serbs believed their country was a democracy. So in a sense, the vast majority of citizens agreed with Djindjic that power would have to be pried from Milošević's grasp.

That was the good news. The bad news was that people were not convinced that the coalition of opposition parties that Panic had worked so hard to wield together amounted to a credible alternative to Milošević. A majority felt that the parties in opposition were "self-interested, uncooperative, likely to fall apart, temporary." If it did collapse, many Serbs would simply stay at home. The poll also showed that none of the current crop of opposition leaders was a strong candidate. Draskovic and Djindjic had both seen their popularity plummet in the wake of the war in Kosovo; their unfavorability ratings were almost as high as Milošević's. The only prominent opposition figure who might be able to win, Dragoslav Avramovic, was dying.

In October, I flew to Budapest to present the results of the survey to the leading members of the opposition, whose members had slipped across the Hungarian border to attend the meeting. I arrived at the Marriott Hotel in downtown only to find that most of my results had already leaked out. Djindjic was said to be so upset by my findings that he boycotted the meeting. Nevertheless, I went ahead with my presentation. I emphasized that Milošević was vulnerable to a well-organized challenge and that the key to victory was opposition unity and a campaign focused on Serbia's calamitous economy.

As soon as I finished my presentation, the audience broke out in a buzz of conversation and argument. Clearly, the audience was excited by my claim that Milošević was vulnerable—and also skeptical. Some participants seemed to believe that there was no point in having an election since Milošević would always find a way to cheat. Others worried that NATO action meant that Milošević might win. I could feel the audience slipping back into the same defeatist stance that I had encountered seven years earlier from people like Tosha Olic, who were supposed to be trying to oust the regime but never acted like they really wanted to. Frankly, this argument angered me and this time, I said so. Switching from exposition to exhortation, I told the audience that skepticism must not become a justification to avoid doing what needed to be done. Milošević could be toppled, I repeated, but only if the opposition worked through the electoral process and stayed united. Win the election and then take to the streets when Milošević cheats. While the meeting ended without any formal decisions, I felt that my presentation had been a forceful one.

The next morning, I had breakfast with Assistant Secretary of State Jim Dobbins and James O'Brien, Secretary of State Madeleine Albright's special advisers for the Balkans. They seemed impressed by my analysis of the situation in Serbia—so impressed that they offered to arrange for me to brief the United States's most important foreign policy group and the world's most important committee, the Principals' Committee, a group whose members included the National Security Adviser and the Secretaries of State and Defense. I was obviously delighted with this opportunity and agreed on the spot.

The two men were as good as their word. In late 1999, I flew down to Washington to brief the Principals' Committee. After landing at National Airport, I hailed a cab. "1600 Pennsylvania Avenue," I told the driver. By this point, I had been working for President Bill Clinton for almost five years, so the White House should have been familiar territory. But going there, slipping in through the northwest side gate that led into the Oval Office—it still gave me a thrill. This time, however, I was a bit preoccupied. Just a few hours earlier, I had just subjected myself to a cardiological test—something called a dual isotope stress test. I was thinking about other things—my health, my lifestyle—and daydreaming a little. My thoughts would soon be interrupted in a dramatic fashion.

Today, Pennsylvania Avenue is closed to traffic, and the White House can be approached only after going through elaborate security procedures. Back then, things were more casual. I'd jump out of a cab and then walk down the alley between the Treasury and the White House on my way to the Oval Office. As soon as I walked through the doorway, however, my wandering thoughts came to an abrupt end. Flashing lights and a siren went off; Secret Service officers rushed up to halt me at the doorway. Fortunately, a more even-tempered supervisor came over. "Sir, either you're carrying a nuclear device and we have a serious problem," the man said calmly, "or you've been to the doctor this morning." I quickly explained that I had in fact recently been to the doctor's office. It was then that I realized that I had set off the White House nuclear radiation detectors. Needless to say, I did not attend the Principals' Committee meeting that day.

The meeting was never rescheduled. Ultimately, someone at the White House decided that it would be unseemly that a political consultant, particularly Bill Clinton's political consultant, tell the Principals' Committee how to beat Milošević, but my memo was widely circulated and discussed. I met with James Steinberg, the deputy national security adviser, and briefed him on my polling results and on my belief that Milošević could be ousted at the ballot box. I briefed Richard Holbrooke, who was then a special envoy to the Balkans and who later became the U.S. Ambassador to the United Nations. Ultimately, everyone came around to the same question: Can this really work? I told them that as far as I was concerned it was a no-lose proposition: if we win, we win. If Milošević cheats and we "lose," then we've further weakened and isolated him. My argument must have proved persuasive because $40 million was appropriated for USAID to do just that.

White House officials weren't the only people interested in my ideas. Soon after my memo started circulating through the upper reaches of the Clinton administration, I got a call from the CIA. I had gone on a Council on Foreign Relations briefing trip to the CIA. In my memo to Director of Central Intelligence George Tenet, I also mentioned the trip, thanked him for the information, and offered to brief the CIA's Yugoslavia analysts on my findings. Well, not for nothing is there a saying, "No good deed goes unpunished." The CIA took me up on my offer.

In February 2000, radioactive heart tests safely in the past, I reported to the George Herbert Walker Bush building, CIA headquarters, in Langley, Virginia. I was greeted with great—and fake—cordiality. My audience at the CIA wasn't just skeptical. It was hostile. The analysts questioned everything—my assumptions, my methodology, my sample sizes, even my ability to conduct a proper poll. Needless to say, they also took issue with my conclusions. The CIA has its own polling operation to monitor public opinion abroad, and their sampling in Serbia was producing very different results, they told me. They insisted that I was underestimating Milošević's levels of popular support. I wasn't an expert on Serbia— I freely acknowledged that—but neither was I remotely persuaded by their objections. The good news was that the CIA was doing everything right—spending millions to monitor public opinion, placing analysts in the region for long periods, and so on. The bad news was that it was coming up with findings that a reasonably intelligent political consultant—one person—could see were wrong after one week in country. It gave me pause—and still does.

The passive fatalism I had encountered among so many Serbian opponents of the Milošević regime was frustrating but at least it was historically comprehensible. The hostility of the CIA, in contrast, seemed to reflect some kind of professional resentment. Instead of welcoming an unusual source of information, I was treated like an undergrad whose presentation was claiming time that would have been better given over to senior professors. It was as if the analysts couldn't stomach the thought of having a political operative on their turf. It was only a brief encounter, but it gave me a disturbing insight about the U.S. intelligence community's analytic capabilities.

After the briefing, I got a call from Assistant Secretary of State James Dobbins. He was concerned about the objections the CIA had raised to pursuing an electoral strategy and, no doubt, by their portrayal of me as an uninformed amateur. "How confident are you of your data?" he asked me. Completely confident, I told him. The polling in the field was good, I assured him. After a short pause, Dobbins responded, "I agree with you. We're going to stay behind you on this one." He asked me to write a short memo summarizing my briefing that he could use to back our position up. Thanks to Dobbins's unwavering support, my argument carried the day. As the year 2000 unfolded, USAID and European donors intensified their support for Otpor.

However, the key to victory—a united opposition—was still not in place. All of the data showed that 60 percent to 70 percent of the electorate disliked Milošević and his policies. In a fair fight, there could be little doubt that Milošević would lose. However, it was also clear that the public wanted the opposition to unite first. Only when they had an appealing alternative in place would the Serbian public take the risky step of challenging their president.

Then, on July 27, 2000, Milošević did something that caught the world off guard and forced the opposition's hand. He called elections for September 24. The opposition was stunned; most observers had expected a vote in the summer of 2001. Clearly, Milošević believed he could win—or at least come close enough to steal the election again. It was time for the opposition to unite or for Serbia and the world to prepare for four more years of Slobodan Milošević.

In early August, I flew back to Budapest for another meeting with the opposition. The ostensible purpose was to present the results of our most recent poll, which had been fielded just before Milošević's surprising announcement. It was a critical moment. Polls showed that Milošević continued to be highly unpopular: our July polls showed that he had a 59 percent unfavorability rating. While that was pretty bad, it was a marked improvement from the preceding fall, when his unfavorability rating stood at 70 percent. In other words, Milošević's standing with the electorate was improving. In contrast, Djindjic's fledgling opposition movement was declining in popularity.

Even more disturbing, dissatisfaction with Milošević was not translating into support for the opposition. There was a clear desire for change: seventy percent of voters believed that Milošević "has had his chance and now it's time for someone new." Yet only about 35 percent of voters supported the Alliance for Democratic Change, the only viable opposition. Fully 29 percent of those with unfavorable opinions of Milošević were undecided about whether they would vote or whom they would vote for. Many voters seemed to suffer from a kind of apathy or demoralization. Part of this apathy reflected the belief that Milošević would steal the election whatever the results. (Fifty-one percent of Serbs expected election fraud.) However, most of it reflected ordinary Serbian's dissatisfaction with and lack of confidence in opposition leaders who were seen as petty bickerers.

It was critically important that we reverse these trends. The best way to do that was to unite the opposition behind a candidate who could win. My task in Budapest was to convince eighteen-odd opposition parties to join forces. However, my most important audience would be one man—Zoran Djindjic.

Djindic was the key to the whole effort. His Alliance for Democratic Change would clearly have to be the core of the opposition; I had already heard rumors that Djindjic might serve as the campaign manager for the opposition. If a truly unified opposition coalition were to come into existence, it would do so at Djindjic's behest. Unfortunately, Djindjic was also someone with whom my relations were at best uncertain. After all, the results of my last major poll had showed that he didn't have the support to ever become president—a finding that so enraged him that he had boycotted my presentation altogether. That finding had not changed. The question was: had Djindjic changed? Or was I heading into a minefield?

My presentation began that morning. As the lights dimmed and my PowerPoint slides went up, I drove home my points. Our polls continued to show that a truly united opposition would beat Milošević's ruling Socialist Party (SPS) by a thumping 46 percent to 26 percent. It was crystal clear that Serbians wanted change. The real question was why, if Milošević and the SPS were so unpopular, the opposition was not doing even better. The answer, unfortunately, was that the opposition seemed less unified to voters today than it had several months ago. The public was fed up with their bickering and that sentiment had started to show up in the polls. During the first half of 2000, Milošević's favorability ratings had slid while the opposition's had increased. Now, however, Milošević's approval ratings were increasing and the opposition's numbers were falling. The trend was small enough to be within the margin of error, but noticeable nonetheless. It was time to unify behind a common candidate or risk four more years of Milošević as president.

In fact, the conference itself offered a vivid illustration of the problem: Zoran Djindjic was not there. However, I was assured that Djindjic would arrive at some point—after my presentation. First, I was told to expect him at 2:00 PM, then at 4:00 PM. But 4:00 PM came and went. No one knew how to interpret his non-arrival.

Finally, at a little after 5:00 PM, Djindjic swept in to the hotel accompanied by five burly bodyguards. Word of his arrival raced through the building. There was a visible sense of expectation. Everyone in the hotel stopped; everyone looked. It was clear that even the other opposition leaders viewed Djindjic as a figure more significant than themselves. An aide informed me that Djindjic wanted to speak with me, privately. The two of us stepped into a barren little function room piled high with unused chairs. With a cool look that divulged nothing, he gestured for me to sit down.

When I had last seen Djindjic, in Washington in 1997, he had spoken only very limited English. Now, he began our conversation with a question in flawless English. "I can't win, can I?" he asked me. "No," I said. During NATO's aerial bombardment of Serbia, Djindic had left Belgrade for the safer environs of Montenegro. Although he claimed that he had left after receiving a credible assassination threat, Milošević's propaganda machine subjected him to a fierce smear campaign, accusing him (among other things) of being a NATO spy. These attacks sent his unfavorability ratings up to nearly 60 percent. People just weren't prepared to believe that his concern about assassination, rather than fear of NATO warplanes, was the cause of the decision to sit out the war.

Politicians rarely like to hear that they can't win, but Djindjic received the news very calmly. There was nothing overwrought or emotional about the man sitting in front of me. He seemed the epitome of sangfroid.

I pressed on. I told him that we had to find someone who could carry the opposition banner but who didn't have high negatives. He needed to have strong nationalist credentials, but he couldn't be complicit in the crimes of the Milošević regime. Then Djindjic said something that surprised me. "What about Kostunica?" Djindjic asked.

It was an intriguing question—and a surprising one. Vojislav Kostunica was the leader of a minor opposition party, the Democratic Party of Serbia, a law professor who had been fired from Belgrade University in 1974 for criticizing the government. Unlike Djindjic, he was not in the first order of opposition leaders. Most people knew very little about him. He was, however, an ardent nationalist—someone who was critical of both Milošević and NATO. Moreover, despite the fact that Kostunica was a fairly minor figure,

I had included him in my Serbia polling. My results showed that people didn't know much about him—and that worked to his advantage. Unlike the other opposition politicians, Kostunica's unfavorability ratings were quite low.

However, Kostunica was also a prickly figure, not a coalition builder. Would Djindjic and other more prominent opposition figures really defer to him?

"He's the most popular non-regime politician," I told Djindjic. "If you get him, you get the nationalist votes, plus you get a clean new face." "But," I asked, "how are your relations with him?"

"Not good," Djindjic responded candidly, "and I don't think they will last. But I can serve my purposes by making him the candidate—and of course it will serve his purposes as well."

"I will control the campaign," Djindjic continued, addressing one of my biggest concerns about Kostunica, "but I will need your messages and advice to do this."

I was amazed and delighted. Gone was the politician who had stayed away from Budapest almost a year earlier out of personal pique. The man sitting across the table was completely pragmatic and Machiavellian in the best sense of the word. He understood that he couldn't win; he understood what needed to be done; and he was obviously ready to do it. I was startled by the extent to which our thinking was now in sync. Indeed, Kostunica proved to be the perfect choice. For one thing, he refused to have anything to do with me or with the other Western advisers working on the campaign. That made it difficult for Milošević to describe his opponent as a lackey of the West. But Kostunica's campaign did take Western money and advice. Ultimately, he would also let Djindjic do what needed to be done to win.

Both Djindjic and I agreed that we could win the election, if we played our cards right. We also agreed that this time, unlike in 1992, we had to make sure the election stayed won. One powerful tool that could serve that purpose was the exit poll. However, after his near-death experience with Panic, Milošević had pushed through legislation that banned exit polls, a move that was clearly designed to make it harder for us to conduct a reality check on Election Day.

Milošević's concern was well founded. Just a few weeks earlier, I had used exit polls in Mexico's presidential elections to ensure that the ruling party, PRI, didn't try to manipulate the election results. By banning exit polls, Milošević was trying to preclude similar scrutiny in Serbia, but Milošević had not completely foreclosed every monitoring option. We might not be able to do exit polls, but there was no legal obstacle to doing what is called a "quick count"—to deploying people to observe voting and tabulation and then report the results back to campaign headquarters in real time. It would take an enormous effort, since it would require volunteers at every polling station in Serbia and a sophisticated organization to collate the results, but in the absence of exit polling, it was the only way to guard against fraud. Djindjic agreed that it was an idea worth pursuing and it soon became one of USAID's most important projects. In the course of the next two months, USAID and Democratic Opposition of Serbia (DOS), the broad anti-Milošević coalition, would train nearly thirty thousand poll watchers and fund a computer center in Belgrade to tabulate the results almost instantaneously—to prevent Milošević from doing what he had done in 1992 to fraudulently hold on to power.

Djindjic also recognized that it in order to actually take power, the opposition would have to find a way to foil Milošević's efforts to steal the election. He said that he was building ties to "different segments" in the society—without specifying exactly who these forces were. It was clear to me that Djindjic was reaching out to Serbia's powerful but shadowy interests—to the gangsters, to the police, to the army—to let them know that this time, if the opposition won the election, they intended to seize the power that was rightfully theirs. This was an important but dangerous move. Just a few weeks earlier, the Milošević regime had begun a particularly harsh crackdown on the student group Otpor, which was fast becoming the largest mass movement in Serbia. In addition, the government had also seized control of two of the remaining independent radio stations in Belgrade, as well as an opposition television station and newspaper. When crowds came out to protest these seizures, the police moved in and forcibly dispersed the crowd with tear gas. The Milošević apparatus, which had long been a "soft" authoritarian regime, seemed to be flirting with the brutal methods of a true dictatorship.

My meeting with Djindjic was businesslike and brief. In a matter of minutes, we had agreed on a plan to take control of Serbia. Kostunica would be the candidate; Djindjic would run the campaign—and lay the groundwork to seize power if necessary. My job would be to pinpoint Milošević's vulnerabilities and develop messages that exploited them.

That task proved easy enough. Polls showed that Serbs viewed poverty, economic sanctions, and "the regime of Slobodan Milošević" as the country's biggest problems. A large majority of Serbs thought it was time for Milošević to go; however, many undecided swing voters were uncertain that the opposition was really a credible alternative. As a result, I recommended a campaign that focused on several basic themes:

- The importance of change and getting rid of Milošević—and the fact that it is an attainable goal.
- The economic revitalization of Serbia and the reintegration of Serbia into Europe.
- The possibility of democratic reform.
- The existence of a united opposition, working together to strengthen the Serbian homeland.

I also helped develop the slogan the opposition would soon emblazon across Serbia—"Gotov je" meaning, "He's finished." Slightly slangy and wonderfully dismissive, it was the perfect phrase for the job at hand. At a time when many Serbs feared that they couldn't get rid of Milošević, we were saying that it was essentially already done.

First, we had to get the message out. Our polling showed that 54 percent of Serbs got their news mostly from television, which was firmly under Milošević's thumb. (Newspapers came in second with 20 percent.) Given the difficulties of accessing the state-controlled media in Serbia, it was critical that the opposition reiterate these messages repeatedly, throughout whatever media they could. Otpor plastered cities and towns throughout Serbia with our "Gotov je" stickers. There was no need to tell Serb voters who "he" was.

We also advised Kostunica to get out of Belgrade and concentrate the campaign on the provinces. The election would be won or lost in second-tier industrial cities like Nis and Novi Sad. At our suggestion, Kostunica circulated throughout small-town Serbia, pressing the flesh and talking to local media outlets. So did Djindjic and other opposition leaders and groups. The goal was to reach as many voters as possible and to distribute the campaign in a way that would make it difficult for the regime to move against its leaders. Our polls, which were now being fielded on almost a weekly basis, showed that our campaign strategy was working. Kostunica and the DOS's favorability continued to climb. By mid-August, Kostunica had the support of roughly 40 percent of the electorate versus Milošević's 25 percent. Moreover, the number of people who said that they didn't plan to vote or didn't know who they would vote for was diminishing. Most of these newly enthusiastic voters were trending toward Kostunica.

Milošević, in contrast, seemed to have learned nothing in the seven years since I last ran a campaign against him. As in 1992, he scarcely campaigned at all. His first campaign rally took place two weeks before the election, and it was an old-style communist party event. Lackeys were paid to attend; Milošević's limousine convoy arrived; he spoke to tepid applause; and then he left. To the extent that Milošević had a theme, it was this: the opposition was a tool the United States was using in an attempt to destroy the Serbian nation. It was a complete canard, but it had the potential to be an effective line of attack. Although Kostunica maintained his distance, his campaign (guided ably by Djindjic) was awash in Western cash and in constant contact with American media consultants and advisers like myself.

Fortunately, however, this line of criticism never really caught on. Serb voters blamed President Clinton and NATO and President Milošević equally for the air attacks on Serbia. Only around 8 percent of Serb voters opposed the opposition because of its "subordination to the West, treason, neglect of the Serbian national interest." On the contrary, Serbs by a margin of 33 percent to 26 percent believed that Kostunica was actually *more* likely to "stand up for Serbia's national interests" than Milošević. The fact of the matter was that a clear majority of Serbs were desperate to end their country's isolation and rejoin Europe. Forty percent believed that

Kostunica would "reintegrate Serbia and Yugoslavia into Europe." Only 16 percent believed the same of Milošević.

Between September 15 and 17, my colleague Nick Goldberg and I conducted our last large-scale preelection poll, the final effort in what had been an impressive push to use public opinion research tools and modern communications techniques to understand Serbian opinion and persuade Serbs to reject their dictatorial leader in favor of alignment with the West.[6] We found that Kostunica now had a twenty-one-point advantage over Milošević. Moreover, Milošević's job approval rating was falling; Kostunica's favorability rating was inching upward. Seventy-two percent of Serbs agreed with the statement "Slobodan Milošević has been president for many years, but now his time has passed and it is time for someone new." Our results showed Kostunica winning by 56 percent to 28 percent. We never really believed Kostunica would actually win by that large a margin. Still, the final polling results did convince us that Milošević could not possibly win anything approaching a fair election.

Sunday, September 24 was Election Day. To preempt any effort by Milošević to claim nonexistent levels of support, the opposition started the day by issuing a press release that announced the results of our poll. It predicted that the Kostunica would win at least 2.2 million votes, Slobodan Milošević—1.3 million (with two other candidates bringing in another million votes). Opposition volunteers stationed at every polling station kept a running tally of the voting. Results from every polling station were emailed back to the USAID-funded tabulation center in Belgrade and fed into the computer bank. Just after midnight, after the polls were closed and the results had been double-checked, Kostunica called a press conference to announce that he had won a clear majority. The question now became, "How would Milošević respond?"

The government seemed caught off guard by this announcement. Clearly, it had never anticipated that the opposition would conduct its own parallel vote count and announce the results just hours after the last poll closed. This time, claiming victory was not an option. At first, the government was simply silent. Then, several days later the Milošević-controlled central election commission asserted that neither candidate had won more than 50 percent of the vote and that a runoff election would therefore be necessary.

Unlike in 1992, this time Milošević's opponents were ready. Kostunica and the DOS rejected plans for a runoff and instead announced plans for a general strike. Three days before the nationwide walkout was scheduled to begin, more than seventeen thousand coal miners throughout the Kolubara region walked off their jobs. The security forces moved into place, but confronted with thousands of angry miners who controlled 70 percent of Serbia's energy resources, they did not act. A dangerous and uneasy standoff had begun.

The opposition now faced a fateful choice: should it accede to fraud and agree to a runoff, a course of action that would give Milošević two more weeks to maneuver for an advantage, or should it risk an open confrontation with the government? Less than a week after the election, we put a poll into the field to find out.

What we found was an electorate transformed. One year earlier, most of the public had rejected mass action against the government: the public wanted an election, not an uprising. After September 24, however, we found that public opinion had shifted. Thanks to polls, publicity, and the quick count, the public saw Kostunica as the rightful president. More than two-thirds of respondents believed he had won in the first round, and his popularity had risen fourteen points over the course of the preceding week. Seventy-one percent believed that the opposition had at last come together for the good of the country. In contrast, Milošević's favorability rating had fallen by 8 percent. A clear majority now saw Milošević as an illegitimate president—a politician who was thwarting the will of the public. In short, our poll gave the opposition the green light to act now instead of risking a runoff election that would only give Milošević more time to dream up new ways to hang on to power.

Fortunately, Djindjic had organized for just such a confrontation. Miners, workers from idled factories, provincial mayors, and others from across Serbia had already pledged their allegiance to the opposition. Djindjic had developed a plan to remove the regime—by force, if necessary. On Thursday, October 5, the plan to drive Milošević from power went into action. That morning, hours before daybreak, convoys of anti-Milošević protestors from across Serbia set off for Belgrade. The regime ordered the police and security forces to block their passage, but time and time again, police officers confronted by an angry—and sometimes armed—mob stepped back from confrontation. It later emerged that Djindjic had also reached out to sympathetic figures in the various "power ministries," as well as to various underworld figures, to reassure them that a new government would view their situation with sympathy—if they stayed out of the final showdown with Milošević. This overture to the men with guns was critical to Otpor's success, but it would later lead to tragedy for Djindjic.

By the early afternoon, a crowd of tens if not hundreds of thousands of people had converged in front of the Federal Parliament. These were not just student protestors. They were hard-boiled politicians like Mayor Velimir Ilic of Cecak, a gimlet-eyed veteran dressed for action in an Adidas tracksuit. He had come to Belgrade with a retinue of roughly a thousand men, many of them former members of the army or the security services. These were men prepared to fight. Indeed, Milan Protic, the mayor of Belgrade, at that very moment was overrunning a police station and dispensing weapons to his supporters.

At approximately 3:00 PM, a protestor drove a bulldozer up toward the parliament building doors. With that, the security forces faded away. The crowd quickly moved on to seize the television station. The next day, Milošević conceded the election. On Saturday, Vojislav Kostunica was sworn in as the new president of Yugoslavia. The man known as "the Butcher of the Balkans" was no more. The opposition had taken a huge gamble based on a single poll and won. Public opinion research, combined with brave and sometimes ruthless opposition leaders who were willing to act on it, had achieved what untold diplomatic missions, multiple international peace conferences, and thousands of bombs had not—it had ended the Balkan Wars. '

The 2000 Serbian campaign marked a historic development in the practice of foreign policy. The Clinton administration had made a courageous decision to support a poll-driven, grassroots electoral effort to oust Slobodan Milošević and won, big-time. Public opinion research had done what military pressure could not. Moreover, it had done it at a fraction of the cost of a military intervention. The *Washington Post*'s diplomatic correspondent, Michael Dobbs, called it "the first poll-driven focus-group revolution."

Our campaign in Serbia showed for the first time that public opinion research could be a potent weapon in promoting regime change. Two weeks after Milošević's ouster, Secretary of State Madeleine Albright wrote me a letter thanking me for my unexpected contribution to peace in the Balkans. "Polling has long been an important part in domestic policy development," she noted. "This may be the first time polling has played such an important role in setting and securing foreign policy objectives."

But regime change is a process, not an event. On October 7, Vojislav Kostunica was sworn in as president of Yugoslavia. In January, Zoran Djindjic became the country's prime minister, however, as Djindjic had predicted at the Budapest Marriot in the fall of 1999, the two men's relationship quickly soured. During May 2001, Djindjic, acting in defiance of President Kostunica and the Yugoslav constitutional court, authorized the extradition of former President Slobodan Milošević to The Hague. In February 2003, Djindjic won the power struggle when Serbia and Montenegro approved a new federal arrangement between Serbia and Montenegro that eliminated Kostunica's position as president of Yugoslavia.

Djindjic's triumph, however, proved sadly short-lived. On Wednesday, March 12, 2003, Milošević supporters got their revenge. At 1:00 PM, Djindjic walked out of the government building in central Belgrade—and was shot twice, once in the stomach, once in the back by an assassin. He died of gunshot wounds soon thereafter in the hospital. The assassination threw Serbia into a panic. In the aftermath, more than a thousand people were detained. Police speculated that the assassination organizers were most likely organized crime figures with ties to a special Milošević security force known as the Red Berets. "Regime change" in Serbia is still an ongoing process with an uncertain future.

During the 1830s, the Prussian military strategist Karl Von Clausewitz famously wrote that war should be "regarded as nothing but the continuation of state policy with other means." Our work in Serbia inverted that maxim. We had shown that the tools of politics could—in some circumstances—be used as a continuation of warfare. We were creating an entirely new art—the art of campaigning against authoritarian regimes.

From my work in Serbia, I developed seven principles for toppling dictators at the polls:

- *Unity is critical.* A unified opposition is the key to facing down a dictator and needs to be in place before an election begins.
- *Harness discontent.* It's not enough that a majority of the public is dissatisfied with a dictator. Opposition leaders must find ways to translate dissatisfaction into active support for the opposition with messages that underscore the discontent and offer hope for the future.
- *Use public opinion polls to identify key issues and then craft simple messages.* Use focus groups to make sure they appeal to critical constituencies. Then repeat them again and again!
- *Expect to be identified with a nefarious outsider.* Deflect this attack by focusing on the issues voters care about most (and that made the regime vulnerable in the first place).
- *Use tracking polls.* It's the only way to monitor whether your message can get through. Moreover, tracking polls can be an important tool in maintaining opposition unity. In Serbia, positive poll numbers helped keep most opposition leaders in line.
- *International pressure is critical.* Play to the media, particularly the international media. In today's interconnected world, external pressure can be the decisive factor in how the government in power behaves. And remember, a unified opposition with a coherent, consistent message is much more likely to be taken seriously by the international community.
- *Prepare for theft.* Winning the election doesn't necessarily mean taking power. Against an opponent who is prepared to steal an election, it's necessary to develop a plan of resistance

that mobilizes the opposition and neutralizes at least some of the people with guns.

Of course, the political context of every campaign is unique. Milošević was undoubtedly an authoritarian leader, but he was not an absolute dictator. Faced with a mobilized opposition that enjoyed the support of a clear majority of the population and that had a plan to take power, his regime did not have the power to forcibly retain control. When hundreds of thousands of fellow citizens took over the streets, the regime's foot soldiers refused to fire, and the regime crumbled.

Public opinion research is destined to become an increasingly important tool of foreign policy. As a tool for orchestrating and/or securing regime change, it's simply too powerful to ignore. In Serbia, public opinion research selected the candidate and crafted the message that removed one of the world's most dangerous and destructive leaders from power. I was proud of what we had accomplished after nearly eight years of work and absolutely convinced that the tools of my profession had played a crucial role in our ultimate victory.

Yet there are also undeniable limits, as I found three years later in Asia. There on the Korean peninsula, I would find myself working on a campaign on the Cold War's last battlefield where the fate of one of the world's most important democracies would be determined as much by a flawed public opinion poll as by the natural workings of democracy.

CHAPTER SIX

BROADBAND POLITICS, UGLY AMERICANS, AND POLITICS IN THE SHADOW OF THE WORLD'S MOST DANGEROUS STATE

When I first started working abroad in the early 1980s, the foreign policies of most countries rarely changed. Save for a handful of nonaligned nations such as India, the world was divided into two blocs—a capitalist bloc led by the United States and a communist bloc led by the Soviet Union. While the characters of these coalitions were quite different, their objectives were basically the same: contain the enemy and (where possible) expand their own influence. Diplomats, generals, and foreign policy experts determined how best to achieve those goals. In this Cold War world, political campaigns played only a minor role in setting foreign policy.

Today, we live in a very different world. Communism's days as a broadly compelling ideology are over: the capitalist, authoritarian regimes that the United States once saw as valuable allies are likewise extinct. South Korea exemplifies the change. When I first went to South Korea in 1983 with then-Congressman Steven Solarz, I saw an industrious but poor nation run by an iron-fisted military junta that enjoyed the support of the United States. Today, South Korea—and the world as a whole—is a much more democratic place. According to Freedom House, in the early 1980s there were approximately fifty-four democracies in the world. Now, 121 of the world's 192 countries are democracies. As democratization has gathered momentum, politics—electoral politics—have become an important part of foreign policy. The political campaign, as well as the cabinet meeting, has become a venue where fundamental foreign policy positions are formulated. Campaign tactics have become as important as effective diplomacy in determining the course of events overseas. Nowhere is this more evident than on the last battlefield of the Cold War, the Korean peninsula.

In North Korea, the United States confronts an impoverished Stalinist dictatorship armed with nuclear weapons that has threatened to turn the world's two largest economies, the United States and Japan, into "a sea of fire." In this dangerous confrontation with the North, we are missing a valuable ally—the South. For years, South Korea was one of America's most dependable supporters. Today, younger South Koreans are routinely critical of, even hostile to, the United States. The story of how our alliance with South Korea has deteriorated is first and foremost a story about electoral politics and public opinion research. It is also a story that provides a rare glimpse into the future of public opinion research, a future where the Internet and cell phones are as important as television ads. Indeed, South Korea even offers insights into a postelection democracy—a world where polls cease to be a tool during elections and instead become a replacement for them.

My work in South Korea came about as the result of our involvement in President Bill Clinton's 1996 reelection campaign.

Winning a presidential campaign solidified our standing as one of the world's "blue chip" political consulting firms; clients from all over the world now wanted us working on their campaigns. Taking off from Kennedy International Airport soon became as familiar to me as taking the shuttle down to Washington, D.C. So it was hardly unusual that, in the spring of 1997, work took me to the Philippines.

My client in Manila was Philippines Defense Minister Renato de Villa. De Villa wanted to succeed Fidel Ramos (another former defense minister) as president; I had been engaged by a wealthy Philippine businessman to help him develop a strategy. Or so I thought when I took off from New York. When I arrived in Manila, it quickly became clear that de Villa's situation was a difficult one. Despite his prominent position, de Villa's name recognition hovered around 4 percent. Since television ads were not allowed during the campaign, there was no way to improve his name recognition: the only way he could win would be with President Ramos's support. But without a higher profile, he had no real chance of being "anointed" by the president. Even if the president were to throw his support behind his defense minister, de Villa faced a formidable opponent in the person of Vice President Joseph Estrada, a college dropout-turned-immensely-popular actor. It also became clear that while de Villa and his supporters were going through the motions, they were not ready to make the effort necessary to mount a real campaign. I spent most of the week in my hotel room in downtown Manila waiting for phone calls and meetings that never seemed to happen. Occasionally, I was summoned forth to meet with the candidate and/or his various advisers at odd hours of the night. When the week was up, I left. But instead of heading home to New York, I decided to first visit a more interesting destination—Seoul.

Politically, South Korea in 1997 was like the United States two hundred years earlier—a new democracy with a fractious party system that was preparing for its first legal transfer of power under a new constitution. Five years earlier, the country had elected its first civilian president, Kim Young Sam. I knew that a presidential election was coming up and was curious to see how politics in his volatile region of the world had changed since my previous visit in 1983. And while I didn't know it at the time, I was about to begin the first in a series of campaigns that would bring me face to face with nuclear blackmail, the next generation of high-tech political warfare, and post-9/11 anti-Americanism.

Several weeks after returning to New York, I got a call from a partner at the high-powered law firm Akin, Gump, Strauss, Hauer & Feld named Sukhan Kim. Sukhan Kim was an amazing success story. As a teenager, he had been sent from South Korea to a military boarding school in Kentucky, where he had become a star halfback and a champion weightlifter, eventually winning the title of Mr. Kentucky. He went on to graduate summa cum laude from Guilford College, attended graduate school at Columbia and then law school at Georgetown, and eventually landed a job at one of America's most powerful law firms (which didn't stop him from occasionally dropping in at gyms, partnering for workouts with bigger men, and then humiliating them by bench-pressing far more than they could). As he rose to the top of the legal profession, Kim developed close ties to the New Korea Party, the party of incumbent President Kim Young Sam. In fact, ten years earlier, Sukhan had actually contacted Penn & Schoen on the party's behalf and asked for information about our firm. Now he wanted to know if I was interested in assisting the ruling party in its bid to retain power.

Thanks to my quick trip to Seoul, I was familiar with the political landscape there. I knew that just a few weeks earlier, the New Korea Party had selected Lee Hoi Chang as its nominee. It seemed an inspired choice. At first glance, Lee appeared to be that rarest of political animals—a politician who was wildly popular, unquestionably honest, and superbly qualified. As an undergraduate, he had attended South Korea's most prestigious university. He had gone on to become a judge at the tender age of twenty-five. In 1978, at the age of forty-six, he became the youngest Supreme Court justice in South Korean history. As a member of the Court, he won the admiration of the public (if not the ruling party) by consistently resisting attempts by the armed forces to try civilians in military courts. In the late 1980s, he stepped down from the Court to head the state election committee, where he delighted the public (and once again alienated many of his fellow party members) by investigating corruption in the ruling party. In the early 1990s, Lee was tapped to head the country's anticorruption drive, a position that soon made him into a well-known public figure. He served briefly as prime minister before his public criticisms of the president led to a hasty resignation.

Lee's greatest asset was his reputation for integrity. As a young judge, he was so upright that he was nicknamed "Bamboo." This unbending honesty played well with the public, but it annoyed many of his fellow politicians, particularly President Kim, whose tenure had been sullied by a series of corruption scandals. In fact, it was common knowledge that Lee viewed President Kim as corrupt and immoral and that the president and many other party leaders despised the rectitudinous Lee. However, to a public disillusioned with the cozy relationships between the government and Korea's powerful business groups, Lee seemed like a breath of fresh air. In short, he looked like a winner, and so the party hierarchy held its nose and nominated him. Almost at once, Lee's lead over his closest rivals soared to over 20 percent.

Unfortunately, the Bamboo Pole had a hidden weakness—his seedlings. Just days after his nomination, Lee's reputation for integrity was called into question by an incident concerning his two sons. It turned out that both had managed to avoid mandatory military service—the bane of every young Korean male—by reporting for induction in such physically emaciated conditions that they flunked their army physicals.

In the United States, stories of politicians' children misbehaving are about as surprising as reports of back room deals in Louisiana. Americans almost expect their politicians to have dysfunctional families and stories of substance abuse among political children are quite commonplace. When the Bush daughters were apprehended for underage drinking, it was not a big deal. Similarly, when Patti Davis posed naked in Playboy, it was simply dismissed as juvenile. Far from being a major detriment, stories of politicians' children behaving badly can sometimes give rise to a certain sense of sympathy. When a young Kennedy goes off the rails, people shake their heads and think, "Poor Ethel."

In Korea, family scandals play out very differently. There, the sense of family shame was still very much alive. When the news broke that Lee's children had avoided the draft, the Korean public turned on him with a vengeance. Older voters concluded that Lee was responsible for his children's bad conduct. Young voters, many of whom had only recently completed their terms of service in the military, were even more upset with the candidate. No one seemed to believe Lee's explanation that his children had always been slight and that his oldest son had been nicknamed "the Skeleton." Nothing could appease the angry electorate: even the decision by Lee's oldest son to volunteer in a local leper colony failed to reverse the candidate's slide. By mid-August, opinion polls showed Lee even with or behind his opponents. By September, his support had fallen to the low teens. The man who had once had a virtual lock on the presidency now looked headed for certain defeat.

From the onset, it was clear that turning around the campaign would not be an easy task. It's easy to make someone unlikable: that's why negative advertising works. It's much more difficult to convince voters to reassess someone they have turned against, and by the end of August, voters had definitely turned against Lee. Moreover, low favorability ratings weren't the only problem I would have to contend with. There was also his opponent—a man who happened to be one of South Korea's most storied figures, Kim Dae Jung.

Kim—D.J. to Koreans—was one of Asia's most celebrated dissidents. In 1971, he had challenged military strongman Park Chung Hee by running for president. He lost and spent several years in and out of prison for his efforts. In 1973, he was kidnapped on a visit to Tokyo by South Korean agents and was taken out of the country on a fishing boat. His kidnappers had already chained heavy concrete blocks to their intended victim in preparation for throwing him overboard, when a last-minute phone call from the CIA saved his life—for a few years, at least. In 1980, a military tribunal sentenced Kim to death for the crime of sedition. Once again, international protests saved Kim's life. Finally, in 1982, he went into exile in the United States. He returned to South Korea three years later and ran in South Korea's first free elections in 1987. Kim lost—and then lost again in 1992.[1] The 1997 election would be his last shot at the Blue House, the presidential mansion in downtown Seoul.

Kim was a fascinating politician. He was reputed to prepare thoroughly and practice extensively before appearing on television or taking the stump, and I believed it. He had the campaign moves of an actor. My candidate, in contrast, was more like the Korean Al Gore—stiff and formal. But I also happened to believe that Lee was a better choice for the country. Kim was undoubtedly a heroic figure, but his virtues were the virtues of a dissident. Moreover, some of his stances were disturbing. Not only was he beholden to Korea's militant and inflexible labor unions, but his views toward North Korea struck me as naive, if not dangerous. Just one year earlier, North Korea had announced its decision to disregard the 1953 armistice that brought the Korean War to a halt. In the following months, it had sent large groups of soldiers into the demilitarized zone, violated South Korean waters, urged South Koreans to overthrow their government, assassinated a prominent defector, probably killed a South Korean diplomat in Vladivostok, and threatened a South Korean newspaper that had criticized the North with "extermination." Evidence of the North's continuing hostility seemed overwhelming, yet Kim dismissed it. He seemed more interested in improving relations with Pyongyang, regardless of its behavior, than in confronting the very real security challenges the North posed. It was a puzzling attitude that I would encounter frequently in the months to come.

I agreed to work with Lee's campaign. That September, I sealed the deal over lunch in New York with Sukhan and with Chairman Lee's nephew, a high-tech investor from New Jersey, Chin Kim, who agreed to finance the effort. An interesting figure, Chin Kim was a messianic Presbyterian who couldn't resist attempting to save people he liked, a skilled martial artist, and Chairman Lee Hoi Chang's devoted nephew. Hiring me was, I suspect, his way of saying thank you to the uncle who had raised him as a young child. I agreed to develop a strategy for the campaign, based on a comprehensive strategy poll. I would go to Seoul in approximately six weeks to present my results. To help me understand the finer points of Korean politics, I turned to Jay Shim, an analyst at Hankook Research, who led me to see the finer points of Korean politics, especially that poll numbers without a model can be worse than nothing. It seemed like a fairly normal business deal at the time. Little did I know that, in a few months time, Asia would confront one of its most serious crises since the Vietnam War.

In mid-October, I flew to Seoul to present the results of my polls and plot strategy with Lee and his top advisers. Unfortunately, the numbers I brought with me were far from encouraging. Rather than rebounding from "starvation-gate," Lee seemed to have settled at a new, permanently lower level of popularity. Before the news about his sons, Lee's favorability rating had been off the charts. By the time I arrived, he was stuck in third place, behind Kim Dae Jung and a maverick conservative governor named Rhee In Je.

The strategy session took place in one of the top-floor conference rooms of the Lotte Hotel, a huge downtown luxury hotel that seemed to be Seoul's center of intrigue. In the United States, the major political parties run fairly lean operations. South Korea is quite different. Political parties there are generally large bureaucracies full of rapacious job seekers, so I didn't know quite whether to expect a party conference or a small huddle when I arrived. It turned out to be a very small group—just Chairman Lee, two or three aides, and a handful of the most prominent men in South Korea, including Hyun Hong Choo, South Korea's former ambassador to the United States, and former Foreign Minister Han Sung Joo.

It was an impressive group; however, it was immediately evident that the candidate himself did not share the general interest in my presentation. Lee seemed inattentive and looked distracted—often I wasn't even certain he spoke English. Clearly, he was not keen to be stuck in a conference room with some American bearing bad news.

I pressed ahead. I quickly went over the (bad) poll numbers and then proceeded to offer my advice. I told the group that it was essential that they make several changes to their campaign. First, Chairman Lee needed to put as much distance as possible between himself and the scandal-ridden incumbent president, Kim Young Sam. Because Lee and President Kim were both members of the ruling New Korea Party, this would be awkward. However, it was also essential (and since Lee and the president were already estranged, I hoped it would also be relatively easy). My second major piece of advice was even more important: Lee had to develop a detailed economic plan.

My suggestions ran against the grain of conventional political wisdom in South Korea. For years, the ruling party had relied on strong economic growth and regional divisions to quell popular unrest and (after South Korea held democratic elections in 1987) to maintain power. For most of the men in the Lotte's conference room, it was a formula that seemed almost fail-proof. They could not believe that the electorate would ever entrust its fate to someone from the small, poor province of Cholla. I thought otherwise. A growing percentage of South Koreans were intent on putting historic regional differences behind them. D.J., a native son of Cholla, might have wider appeal than the old guard thought. Moreover, the second pillar of ruling party dominance—strong economic growth—was starting to look shaky. After years of feverish economic growth, South Korea's economy was slowing down. Most experts expected growth of "only" 6 percent that year. While that may sound great to Western ears, in Korea, a country determined to join the ranks of the world's rich, industrialized nations as quickly as possible, it was a huge disappointment.

In addition, there were faint but ominous signs of more serious trouble down the road. Earlier in the year, Hanbo Steel had become the first Korean *chaebol* (conglomerate) in a decade to go bankrupt. In July, South Korea's third largest automaker, Kia, was forced to request emergency loans from the government. Currency speculators had begun to drive down other Asians currencies like the Thai baht. Although no one suspected what would happen in a few short weeks, even then it was clear that Lee and his party needed to define themselves in terms of an attractive economic philosophy. In other words, they needed to build an appealing brand. I wanted the New Korea Party to emulate President Clinton and put forward a comprehensive economic plan.

Doing so would be an important step toward victory, I told the group, but it wouldn't be enough. Lee was in too much trouble. While support for D.J. and Governor Rhee hovered in the thirties, Lee's numbers had fallen to the mid-teens. He needed to do something dramatic to get his numbers up. I suggested a bold idea: Lee should merge his New Korea party with former Seoul mayor and finance minister Cho Soon's Democratic Party.

Joining forces with Cho was not an obvious move. If you just looked at the polls, Cho might have seemed a marginal figure: nationwide, he was attracting only 3 percent or 4 percent support.

However, I believed that an alliance with Cho would serve several important purposes. It would shore up Lee's economic credentials and garner support for him in the capital region, which accounted for roughly 40 percent of total votes. It would also foreclose the possibility of an alliance between Cho and Rhee. There was only one problem with my idea: Lee and Cho were not exactly fond of one another.

Chairman Lee listened noncommittally to my ideas. At one point during the meeting, Ambassador Hyun asked provocatively, "Mr. Schoen," he began, "you've had a great deal of experience. Your fame is renowned throughout the world. But have you ever seen a situation this bad turn into victory?" The question, though unstated, seemed clear enough: did I think Chairman Lee should drop out of the race? It was a question that I as an American new-comer to Korean politics certainly wasn't going to answer. I politely ducked it. It was clear, however, that Lee would be facing an uphill race in more ways than one.

Despite his apparent indifference to my presentation, Lee took my most concrete piece of advice and agreed to merge the New Korea Party with Cho's party. The merger succeeded beyond my wildest expectations. The resulting Grand National Party (GNP) was an immediate hit with the electorate. Within just days of the announcement, Lee had moved from last place to a position where he was essentially tied for the lead. One plus one had somehow yielded four.

Lee's standings in the polls weren't the only thing that got a boost from the merger. My standing with Lee rose as well. Although he was by no means ready to bring me onto his team, he and his aides were now interested in hearing my advice. Unfortunately, they weren't ready to act on everything that I had to say. For instance, I believed that the best way for Lee to convince voters that he had a coherent economic plan was to highlight the support of Cho. That meant making ads with Cho. Yet Lee insisted on keeping Cho on a short leash. When I pressed Lee's aides on this point, I was told that Lee "didn't want to deal with Cho." "He doesn't have to deal with him personally," I would reply. "You just have to use him—put him on the air." But the chairman was clearly reluctant. The postmerger bounce had restored the campaign's confidence, and unfortunately confidence translated directly into stubbornness.

Lee also suffered from his unwillingness or inability to run a modern, media-centric campaign. While the seventy-one-year-old Kim Dae Jung made every effort to put himself before the television cameras, the much younger Lee was clearly uncomfortable with them. Television rewards candidates who are relaxed, who seem to engage with people, and who can project that ineffable quality of leadership. Despite his many accomplishments and appealing attributes, Chairman Lee did not have these attributes. Fundamentally, he was a shy person. In private, he often displayed a wry sense of humor. In public, however, he was every bit the judge—stiff, rigid, and unable to connect with people. Even when he dressed informally, he wasn't able to act like an ordinary guy.

Working for an American campaign, I would almost certainly have been able to push for a major change to the candidate's image. However, my role in Lee's campaign was quite different. I was working for Chin Kim and Sukhan Kim, not the campaign. I also knew that given the sensitivity of the South Korean electorate to anything that might be perceived as American meddling in the campaign, I needed to be especially cautious about how I met with Lee. As a result, I did everything I could to keep our monthly meetings private. Toward the end of the campaign, rather than risk being seen with Chairman Lee in public, I insisted on conducting all of our meetings in private. I knew it would not be helpful to have an American visibly involved in the campaign.

Even in my private meetings with Chairman Lee, it was hard to tell if I was getting through. After the initial strategy meeting (and my helpful party merger advice), Chairman Lee was unfailingly polite. He now listened attentively to what I had to say. Typically, I would respectfully explain why I believed he needed to actively establish an appealing identity for the new GNP by putting forward a coherent economic policy. My foremost concern was thematic: what would the message of the new party be? Everyone else on the campaign seemed to be more concerned about what job did I now have, who would the party chairman be, and so forth. While I made my points, Lee would sit politely across the table, nodding his head and saying, "Yes, yes." Those were just about the only words he spoke to me during the campaign.

Lee also paid a price for his deepening feud with President Kim Young Sam. The presidential administration denied him the advantages the ruling party candidate typically enjoyed, such as sympathetic treatment from the state-controlled media, while saddling him with the baggage of the current government's various corruption scandals. To make matters worse, the president was quietly encouraging the candidacy of a young, conservative governor, Rhee In Je, who was eating into Lee's base of older, conservative voters. Rhee's challenge was one Lee could ill afford, yet he refused to preempt Rhee or make a deal with President Kim Young Sam.

In November, Lee debated Kim Dae Jung for the first time. It was actually the first live televised presidential debate in Korean history, and Lee came off badly. Like Al Gore, Lee had the unfortunate habit of coming across as aloof. While we were struggling to present an appealing face to the public, D.J. was taking moves from our playbook. In early November, he struck a deal with the ultra-conservative Kim Jong Pil, who held considerable sway over conservative Chungchong province, despite the fact that he was the founder of the very secret intelligence force that had made at least two attempts on D.J.'s life. Evidently, the fact that D.J.'s positions on North Korea were much softer than Lee's mattered not a whit to the ruthless old agent. In South Korea, political parties were first and foremost vehicles for personal power.

Eventually, the GNP decided to follow my advice and make an ad using former finance minister Cho. I explained that the ad needed to be up-beat, inspiring confidence in voters that the new GNP had a viable economic plan. However, when I saw the ad Lee's media team had produced, I was shocked. Cho was barely visible: the producers had failed to use bright klieg lights. The background music was doleful. The whole thing reminded me of a Soviet death announcement. As the campaign went on, the quality of Lee's ads did improve, somewhat, and the campaign eventually managed to get a few good positive advertisements on the air. Nevertheless, they didn't figured out how to produce an effective issue ad, much less a negative ad. Despite months of effort and untold dollars (or rather, the South Korean currency, the won), Lee's team never managed to produce even a single ad as good as the one Bill Press and I assembled in two days in a dilapidated Belgrade soundstage five years earlier.

All in all, our position was a difficult one. For all practical purposes, Chairman Lee was stuck four to five points behind Kim Dae Jung. But unbeknownst to us, the campaign was about to be upended by something bigger than anything we could ever have imagined—the Asian economic crisis.

By early August, it was clear that much of Asia was entering a full-fledged economic meltdown. Stock markets across the region were collapsing, and countries in Southeast Asia such as Thailand were beginning to own up to staggering debts that clearly threatened the regional banking system. Despite the Bank of Korea's interventions, South Korea's currency was falling sharply against the dollar, making it harder for the country's overburdened banks to pay dollar-denominated debts. In Seoul, the question on everyone's mind was, would the contagion spread to us? The answer was not at all clear. As late as that fall, Michael Camdessus, the managing director of the International Monetary Fund (IMF), was expressing confidence that South Korea would escape the spreading panic. It did not. One week after Camdessus's statement, South Korea's currency went into a free fall, and the stock market staggered.

It was a frightening, bewildering moment for the candidates and for the nation. South Korea is a proud, intensely nationalistic country. By 1997, it had enjoyed years of strong economic growth. Unemployment was virtually unknown, and many South Koreans were fast approaching the living standards of citizens of rich, industrialized countries like Great Britain. In short, South Korea was one of the world's great success stories. Now in a matter of weeks, the economy South Koreans had struggled to build for decades was dissolving before the populace's very eyes.

At first, the politicians seemed paralyzed by this unfolding catastrophe. President Kim denounced "speculative foreign press reports" for creating an atmosphere of panic and insisted that an IMF bailout was "unthinkable," yet at the same time, his government failed to pass legislation that might have stabilized the country's tee-tering banks. At the beginning, our candidate shared the general sense of shock. Initially, he hesitated, caught between the urge to de-nounce the government and the *chaebols* for their reckless borrow-ing and the need to find a way out of the crisis. However, Lee soon faced up to the reality of the situation. South Korea was going to have to accept an IMF bailout, on the IMF's terms. When, at the end of the month, President Kim Young Sam announced that negotia-tions were underway, Lee was generally supportive.

Kim Dae Jung was an entirely different matter. Ironically, D.J. was the first candidate to raise the possibility of an IMF bailout. But once negotiations began, he turned against them. Instead of facing up to the frightening reality, D.J. generally preferred to strike angry nationalistic chords, and such criticisms were not made casually. They were crafted to tap into the reservoir of pride and resentment that is so distinctly Korean. (Koreans even have a word, *han,* that de-scribes this unique sense of oppression—a feeling that reflects a cen-tury of war, occupation, and division and occasionally gives rise to outbursts of rage and frustration.) At first, they played extremely well with the public. However, as the magnitude of the problem be-came manifest, popular support for D.J.'s stance dwindled, replaced by a growing recognition that South Korea shouldn't bite the hand that feeds it. So Kim changed his tune. Now he was for an IMF bailout, but only on terms favorable to South Korea. As for the offer on the table, D.J. was still against it.

This was our opening. At my recommendation, the Lee cam-paign made the most of Kim Dae Jung's flip-flops. In ads and public appearances, Lee relentlessly criticized his opponent, describing him as an untrustworthy figure who would bungle South Korea's recov-ery. Many Koreans already viewed D.J.'s native province, the poverty-stricken southwestern region of Cholla, as a shiftless place—not a region to chose a president from. Our attacks played on those preconceptions. Our message was simple: it's too risky to trust the economy to Kim Dae Jung.

On November 21, the government of President Kim Young Sam announced that it would seek the largest bailout in the history of the IMF, a credit line of more than $60 billion. As the South Korean public faced up to the magnitude of the crisis, it began to back Lee. Our message (Lee equals stability) was working. Lee also got a boost from his old nemesis the president when Kim Young Sam finally came out and formally endorsed another presidential candidate Governor Rhee: by this point in the campaign, the president was so unpopular that his endorsement actually cost Rhee support. By the end of the month, Lee had fought back to a virtual draw with Kim Dae Jung in what the *Financial Times* called "one of the most remarkable comebacks in South Korean politics."[2]

The presidential election itself was slated for Thursday, December 18. In the final weeks leading up to the election, Kim Dae Jung staked his candidacy on a wildly unrealistic platform that promised not only to protect jobs but also to raise per capita gross national product from ten thousand dollars to thirty thousand dollars in five years' time. He also continued to criticize the IMF bailout plan. D.J.'s comments so unsettled the markets that the IMF sought written statements of support from the various candidates. Lee complied; D.J. refused, saying that while he supported the idea of a bailout, he wanted to renegotiate the details. Kim Dae Jung was playing to South Koreans' anger, Lee to their anxiety. The outcome of the election would hinge on which emotion prevailed.

South Koreans weren't the only ones following the campaign and listening to the candidates' rhetoric. International investors and bankers were listening too, and they didn't like what they were hearing. Investors continued to sell South Korean assets and drive down the won. Despite the IMF agreement, South Korea faced the real threat of a governmental default—an event that would slam the door on foreign capital and cripple economic growth for years to come. The country was heading toward a truly dangerous crisis, but politically speaking, it was a good crisis. If people believed that voting for D.J. was a risk that could lead to economic ruin, voters would ultimately choose Lee. In order to win, Lee had to maintain the atmosphere of crisis that D.J. had done so much to create. Only then would the message that voters couldn't risk choosing Kim prevail.

Lee seemed to understand this. At my urging, the campaign pounded Kim Dae Jung for recklessly vacillating on the IMF agreement in the face of a looming economic disaster. At last, Lee was on message, driving home the point that he was the only prudent choice, that D.J. was too risky. One week out, we had drawn to within a point or two of front-runner Kim Dae Jung. So when President Kim Young Sam summoned Lee Hoi Chang, Kim Dae Jung, and Rhee In Je to the Blue House (South Korea's White House) for an emergency "unity luncheon" on Saturday, five days before the election, I was immediately alarmed. I sensed a trap, and I even had a pretty good sense of what it was—I suspected that President Kim wanted to take our best issue, the IMF bailout, off the table. The evening before the meeting, I got the chance to share my concerns with Chin Kim, who had moved from New Jersey to Seoul for the final stages of the election.

"This is a 'make or break' event. This is the Saturday before the election," I told Kim. "This unity lunch is a terrible track for us to take. If you come out and say, 'The country is united. We're all in this together,' then the loser is Chairman Lee because it takes the issue of D.J. off the table." Chim understood my point and promised to convey it to his uncle via Ambassador Han.

That Saturday evening I got a chance to see the candidate. When I asked him how the meeting had gone, he replied, in fluent English, that the meeting had gone well and then proceeded to tell me about the luncheon and about how he and the other candidates had agreed to issue a joint statement of support for the IMF bailout deal. Though Lee was in good spirits, I was stunned. By agreeing to a joint statement on the IMF, Lee had at once aligned himself with the unpopular incumbent president and removed our best argument against D.J.—that he was too risky, that a vote for him would jeopardize the IMF deal and by extension the future of the country. Yet as Lee proceeded with his leisurely report of the afternoon, it was clear that he did not even realize he had made a serious and irrevocable strategic mistake. I saw no way to correct it, so I bit my tongue and listened politely. Later that evening, I told Chin Kim, "That's it; we're going to lose." I repeated the same message to Sukhan Kim who sadly had been summoned back to the United States because of an emergency related to South Korea's ongoing economic crisis.

I stayed in Seoul to supervise the final tracking poll on Tuesday morning, two days before the election. The numbers vindicated my fears. They showed Lee two or three points behind. Since undecided voters rarely go with the incumbent (in this case, the incumbent party), I knew that Lee would probably lose though his aides refused to believe it. Two days later, Kim Dae Jung edged Lee out, 40.4 percent to 38.6 percent. Rhee In Je, the conservative governor who had enjoyed the support of President Kim Young Sam, got 19.2 percent of the vote. The GNP had stood by while its natural constituency was split and bungled the one issue that might have brought victory.

Even in defeat, I believed that the advice I had given Lee was good: if he had made Kim Dae Jung's trustworthiness the issue from day one and not let him off the hook on the IMF, I am confident that Lee would have won the election. But I didn't expect Lee or his team to recognize that. I expected them to blame me and take their business elsewhere. When candidates lose, they usually do.

But this time, they didn't. Sukhan Kim and Lee had both been sufficiently impressed by my insights and advice—particularly by my suggestion to merge parties and by my correct prediction of the outcome of the election—that they decided to keep me on board. As a result, five years later, I got a second chance to put Lee into the Blue House. I knew that this time I would not be running against the wily Kim Dae Jung; like his predecessors, D.J. was limited to a single five-year term. But just who our opponent would be was still unclear, because for the first time in its history, South Korea's Millennium Democratic Party was going to select its candidate via a primary election.

The skills and personality of an opponent are always important. In this case, however, the identity of our opponent was secondary. Primary or not, the only successor President Kim Dae Jung would allow his party to select was someone who supported his most cherished legacy—the so-called sunshine policy toward the North.

By the turn of the twenty-first century, North Korea was a country that had been on the brink of mass starvation for at least four years. Disastrous agricultural policies forced on the countryside by its long-time leader, Kim Il Sung, had destroyed the peasant plots that once had fed the population. Even the North's million-man army was going hungry. For the first time since October 1950, when a U.S. counteroffensive drove back North Korean forces and captured Pyongyang, the very existence of the North Korean regime was at stake. The North's response to this unprecedented crisis was to enact legislation making Kim Il Sung "President for Eternity." This despite the fact that the leader North Koreans called "the perfect brain" had died in 1994.

To most of the outside world, responses of this sort seemed almost insane, yet there was a method to this madness. The North's new leader, Kim Jong Il, son of "the perfect brain," was in a tenuous position. His father had been a leader of the anti-Japanese resistance in the years leading up to World War II. While his only real success during this period was staying alive, North Korean propaganda had succeeded in making him into a heroic figure in the eyes of North Koreans and a figure of sneaking admiration among some in the South. In contrast, the bouffant-haired younger Kim had spent his youth watching movies and occasionally kidnapping a South Korean director or actress. Strengthening his father's claim on power was a way to strengthen his own power. It was a mistake (albeit a tempting one) to dismiss the younger Kim as a madman and a clown. Indeed, the young North Korean dictator would soon show himself to be a shrewd manipulator of the South Korean electorate. In South Korean president Kim Dae Jung, he would find an eager and endlessly credulous partner.

As a candidate, Kim Dae Jung had made no secret of his desire to improve South Korea's relations with the North. His goal, however, was not to ameliorate the misery of ordinary North Koreans or to accelerate the collapse of one of the world's most evil regimes—quite the contrary. D.J. and his circle actually *worried* about the collapse of the North Korean regime: they feared the gulag to the north might disintegrate and send millions of impoverished North Koreans streaming into the prosperous South. To forestall this "catastrophe," D.J. decided to offer economic incentives and aid to the North Korean dictator. Of course, they didn't present it to the public in quite those terms. The purported rationale for the policy was that greater trade with the South would open the secretive North Korean regime to the outside world and lead it to gradually abandon such unpleasant habits as attacking South Korean vessels and soldiers, smuggling drugs, building nuclear weapons, and threatening Seoul with "unspeakable disaster."

From the beginning, it seemed like a dubious strategy. Nevertheless, two years after it was unveiled, D.J.'s sunshine policy produced what at the time looked like an historic breakthrough. In the summer of 2000, Kim delighted the world by traveling to Pyongyang for an historic summit meeting with "the Dear Leader" Kim Jong Il. Schoolchildren sang; Kim Dae Jung and Kim Jong Il held hands; and commentators confidently predicted a warm ending to the world's last Cold War confrontation. Later that year, D.J. even won the Nobel Peace Prize. If the truth about his overture to North Korea (and the massive, illegal cash kickbacks from South Korea's most prominent company that had lubricated it) had come out, the president might well have received a trial summons rather than a ticket to Oslo, but those revelations would not surface until three months after the 2002 election.

Still, even before the public learned the full price of the sunshine policy, it was becoming clear that South Korea's bargain had not been a good one. By that fall, most South Koreans understood that Kim Jong Il was using economic and food aid to feed his soldiers and shore up his regime, not soften it. Disquieting rumors of clandestine nuclear weapons facilities continued to circulate. North Korea continued to export advanced military technology to some of the world's nastiest regimes and to conduct tests of ballistic missiles that could strike Japan and eventually parts of the United States as well.

After George W. Bush became President in early 2001, rhetoric on both sides got tougher. Bush likened the younger Kim to "a pygmy" and North Korea to "a Gulag the size of Houston."[3] The North Korean media responded by branding Bush "a moral leper" and, in a rare failure of vituperative creativity, described the United States, as "an empire of evil." Despite ample evidence of Kim Jong Il's bad intentions, South Korean attitudes toward the North were surprisingly ambivalent. No one could deny the barbarity of the regime in Pyongyang, yet sometimes it seemed as if many South Koreans, particularly South Koreans under the age of fifty, wanted to overlook it. Younger South Koreans felt more secure than older voters and viewed the North as less of a priority and threat. When they did take notice of the North's actions, some South Koreans blamed Bush for provoking the mercurial dictator. Occasionally, I almost got the sense that some people took a kind of perverse pride in the North for standing up to their country's historic enemy (Japan) and to the world's sole superpower (the United States). While I didn't fully understand these attitudes, it was clear that North Korea's relations toward the world—and South Koreans' ambivalent feelings toward their kinspeople north of the thirty-eighth parallel—would be the wildly unpredictable backdrop to this unusual campaign.

Successful American consultants have always found clients abroad because political campaigns in the United States are seen as cutting edge. I returned to South Korea confident that if we could keep the conservative party united and run a better media campaign, Lee would probably win. In many ways, I expected to run a fairly conventional campaign in a country one step behind the United States. Instead, I found a country that was more technologically sophisticated than my own. The "Internet insurgency" approach to political campaigning originated not with Howard Dean in Vermont but in South Korea. Dean and his supporters never used the Internet as effectively as our South Korean opponents did.

High-tech wizardry wasn't the only intriguing feature of this race. The 2002 Korean presidential race would also raise fundamental questions about the goals—and the limits—of public opinion research. For the first time, I would come face to face with the possibility that the techniques I had spent my life perfecting might one day cease to be tools of politics and instead become its replacement. I would also find myself working in an environment defined by virulent anti-Americanism skillfully fanned by a fellow political consultant.

My first surprise came early. Like most other observers, I had expected the Millennium Democratic Party to choose as its nominee Rhee In Je, the upstart governor who had probably cost Lee the election five years earlier. Rhee would have been a good opponent for us because he was every bit as unpopular as Lee. However, when members went to the polls in April 2002, they instead selected Roh Moo Hyun as their party's presidential candidate.

The selection of Roh (pronounced "No") surprised almost everyone. Roh was a maverick, a former human rights lawyer who had bucked the Korean political system at every turn, and the contrast with Lee could not have been more striking. While Lee was a cosmopolitan technocrat who had lived and taught abroad, Roh was the son of a poor fisherman who hadn't gone to college and who had never visited South Korea's most important ally, the United States. In fact, Roh's only previous experience at the national level had been as the head of the administration of maritime affairs and fisheries. For a country that valued expertise and education, he was a startling choice.

But Roh did have one qualification that trumped these deficiencies, at least in President Kim Dae Jung's eyes: he was a strong supporter of the sunshine policy. Indeed, Roh made D.J. seem like a hawk in a comparison. When confronted with the sunshine policy's total failure to moderate North Korean behavior, Roh responded that South Korea needed to *intensify* its overtures to the North. Moreover, unlike D.J., Roh was an outspoken critic of the United States. During his career as a member of parliament, he had even questioned the need to have American troops on South Korean soil at all. Though President Kim's sunshine policy was wildly optimistic, he at least understood the importance of maintaining the U.S.-South Korean alliance that shielded the South from both the madman to the north and from the pressure of China to the east. Roh, in contrast, seemed to have a limited understanding of geopolitical realities.

With ten thousand North Korean rocket launchers and pieces of field artillery just thirty-seven miles from Seoul, Roh's rhetoric and platform was reckless in the extreme. His supporters, however, seemed to trust the million-man army to the north. They were more upset with President George W. Bush than with North Korean dictator Kim Jong Il. Earlier that year, Bush had delivered a speech on

terrorism that propounded the notion of "an axis of evil" and identified North Korea as one of its charter members. (The others were
Iraq and Iran.) To many South Koreans, such rhetoric seemed reckless. It also served as a way to blame Bush, not Kim Jong Il, for the
broken-down peace process. Unfortunately, in the coming months,
the Bush administration would do nothing to change this attitude.
The administration's indifference to South Korean public opinion
and to the campaign getting underway would cost my candidate—
and ultimately all of northeast Asia—dearly.

Although Roh came from nowhere, had none of the experience
necessary to govern a country, and advocated a policy toward the
North that was at best wildly optimistic. His standing with the public soared as soon as he was selected as his party's nominee. In the
days following his primary victory, Roh moved ahead by thirty
points. By then, I knew enough not to be too concerned by such
movements. After all, five years earlier, I had seen Lee's twenty-point
bounce fade quickly as well. (In the United States, in contrast, presidential nominees typically see a ten-point bounce that usually evap-
orates within a week.) I assured Lee that Roh's rise wouldn't last—if
we responded with the right strategy.

The first step to victory was linking Roh to Kim Dae Jung. By
2002, President Kim Dae Jung, like his predecessors, was a man en-
meshed in kickback scandals. As with President Kim Young Sam,
voters had responded by angrily turning against the president. Our
job was to try to convince voters to turn their backs on Roh as well,
and so throughout the spring and summer, we sought to link Roh to
the now unpopular Kim Dae Jung at every opportunity. This time, I
was relieved to find that Lee's Korean local advisers were capable of
developing effective counterattacks on the media, getting our counterattack off to a good start.

Roh did not take our attacks lying down. Indeed, his counter-
moves showed him to be a shrewd tactician. While we pounded him
on the airwaves, the Roh campaign took refuge on the Internet.

South Korea is a technology-loving nation of apartment dwellers.
As a result, it has the highest broadband penetration rate of any place
in the world. Roh was a politician who had found a way to use the In-

ternet like no one before him. His Internet operations had begun as an afterthought. Two years earlier, Roh had lost a hard-fought campaign for parliament. After his loss, Roh supporters had flooded Roh's homepage with emails expressing their continuing support. A handful of supporters then decided to start a cyber fan club, which they dubbed "Nosamo" (an acronym that stands for "People Who Love Roh"). Two years later, Nosamo had thirty-seven thousand members; by election day, that number would swell to the hundreds of thousands. Many of these supporters were members of South Korea's "386 generation"—people in their thirties who were educated in the 1980s and born in the 1960s. Many of these voters had spent their student years protesting South Korea's military dictatorship and its American supporters. As a result, they tended to be skeptical of the old ruling party, distrustful of the United States, and technologically savvy.[4] Like Howard Dean, Roh was building support in a way that was largely invisible to his opponents. This Internet approach to organizing would prove invaluable for him in the months to come.

Still, the momentum was on our side. Roh was sinking, and we were moving up. The lack of results from the sunshine policy, combined with allegations of corruption now enveloping D.J.'s sons, put Lee and his supporters in a steadily improving position. Lee had lost the previous election only because his natural constituency had been split when Rhee In Je entered the field. If Lee could keep his base united, I thought he would win. Unfortunately, I had a blind spot—the classic blind spot, in fact, of the American abroad. I forgot about soccer.

Soccer is South Korea's most popular sport. That summer, for the first time ever, the World Cup was being played in Asia. The hosts of the games were Tokyo and Seoul. I knew that South Koreans were proud to be hosting soccer's world championship; what I never expected is that they would actually win a game. Yet on June 4, before more than fifty thousand wildly cheering fans at Asian Stadium in Pusan, the South Korean team did just that, overwhelming a stunned Polish team by the score of 2 to 0. A week later, South Korea fought the American team to a tie, keeping the country's World Cup hopes alive before nearly seventy thousand fans in

Seoul. After a win against Portugal, South Korea beat three-time champion Italy to advance to the quarterfinals against Spain. Suddenly, South Korea had a new hero—Chung Mong Joon, the head of the South Korean football association and the man responsible for bringing the World Cup to Korea. Chung was comparatively young (fifty years old), handsome, and rich. (His father was the billionaire founder of Hyundai, and Chung himself controlled its shipbuilding division and a fortune estimated at more than $500 million.) He was also a member of parliament with a long-standing interest in the presidency. It was even rumored that Dick Morris was masterminding Chung's candidacy. I never could ascertain if that was true, but Chung's "outsider" rhetoric certainly had a Morrisonian touch.

As soon as Chung let it be known that he was considering entering the race for the presidency as an Independent, he shot to the front of the opinion polls. The person who suffered the most from his appearance on the political stage was Roh. While the youthful, soccer-loving multimillionaire flirted with a smitten public, Roh's standing with the electorate was falling fast. The culprit was that familiar Korean combination of corruption and children—not Roh's but the children of the incumbent president, Kim Dae Jung. The day after South Korea's unexpected win over Poland, prosecutors arrested President Kim's youngest son on charges that he had evaded taxes and solicited $3 million in kickbacks. D.J. was looking worse every day; our message, which connected Roh and the president, was working very effectively. By June, Lee and Roh were virtually neck and neck in the polls. In fact, I became concerned that our strategy was working too well. Inside the ruling Millennium Democratic Party, a growing number of politicians were agitating to replace Roh with Chung. If they succeeded, Lee's situation would be much tougher. If Chung maintained his current momentum, he would almost certainly beat Lee. If, however, Chung could be persuaded to enter the race as an independent, then he would split Roh's vote and potentially ensure Lee's election.

Fortunately for us, Roh stubbornly refused to give up the fight. All of the conditions for a divided election seemed to be falling into place. In August, disgusted voters opted overwhelmingly for the

GNP in parliamentary elections. Once more Roh came under pressure to resign. Once more, he refused to step down.

Yet as heartening as these trends were, Lee's path to the presidency was not a completely clear one. For one thing, his favorability rating continued to be alarming low. Only 29 percent of voters had a favorable opinion of Lee; 56 percent had an unfavorable opinion. Those are bad numbers: Richard Nixon enjoyed a similar level of support, *after* the Watergate scandal had broken. They showed that Lee had no reservoir of goodwill. Even his supporters were voting more for the GNP and for change rather than for Lee personally. We had no room for mistakes. In contrast, Roh's favorable/unfavorable ratings stood at a much more encouraging 44/40 percent, despite months of negative communications aimed at lowering his numbers. One on one, Roh was still the stronger candidate.

Lee's most serious problem was with the "386 generation"—a group that had grown up in an affluent, high-tech society. Its members had no recollection of the devastation wrought by the North's invasion of the South in 1950. As a result, they tended to worry less about security in general and the North in particular than did voters over the age of fifty did. However, they did remember battling the U.S.-backed South Korean military during the 1980s. They tended to view the conservative establishment as South Korea's most serious problem, while dismissing North Korean dictator Kim Jong Il as a blowhard who would never go so far as to actually attack his brothers to the South. Despite its mixed results, President Kim Dae Jung's sunshine policy was popular among the 386 generation.

We knew that Lee would have a hard time connecting with most of these voters. Young men were particularly hostile: they had never forgiven Lee for his sons' failure to serve in the military. Nevertheless, in order to win, Lee had to make some inroads with this group. I decided that our best bet was to target families with young children and, to a lesser extent, women. Toward that end, I pushed hard for Lee to develop a "young families" initiative and promote other ideas that would demonstrate his commitment to the younger generation. "You've got to put forward a more populist economic campaign," I told the candidate. "You've got to talk about closing the gap between the rich and poor and propose programs that will address people's concerns about affordable housing, day care, and health care."

By this point in our relationship, Lee and I were on speaking terms—no more nodding and just saying, "yes, yes." But while Lee agreed with me in principle, in practice, he simply wasn't willing to run the Clintonesque campaign that I was pushing for. Lee was a decent fellow with an acerbic sense of humor, but as I witnessed during his televised debate in 1997, if you put him before an audience, he became cold and rigid. Lee had also become a touch fatalistic. Sometimes when I would urge his team to come up with a package of housing or health care initiatives, Lee would respond in a way that suggested he was bemused by my enthusiasm. "Mr. Schoen," he would say to me, "solve this problem for me. You tell me I must appeal to young people, yet the media only wants to report on my fights in parliament. What use is it to propose initiatives when the press only covers controversies?"

Lee's problems did not lie only with the media. His campaign also suffered from tactical problems. Simply put, Lee proved unwilling to make the deals necessary to secure support that was still in play. In Korean politics, making alliances and managing competition is often the key to victory. In 1997, Lee lost in part because Kim Dae Jung struck an alliance with the reactionary Kim Jong Pil—a man who had twice ordered his assassination—and because Lee's rivals had maneuvered Rhee In Je into the race. Building the right alliances would be the key to victory this time around as well. Unfortunately, making alliances and protecting his base was something that Lee refused to do, as his handling of Park Geun Hye would demonstrate.

Park was the daughter of General Park Chung Hee, one of the most controversial leaders in South Korean history. In 1961, General Park had seized control of the government in a coup d'etat. Once in charge, however, his reforms laid the foundation for South Korea's economic miracle. Even four decades later, he remained a popular but polarizing figure. However, a few things about his political presence were clear. Ideologically, supporters of the young Ms. Park should have been Lee backers. Like all Korean politicians, however, Park wanted to negotiate the best deal she could for herself. For her support, the younger Park was holding out for nothing less than the position of prime minister. It was a high price, but I thought Lee should pay it.

That winter, Lee and I had met up for a strategy session in New York. At the meeting, I urged him to seal the deal with Park. He refused. "She's not qualified; I won't do it," he told me firmly. At that moment, I realized with dismay that Lee simply wasn't willing to make the deals that were necessary to win. (At the end of the campaign, Lee would belatedly change his mind, but by then it was too late.) Such principle was very admirable, in the abstract, but in practice, however, Lee's unwillingness to deal coupled with an almost unwavering belief in his own righteousness had turned nearly every significant political figure in South Korea against him.

Still, all was not lost. If soccer organizer and Hyundai heir Chung Mong Joon did in fact enter the race as an independent, he might split the vote with Roh and allow Lee to eke out a narrow win. And in September, Chung did just that, ending his long flirtation and announcing his entry into the presidential race. I was delighted. In a two-way race, Chung enjoyed more than a ten-point lead over Lee, but with Roh and other marginal candidates in the race, our polls showed Lee with a narrow lead. Even if Chung eventually drove Roh out of the race, I believed Lee would have a good shot. Allegations from former Hyundai executives that Chung had illegally manipulated Hyundai share prices four years earlier were already starting to leak out.[5] Increasingly, I saw Chung as a Ross Perot figure—someone with broad but shallow support, who looked appealing now but whom voters would eventually recognize could never govern. That assessment would soon prove correct. What I didn't anticipate was a terrible accident, one of the most bizarre uses of a political poll in the history of electoral politics, and an orchestrated campaign of anti-Americanism.

The accident came first. In June, at the peak of South Korea's soccer craze, at a moment when South Koreans were flush with feelings of nationalism and pride, an American armored personnel carrier out on a training exercise ran over and killed two young Korean girls who were walking along the side of the road on the way to a friend's birthday party. The accident enraged the South Korean public. The presence of American forces in South Korea had long been a sore point—particularly in Seoul, where the enormous Yongsan military base occupied much of the downtown area. During the tense decades of confrontation with the North, the annoyance of having American forces occupying an important part of the capital had been offset by the comforting knowledge that an attack on Seoul would be an attack on the United States. By 2002, the mood had changed, and in light of D.J.'s overtures to the North, the fear of attack had subsided. Gratitude for U.S. support had given way to resentment—resentment at the large U.S. military footprint, resentment at the years of support for South Korean military dictators, resentment that the American soldiers who drove too fast and killed two Korean girls would be tried before an American military tribunal instead of before a Korean court. So it was no surprise that after the accident, South Korea erupted in protest.

At first, these anti-American demonstrations had no discernible effect on the campaign. Roh had attempted to capitalize on anti-American feelings in the past, and in the wake of the killing, he launched a few more provocative sallies, gaining little traction. Indeed, his tough-on-America, soft-on-North Korea stance soon backfired. In October, American officials revealed that North Korean negotiators had admitted the existence of a secret weapons program—a revelation that quickly pushed anti-American protests off the front pages of South Korean newspapers. Although older and younger South Koreans ranked the threat posed by North Korea differently, they largely agreed that a nuclear North was very bad news indeed for the country. By the end of the month, Roh was running a distant third to Chung and Lee. One newspaper noted that Roh "has not so much an uphill struggle as a vertical cliff to climb if he is to have any chance of winning the December 19th poll." At least half of the Millennium Democratic Party lawmakers reportedly wanted Roh to step down so that the party could unite behind Chung Mong Joon.

Then in the third week of November, Roh got his big break: the American soldiers who had run over the two girls were acquitted of negligent homicide by the American military tribunal and quickly flown out of the country. The verdict convulsed South Korean society. Koreans were stunned by the acquittal and upset that there was no possibility of some other form of punishment. The verdict surprised me as well, since given the obvious political volatility, it seemed clear to me that the military should have made a major effort to postpone the trial or in some way prepare for the aftershocks that would result from an acquittal. Instead, Washington had stood by while a military tribunal effectively struck a match in the tinderbox of South Korean public opinion.

The folly of this indifference soon became apparent. The sprawling Yongsan base was besieged by protestors. "Yankee go home" signs appeared in Korean storefronts. Worst of all, the verdict also resurrected Roh's campaign. For months, Lee had been trending steadily upwards. In the days after the acquittal, the upswing paused. Within a week, a new trend was unmistakable: Lee's ratings had started to slide while Roh's numbers were beginning to rise.

Roh saw his opportunity, and he seized it. For years, he had toyed with anti-American sentiment. Now he moved even closer to the edge. In a televised debate, Roh declared that as president, "I have no intention of kowtowing to the United States"—unprecedented language from the president of a country whose fate would have been starvation under a Communist dictatorship but for the sacrifice of fifty-four thousand American lives five decades earlier. At one point, he went even further. In its negotiations with the United States, North Korea at one point had demanded a "security guarantee" against American aggression. Coming from the country that had started the Korean War, it was a ludicrous demand. Yet instead of dismissing it as nonsense, Roh suggested that South Korea would be willing to extend such a guarantee—presumably to protect North Korea from the United States. In the face of vociferous protests from Washington, Roh quickly withdrew this suggestion; however, it was clear that his overall strategy was working. With the enthusiastic support of his "Nosamo" Internet club members, Roh had managed to ride out the Chung challenge and maintain a strong base of highly committed supporters in much the same way that Howard Dean maintained support during the Democratic primaries. Now he was making a strong appeal to the 386 generation both in style and substance. After languishing behind both Chung and Lee for months, Roh was surging. By late November, he was neck and neck with Chung.

Despite the revelations that North Korea had been secretly pursuing a nuclear weapons program, both Chung and Roh continued to support Kim Dae Jung's sunshine policy. As a result, the two men were effectively competing for the same vote. By then, it was clear that if they both stayed in the race, Lee would win, a result neither candidate wanted. So Chung and Roh struck what was surely one of the strangest bargains in the history of modern politics: they decided to let a poll determine who Lee's opponent would be. Under the terms of the deal, the two candidates agreed to face off in a televised debate. At the end of the debate, a Korean research company would poll two thousand viewers. Whoever came out ahead would then become "the unified candidate."

I was astonished by this arrangement. Chung might well know a lot about soccer and shipbuilding, but when it came to politics, he was clearly out to sea. In exchange for agreeing to drop out of the race, he could have extracted any concession he wanted from Roh. The prime ministership could easily have been his. Instead, he had agreed to step before the television cameras and let whoever tuned in decide his political fate and determine the future of the country. It was an amazing and disquieting decision: just imagine if in 1992 Ross Perot and Bill Clinton had decided to debate each other on *Larry King Live* and then field a poll to determine who would challenge George Bush. The idea seems ludicrous. Yet that is precisely what Chung and Roh were agreeing to do.

On Friday, November 23, Chung and Roh met for this most unusual of showdowns. It was hard to imagine a more high-stakes debate, yet Chung arrived unprepared. Given the many similarities between his platform and Roh's, Chung needed to present a coherent rationale for his candidacy, but by the end of the encounter, it was clear that he had none. After the debate was over, a private research firm called two thousand people to gauge their reaction. Thirty-two percent of the respondents expressed a preference for

Lee; those preferences were disregarded. Of the remaining 68 percent, 48.6 percent supported Roh versus 42.2 for Chung with the remaining 10 percent undecided. The "outcome" was a close one; it was not even entirely clear that the questionnaire had been put together properly. Yet incredibly, Chung agreed to end his candidacy.

Chung's decision to let polls determine his fate wasn't just a novice's error. It also reflected an alarming, seemingly logical, conclusion: if polls are accurate, why bother with elections anyway? Though it might sound surprising, I had never before thought of polls as a substitute for an election. As a professional political consultant, I view polls as a starting point, a way to understand the public, and move them in the direction of my candidate. Chung, in contrast, had let polling replace voting. For him, it was a major mistake, and yet I wonder if it is a mistake that won't be repeated in the future.

The scenario that we had feared most had now come to pass. Roh and Lee would be going head-to-head against each other. The first poll numbers after the bizarre television showdown were even worse than I had feared: they showed Roh with a startling twenty-point lead over our candidate. The only way we could win was if we drove Roh's favorability ratings down. That meant a relentless negative attack on Roh for the final weeks of the campaign.

Our polling suggested three lines of attack. The first two were personal: "Roh has no clear/workable policies," and "Roh is not able to reform South Korea." The numbers also showed that Roh was vulnerable on the issue closest to his heart—the sunshine policy toward the north. The revelation that North Korea had been secretly pursuing atomic bombs while receiving aid from the United States and South Korea was evidently enough to give at least some South Koreans pause. Fifty-five percent of the public now believed that Roh's pledge to respond to North Korea's treachery by continuing and even strengthening the sunshine policy would put South Korea at risk. It wasn't the biggest margin to work with, but it was the best shot we had. Consequently, driving home the message that "Roh's North Korea policy will be a failure" became a major part of the final push.

Unfortunately, despite this strong message, the anti-American protests just wouldn't go away. Early December saw the largest anti-American protests in South Korea's history. A mob broke through police barricades in front of the U.S. embassy and briefly threatened the embassy itself. An American soldier was attacked by three Korean men and stabbed on his way back to his post. Both the Korean and the international media treated these rallies as spontaneous outbursts of Korean anger. Sociologists and historians took to the air and the op-ed pages to explain the roots of Korean rage. No doubt there were historical grievances; no doubt many people were angry, but on the eve of the election, I would learn that those factors hadn't been the only thing behind the protests.

The final stage of the campaign was dominated by the candidates' differences over North Korea. In his debates with Roh, Lee stayed on message, hammering his opponent for supporting Kim Dae Jung's sunshine policy. Time and time again, Lee drove home a simple point: "We've had five years of sunshine policy, while behind our backs North Korea has been working on nuclear weapons." As president, Lee vowed to cut off economic aid to the impoverished North unless Kim Jong Il agreed to reopen all of his nuclear facilities to inspection by the International Atomic Energy Agency and cease all work on nuclear weapons.

Roh responded by defending the sunshine policy and attacking George W. Bush. He criticized Bush for refusing to negotiate with the North. "I do not think Bush's hardline policy will help North Korea to eliminate its nuclear program," Roh told foreign reporters—as if all North Korea needed was "help" to deal with the problem. It was Kim Jong Il versus George W. Bush—and the North Korean dictator was winning.

The situation for Lee was not improved by U.S. policy. The Bush administration had originally responded to the death of the two young Korean girls with a desultory apology. After the acquittal of the U.S. soldiers, President Bush delivered another official message of regret through U.S. Ambassador Thomas Hubbard that many in South Korea saw as insultingly inadequate. Eventually, through his contacts in Washington, my lawyer friend Sukhan Kim managed to pry forth another, more sincere apology—a direct phone call from President Bush to President Kim. But by then it was too little, too late. Lee's victory or defeat would have a huge impact on the U.S. position in Asia—it would determine whether we would have a stalwart ally ready to stand up to the North or a regime committed to a policy of appeasement—yet Washington seemed unaware of just how much its inept response was hurting the man who might be a valuable ally. The U.S. government's halting response was now another hurdle we would have to overcome.

While South Koreans raged against U.S. forces, North Korea suddenly made a few efforts to be nice. It agreed to another round of family reunions and announced that it had finished clearing land mines on its side of the demilitarized zone, thus laying the groundwork for the resumption of rail service. It didn't always sustain the effort (for instance, it failed to open a road through the demilitarized zone that had been under discussion for some time); however, it was clear that Kim Jong Il was far more closely attuned to South Korean public opinion than the Bush administration. Certainly, we would have welcomed similar gestures from the U.S. government because all indications were that the race was very tight. Roh enjoyed a clear lead among younger voters, and voters under the age of forty made up 48 percent of the electorate. That suggested Roh had a clear lead. While public polls were banned in the final stages of the campaign, our numbers showed Lee trailing Roh by the narrowest of margins. Our only hope was that young voters wouldn't mobilize to vote—and that when voters did step into the voting booth, they would think twice about casting their ballot for Roh and his policy of appeasement.

One afternoon about four days before the election as I was walking across the lobby of the Lotte Hotel, I heard a loud, "Hey, Doug!" It was David Morey, an American consultant who seemed to be something of a fixture in quasi-authoritarian East Asian countries, and his business partner, Bob Armao, former personal adviser to the Shah of Iran. No one knew much about exactly what David did. However, I did know that he had long worked for President Kim Dae Jung. D.J. was theoretically neutral, but it was obvious he was pulling for Roh. I had run into Morey in the fall and over lunch he claimed to be out of the electoral game—at least as far as South Korea was concerned. Now, Morey and Armao seemed delighted to see me, and frankly, I was glad to see them as well. Anti-American sentiments were running so high in Seoul that the Lee campaign preferred to keep me away from campaign headquarters. I had been identified in the South Korean press as an American "smear master," and they wanted to make sure that no one thought I was orchestrating the campaign. So when Morey proposed dinner for the night before the election, I accepted at once.

We met at the Hyatt Hotel. Over drinks, Morey filled me in on his recent activities. He had been in Seoul for two weeks working for his old client, President Kim Dae Jung during the recent period of demonstrations. I could not help but wonder if American political consultants had some involvement in orchestrating the anti-American demonstrations that had rocked the capital for the past month, making sure they had the right touch. Could the apparently spontaneous anti-American demonstrations have been stage-managed by an American political consultant? Those were the same demonstations that supposedly reflected generations of frustration. It was a possibility that took my breath away.

That wasn't the only startling revelation of the evening. That same night in another part of town, Hyundai ship-building heir and former presidential candidate Chung announced that he was withdrawing his support for Roh and coming out in favor of Lee. If the announcement had come a few days earlier, it might have given an important boost to Lee's campaign: Chung would have been seen as a statesman making a tough decision for the good of the country, and the Lee campaign would have had time to publicize his decision. Announcing his decision the night before the election made Chung look erratic instead. The Hyundai heir had bungled things again.

Thursday, December 19 was Election Day. Lee's fate would hinge on whether South Koreans stepped into the voting booth and thought twice about committing the country to Roh's "see no evil" policy toward the North. The first wave of exit polls gave me reason for modest optimism. Morning voters were leaning toward Lee. Many of these voters were undoubtedly older Lee supporters. Still, it was a hopeful sign. If Roh's younger supporters turned out in light numbers, Lee just might eke it out, I thought, particularly if the undecideds broke our way. Unfortunately, neither I nor anyone else in the Lee camp was prepared for what happened next.

It turned out I wasn't the only political consultant conducting exit polls. Roh's strategists were watching the pro-Lee trend too.

That afternoon, they resorted to a novel weapon: they sent out approximately one million text messages to young voters who had registered on "Go Roh" websites and urged them to go to the polls.

These last minute efforts apparently worked. Lee lost again, this time by 2.3 percent.

In the end, Lee's campaign had floundered on a generational rift. While he won easily among voters ages fifty and over, younger voters, particularly members of the 386 generation, went overwhelmingly for the fifty-seven-year-old Roh. Young voters wanted change, not counsels of caution from sixty-seven-year-old Lee. We had gone up against a generational divide and an opponent who had skillfully played on anti-American sentiments and lost. I was disappointed. My job was to win, even against opponents who played dirty. However, a few months after the campaign, I found out that I had once again underestimated my South Korean opponents. Roh's win was not the result just of a generational desire for change and dirty tricks. He was also the beneficiary of a fraud perpetrated during President Kim Dae Jung's presidency—a fraud of truly shocking proportions.

Two months after Roh Moo Hyun took office as president, it was revealed that the summit meeting between North Korean tyrant Kim Jong Il and then-President Kim Dae Jung—the summit meeting that had launched the so-called sunshine policy—had been bought and paid for. The price of the meeting was a payment to North Korea of somewhere between $100 million and $500 million from the *chaebol* Hyundai. The world may never know just how much Hyundai paid: soon after the scandal broke, the chairman of the Hyundai Group committed suicide by leaping out of his office window. Far from being the result of a successful diplomatic overture, dialogue with North Korea had been purchased with a bribe. Of course, these revelations came too late for Lee Hoi Chang. Two days after the election, he had announced he would resign from politics. Instead of a strong leader who was prepared to confront North Korean provocations head on, South Korea—and the world—face the future with a leader whose election was based on a massive bribe to the world's last Stalinist dictator.

PART THREE

THE BIG TIME

CHAPTER SEVEN

NEW DEMOCRATS, INOCULATION, BILL CLINTON, AND MA BELL

T hough our work abroad was a crucial component to our business, consulting on the domestic front was always our primary focus. At a very early stage in our business, we discovered that we were particularly effective at helping to elect Democrats in states or districts where the cultural and political climate was unfriendly—if not outright hostile—to liberal policies. Such was the case in the early 1980s, when being a Democrat was no easy task. Ronald Reagan's sweeping win in 1980 marked the eclipse of the old New Deal coalition by a new and potentially long-lasting Republican coalition.

Lulled into complacency by decades of power, many of the Democratic congressional barons in Congress hadn't seen the threat coming, but as two young pollsters working to make a living, Mark and I had. Outside of Democratic enclaves like New York City,

Democrats had a near-fatal image problem with the electorate. Democrats were seen as tax-and-spenders who were soft on defense and overly permissive when it came to crime and welfare. For Democrats to succeed in the increasingly rough-and-tumble world of national politics, we needed to change the rules. It was time to put aside failed orthodoxies and start paying attention to what swing voters actually wanted. It was also time to start playing tough.

In the summer of 1982, we got just that chance in the form of a call from a long-shot senatorial candidate in New Jersey named Frank Lautenberg. Lautenberg, a wealthy (he had founded the payroll-processing company Automatic Data Processing) but little-known businessman, was running against the doyenne of New Jersey politics, Millicent Fenwick, a Republican known for her support of civil rights, her dashing looks, and her eccentric behavior. (When her doctor insisted that she stop smoking cigarettes, she switched to pipes and inspired Garry Trudeau to create the pipe-smoking grandmother *Doonesbury* character, Lacey Davenport.)

When we joined the campaign, Lautenberg was trailing badly. Our benchmark poll showed Fenwick had a twenty-five-point lead. Lautenberg was recognized by a mere 28 percent of New Jersey voters.[1] The *Wall Street Journal* referred to Fenwick as the most formidable GOP candidate in the entire fall election.[2] However, a deeper look at the polling numbers revealed glaring vulnerabilities. Fenwick's visibility was high, but on many of the most important issues she was perceived as being out of touch. Nor was she fully trusted by voters worried about the state of the economy.

We mapped out a two-tier strategy for Lautenberg. First, we set out to win suburban New Jersey voters, especially women. Fourteen years before the "soccer mom" was identified as the reigning force in American politics, we realized that suburban, female voters were the key to winning national elections. The campaign began running a series of positive spots highlighting Lautenberg's business background and identifying him with the middle class that proved extremely effective in raising Lautenberg's profile with New Jersey voters. When these first rounds of ads had finished running, the electorate was willing to consider him as a viable candidate. However, we knew that positive ads alone would not win the race against a relatively popular opponent. Once Lautenberg's favorabilities began to crest, it was time to go on the attack.

Fenwick's, and indeed the Republican Party's, great vulnerability was the economy. With an unemployment rate of nearly 10 percent, we had a huge opening. Those numbers helped to move the polls in Lautenberg's favor. In 1982, many people were fed up with the GOP, who seemed to be out of touch with middle-class values and concerns. We turned these worries against Fenwick, highlighting votes in Congress that had cost jobs in New Jersey.

We also played hardball, indirectly attacking her as an eccentric old woman. Many in the campaign felt that going negative against a female candidate would boomerang against Lautenberg, but my partner Mark Penn insisted on it. Both Mark and I thought voters could handle the truth, and in fact, our polls soon showed that only hard-core Fenwick voters were troubled by Lautenberg's attacks. Most voters wanted to hear what we had to say, and when they did, they shifted their support to Lautenberg. In the end, Lautenberg won a narrow victory with 52 percent of the vote.

Nationwide, Democrats picked up twenty-six seats in the House during that election year. Some saw this as a triumph but Mark and I were bitterly disappointed. Considering the economic situation in the country, the numbers should have been much higher. Too many Democrats had run on an old-fashioned platform calling for greater "fairness" and more distribution—a stance that fundamentally misread the American electorate. As we noted in a *New York Times* op-ed written only a few months before the November election, "the electorate is far more concerned with results than with fairness." Voters were bothered by the economic downturn, but unless Democrats could come up with an alternative that cut across class and income lines, they would stick with Reagan. "Continued resort to the politics of class could be a waste of their precious campaign time," we warned.

Contributing to the Democrats' malaise was their failure to recognize the potency of the tax issue. Voters from across the political spectrum saw Democrats as liberal, big-spenders who were taking their hard-earned tax dollars and putting them into a bottomless pit of social programs. In the late 1970s and 1980s, many voters saw this as the number one issue in the country. Of course, the reality was far more nuanced, but in politics, perception is key, and the perception in the 1980s was that the GOP represented the middle class—and that Democrats had become the party of the lower class. As long as they were unable to show empathy for the struggling middle class, electoral victory would, in our view, remain out of reach.

Yet few of the politicians running for office in 1982 or even 1984 seemed to understand this. Democrats were still far too tied to the special interests (particularly the labor movement), which had come to define the Democratic coalition, making them unable to get away from a message that had brought success in the 1960s but by the mid-1980s was completely outdated. The tunnel vision of Democrats was clearest during the 1984 Democratic convention when the party's candidate Walter Mondale declared he would raise taxes if elected—a proclamation that played right into the hands of the Republicans tax-and-spend attacks and the electorate's negative perceptions of the Democratic Party. Reagan's popularity was matched only by the unpopularity of Democrats. We were hardly surprised when Mondale was swept in a forty-nine-state landslide.

After our experience with Lautenberg, we encountered similar success with Jay Rockefeller in West Virginia during his successful run for Senate in 1984 and with Richard Shelby of Alabama in his 1986 Senate bid, convincing us that Democrats had to move to the Right on fiscal issues. Democrats had to get serious on the deficit, call for spending cuts, and avoid any talk of tax increases. In short, they needed to co-opt Republicans on fiscal matters and take these issues off the table, instead fighting the party's political battles on a battlefield where local and social issues dominated and Democrats had a natural advantage.

The election of Richard Shelby to the U.S. Senate was a case study of this approach. With little more than myself, a press secretary, and media strategist Bob Squier, Shelby, a four-term Democratic congressman, set out to unseat Republican Senator Jeremiah Denton, a retired admiral and Vietnam War vet who had been one of the longest serving POWs in American history. Both men were seen as conservatives; our polls showed that the electorate did not see a huge difference between the two candidates. However, Denton was vulnerable to attacks on his lukewarm support for Social Security, his vote in the Senate to increase his salary, and his lifestyle. (He owned two Mercedes, was seen infrequently in Alabama after his election, and made a crude remark in public about spousal abuse.) Under our advertising onslaught, his twenty-five-point lead quickly evaporated, and on Election Day, Shelby eked out a narrow victory.

As we wrote in another *New York Times* op-ed, days after the 1986 midterm election, the Democrats were starting to get the message, and it helped them take back control of the Senate:

> Indeed, virtually every successful Democratic candidate for the Senate made it clear from the start that he opposed wasteful spending on social programs, opposed using tax reform as a way to raise taxes and supported government policies to encourage economic growth and traditional American family values. . . . In this year's Senate contest, it was only after Democratic candidates succeeded in demonstrating that they shared the Republicans' basic positions on fiscal issues that they could turn to the local issues and personality differences that led to such a sweeping Democratic victory.[3]

What we realized in 1986 was that when Democrats neutralized the differences between them and Republicans on fiscal issues, it became almost impossible for GOP charges that Democrats were liberal, big-spenders to stick. On local issues—whether it was nuclear waste in Nevada and Washington, support for textile workers in North Carolina, or Social Security in Alabama—the Democrats trumped Republicans. The key was to take fiscal matters off the table.

What we had put forward in November 1986 was not just a blueprint for electoral victory: it was also a road map for remaking the Democratic Party. We were certainly not the only strategists groping for a new approach; however, our efforts to move the Democratic Party beyond class warfare and toward more centrist positions—along with a record of success that demonstrated that candidates who followed this counsel could win—played a critical role in bolstering the New Democratic movement that Al From organized nationally as the Democratic Leadership Council. Nowhere was the strength of our argument demonstrated more forcefully than in the 1988 election of Evan Bayh as governor of Indiana, a rock-solid Republican state. Bayh's election demonstrated just how potent this approach could be.

Bayh entered the race with the advantage of fame—his father, Birch, had been a U.S. Senator and had run unsuccessfully for the Democratic presidential nomination in 1976—and the disadvantage of running in an overwhelming Republican state. His father had been defeated in 1980, and it had been twenty years since Indiana had selected a Democratic governor. To beat his opponent, the state's lieutenant governor, John Mutz, we quickly realized that Bayh would have to inoculate himself from traditional GOP attacks.

We found our opportunity in the most unusual of places—taxes. Mutz was lieutenant governor during a time when Indiana levied the two largest tax increases in state history. According to our poll results, he was vulnerable on the issue. Indianans believed that higher taxes could have been avoided if politicians had shown greater fiscal restraint. Even though the increases went to fund education programs and avert a budget deficit, Mutz's explanations fell on deaf ears. More consequently, his attempts to portray Bayh as a big-spending, tax-raising liberal Democrat simply didn't resonate. Quite simply, Indiana voters viewed Bayh as the fiscal conservative.

Once the tax issue was off the table, Bayh was able to fight Mutz on more favorable turf. He took the Republican to task for his administration's lack of support for economic development in the state. One of the ironies of that campaign was that Mutz spent much of his time taking credit for attracting a $500 million Subaru-Isuzu plant to the state. What he failed to understand was that this issue actually hurt him: at our behest, Bayh successfully accused Mutz of focusing more attention on overseas investment rather than Indiana's own business community.

Our strategy was extraordinarily successful. An October article in the *Wall Street Journal* noted that Bayh "often sounds more Republican than his Republican opponent."[4] A statewide poll in early

October indicated that Indiana voters thought Bayh would be more effective than his opponent in keeping taxes down, creating jobs, and spending money wisely. For a Democrat in 1988 to have that kind of image in a state as conservative and hostile to Democrats as Indiana was nothing short of a political miracle. Even more extraordinary was the fact that 1988 was a presidential election and the GOP's vice presidential candidate was an Indiana native son—Dan Quayle. Yet on November 8, the thirty-two-year-old Evan Bayh was elected governor by a comfortable six-point margin.

Something remarkable had occurred on the Indiana plains: a new type of Democrat had been born, one that could neutralize Republican strengths and appeal to more conservative voters. It was a victory that helped pave the way for a similar triumph four years later—the election of Bill Clinton.

O ur successes during the early 1980s had turned us into a sought-after commodity in the small world of political polling, convincing us that we could build successful careers in the field. While my natural assumption was that we would limit the success of our firm to the world of politics, I soon learned that Mark had another idea: He wanted to enter the corporate marketplace.

I'd worked with corporate clients before. In 1979, Ron Maiorana, a former speechwriter for Nelson Rockefeller, had asked me if we'd be interested in doing a series of polls and focus groups on consumers' attitudes toward gas station convenience stores for Hess Oil. I'd readily agreed, and a meeting had been scheduled with Phil Kramer, the company's president. Somewhat nervously, I'd informed him that the cost of the project would be thirty-five thousand dollars—a princely sum for two young pollsters in 1979. Fine, fine, he'd nodded, somewhat absent-mindedly. I'd wanted to make sure he'd agreed, so I looped back to the cost once or twice in the remainder of the presentation. Each time I'd gotten the same faint nod. Eventually, I decided it was a yes and wrapped up the meeting. Needless to say, I was in high spirits when I left the boardroom so I was surprised when Maiorana turned on me and said, "Schoen, you are a fucking idiot."

"Fucking idiot?" I responded. "I just sold that guy on a thirty-five thousand-dollar poll! Give me a break. How do you figure on that?"

"Look," he said, "all that the guy could have said when you asked for the fourth time was, well, I'll think about it. When you have a sale, leave the room as quickly as you possibly can."

In the years that followed, Mark and I continued to take on the occasional corporate client, but it wasn't until 1984 that Mark got the push that would change the future of our firm. It came in the form of an unexpected rejection from Frank Lautenberg.

During Lautenberg's 1982 campaign, Mark had played an important role—indeed, the critical role—in helping to win the election. In the final days of the campaign, Mark insisted—in the face of considerable internal opposition—that Lautenberg continue to run tough, negative ads against Millicent Fenwick. Those ads almost certainly put Lautenberg over the top, and in the years that followed Mark took justifiable credit for this strategy. That evidently rubbed some on the junior senator from New Jersey's staff the wrong way. Two years later, Mark was abruptly dropped from Lautenberg's team. It was a bruising experience for Mark, one that temporarily soured him on working for politicians and subsequently resulted in his major effort to attract nonpolitical customers building on our experience working for Avis, which had begun with David Garth and Ron Maiorana in early 1982.

For a small political consulting firm—a small *Democratic* political consulting firm—it was audacious idea. Compared to the advertising and public relations leviathans of Madison Avenue, Penn & Schoen was only a small player. Frankly, I was skeptical that we should try to cultivate corporate America as a major part of our client base. My work for Hess had certainly been remunerative, but it had hardly seemed central to my primary interests, or Mark's for that matter. Little did I realize that Mark's push into the corporate world would ultimately throw us into one of corporate America's most contentious battles—the long distance phone wars that pitted industry giant AT&T against upstart MCI. Still less could I foresee that in the course of our work for AT&T, Mark and I would develop a new tool—the mall tests—that would revolutionize the practice of public opinion research and

ultimately play a pivotal role in the reelection of President Bill
Clinton in 1996.

Our first major client was Texaco. In 1985, Texaco had
snatched Getty Oil away from Pennzoil, just days before the
two companies had planned to sign a formal acquisition agreement. Pennzoil responded to the Texaco raid with a lawsuit charging Getty with violating a binding oral agreement. Since Texaco
had indemnified Getty against any lawsuits arising from the deal,
Texaco was at risk. Later that year a Texas jury delivered a stunning blow to the company, ruling in favor of Pennzoil and fining
Texaco more than $10 *billion*—a ruling that forced the oil giant to
file for bankruptcy.

During this difficult period, I got a call from Mike Rowan, who
had worked for political consultant Joseph Napolitan before joining
Hill & Knowlton. Rowan knew that Texaco needed someone with
political skills to help the company navigate the difficult bankruptcy
proceedings and protect its brand. As we moved deeper into the
business world, we were increasingly struck by the similarities between political campaigns and corporate marketing. The act of voting was quite akin to what consumers did in the marketplace.
Building a candidate's name recognition was like building brand
awareness. Party identification guided people's likely choices in the
voting booth in much the same way that the reputations of parent
companies influenced consumer views of individual products. Both
politicians and companies were eager to attract new voters (or consumers); both competed for the loyalties of swing voters—or, in corporate parlance, occasional buyers.

The mechanics of the two businesses also matched up. Corporate brands, like political brands, are built via expensive marketing
campaigns where the stakes are higher every day. Both candidates
and companies had to compete and project themselves on multiple
fronts. Like candidates, corporations are frequently blindsided by
events outside of the brand's control. Finally, like the best politicians, the best brands are the ones that respond directly and effectively to attacks, manage crises, inoculate themselves against

threats, exploit weaknesses in opponents, and persuade people to believe in them.

Political and Corporate Similarities

Political Campaign		Corporate Campaign
The Vote	⬅➡	Consumer Choices
Party Identification	⬅➡	Brand Loyalty
Job Performance	⬅➡	Customer Satisfaction
Candidate Reputation	⬅➡	Brand Image
Swing Vote	⬅➡	Occasional Buyers
Campaign Promises	⬅➡	Brand Promises
New Voters	⬅➡	New Buyers
Advertising, Events	⬅➡	Advertising, Promotions
Negative Advertising	⬅➡	Comparative Advertising
Multiple Fronts	⬅➡	Multiple Fronts
Reality/Perception Gap	⬅➡	Confusion

Source: Penn, Schoen & Berland Associates, Inc.

We were equally struck by the differences. As political consultants, we were used to polling instantly and making big decisions quickly, often in the course of a single twenty-four-hour period. As campaign veterans, we were accustomed to high-stake, cutthroat environments and to operating with a sense of speed and urgency. Yet this seemed rare in many corporate suites. In part, this reflected the winner-take-all nature of politics, whereas in the marketplace, a company that came in second might still make a decent profit. A politician who comes in second is a loser. At that time, most large corporations were strikingly bureaucratic, plotting their moves for months, even years, often paying little attention to their public images. While campaigns relentlessly focused their efforts on swing voters and maintaining the base, all too often, corporations attempted to target everyone when selling a product. We thought we could do better. What we were attempting was not unprecedented, but still, our foray into the corporate world represented an interesting historical reversal. Four decades earlier, back when President Eisenhower first ran for the presidency, he had relied on skilled advertising executives to shape his campaign. In the years that followed, however, most of Madison Avenue lost touch with the practice of politics, save for the occasional presidential campaign. The campaigns were too short and bare-knuckled, and the pay was inadequate. As a result, the big advertising firms and the corporate research departments had failed to develop the kind of quick research and response capacities that we had honed over the course of the preceding decade. There was an irony here: Mark and I were now coming full circle, bringing skills once thought to reside exclusively on Madison Avenue back to corporate America.

Our basic advice to Texaco was straightforward: present the company's bankruptcy as a story about a blue chip firm with high-quality products and good ethics brought low by a ridiculously large jury award and fight the counter-story that this was the cautionary tale of an oil giant damaged by its own unethical behavior. The strategy we helped formulate was successful, as the torrent of criticism directed at the jury award would eventually attest. It ultimately led to an ongoing arrangement with Texaco which Mark grew substantially, and then with Caltex, Texaco (and Chevron's) international division. That work eventually became so important that Mark and I soon made Mike Berland a partner in the firm in large part to handle that relationship. Penn & Schoen—now Penn, Schoen & Berland—had a new business.

It was not surprising that a significant early corporate client came from the world of politics. What was surprising was that he was a prominent Republican—Mitch Daniels. Today, Daniels is the governor of Indiana, a position he secured after serving as the director of the Office of Management and Budget, one of Washington's most powerful jobs, in President George W. Bush's first term. Even back in 1990, he was a prominent figure in the GOP. A former aide to Senate Foreign Relations Committee Chairman Richard Lugar, Daniels had worked in a variety of positions in the Reagan White House. In 1987, he moved back to Indianapolis to head the Hudson Institute, a conservative think tank, switching again in 1990 to the private sector and taking over North American pharmaceutical operations for the Indiana-based drug giant, Eli Lilly.

I never expected to find myself talking strategy over lunch with the likes of Mitch Daniels, nevertheless that's exactly where I found myself in the summer of 1990. An intermediary had suggested that Eli Lilly work with us on a super-secret new project, and Daniels, looking past our political differences, had agreed to a meeting. Over lunch in a generic Eli Lilly cafeteria, he was frank and disarming. He told me that he had been impressed by our work for then-Governor Evan Bayh and that he wanted us to consult on a similar campaign for a revolutionary new treatment for depression that Eli Lilly was staking its future on—Prozac. Our work on Prozac would not only help usher in a new era of life-saving blockbuster drugs but also introduce a new concept into political strategy—the concept of "inoculation."

Today, Prozac is one of the world's most famous brands, and depression is universally accepted as a treatable illness. If a friend told you tomorrow that he'd been "feeling blue" for weeks, it would be quite natural to say, "Hey, you should see a doctor about that." The situation was very different in the late 1980s. Back then, it was rare indeed for someone, particularly a man, to confess to feelings of depression or haplessness. Mental illness carried a heavy stigma; it was a failing to be hidden, not an illness to be discussed. While psychotherapy offered some relief, only a small percentage of people had the time and money to afford it.

Prozac had the potential to change that—and to make Eli Lilly billions of dollars. According to the National Institute of Mental Health, nearly 10 percent of adults cope with serious depression in the course of a typical year—18.8 million people a year. Our charge was to help Prozac reach this potentially huge market. As I talked with company personnel, I realized that our true task was something far more challenging. We were not trying to convince consumers to purchase Eli Lilly's product instead of a competitor. Prozac was unique, and at that point, it had no real competitor, so instead of trying to beat the competition, we were trying to topple a centuries-old stigma. As a political consultant—a *Democratic* political consultant—it was an intriguing assignment.

Our first task was to help Eli Lilly differentiate Prozac from the older generation of antidepressants. From a biomedical viewpoint, Prozac was a revolutionary product, the first in a generation of drugs known as serotonin reuptake inhibitors (SSRIs)—something different from and better than what had come before. Researchers had long noted that people suffering from depression generally had lower levels of neurotransmitters such as serotonin in their brains than people who did not. Prozac alleviates feelings of depression by prolonging the presence of neurotransmitters in the brain. But despite its novel features, Prozac wasn't the first psychotropic drug. For decades, physicians had been prescribing—in fact, often *mis*-prescribing—a wide array of tricyclic drugs. Tricylcics have potent side effects, and physicians were supposed to take multiple blood tests in order to carefully adjust dosing. Many, however, did not. The result was a class of drugs with a very bad reputation. People on medication were, in the minds of most members of the public "crazy"—not ordinary "people like us" who just happened to be suffering from an illness. It was critical that Eli Lilly inoculate Prozac against any association with these older, less successful products.

We advised Eli Lilly to focus on three audiences in particular for its marketing campaign. First, physicians. To maximize sales, Lilly needed to emphasize Prozac's safety. Roughly sixty-seven hundred people had been involved in clinical trials of the drug—more than for any other antidepressant approved by the Food and Drug Administration (FDA) in the proceeding decade. Prozac was easy to dose, unlike the earlier generation of antidepressants, which often required physicians to do blood work before writing a prescription. It was much less toxic than other antidepressants, and its side effects were much milder and less objectionable (for instance, weight loss rather than weight gain). We helped Eli Lilly develop language for its sales reps to use during their physician visits. We also advised Eli Lilly that a major focus of its campaign should be to persuade ordinary physicians that Prozac's combination of safety, ease, and effectiveness made it a safe prescription choice for a whole range of symptoms.

Our second target audience was potential users. Today, anyone who spends an evening watching television will see a slew of advertisements touting medications aimed at alleviating every conceivable ailment, but back in the late 1980s, the idea of marketing a drug not to doctors but to the public was novel, indeed unprecedented. Doing so with Prozac was not only a smart business move; it was also good from a public health perspective. Although many experts (and television viewers) now decry the proliferation of prescription drug ads, the fact of the matter is that people suffering from depression were not getting the help they needed from the medical profession. We advised Eli Lilly to speak directly to this significant segment of the public. This was particularly true of men, who were much less willing to talk about their feelings than women. Lilly needed to communicate to men that talking about how they felt, about depression, was socially acceptable.

We also sought to inoculate the product against future attacks. Given the high occurrence of suicide among people suffering from depression and other mental illnesses, we knew deaths would be "linked" to Prozac. Since we lived in the most litigious country in the world, we could also predict with perfect confidence that the trial lawyer's bar would come after Lilly. While individual cases would be fought in court, a major part of our effort was to ensure that these legal battles occurred in an environment where public opinion was favorable.

With our guidelines in mind—and with potential pitfalls identified—Eli Lilly went on to craft one of the most effective marketing campaigns in American corporate history. People suffering from depression sought out Prozac by the millions. And when trial lawyers began to come forward with cases accusing Prozac of causing (in addition to suicide), murder, self-mutilation, spending sprees, and nymphomania, the response from the public—and from most juries—was skeptical. In short, inoculation worked. By the end of its first decade in the marketplace, sales of Prozac had earned Eli Lilly an estimated $28 *billion*.

What was most important to us, however, was the idea of inoculation. We quickly realized that companies who were unable to inoculate products with perceived risks early on did so at some considerable risk. A classic case in point was Procter & Gamble's launch of the fat substitute olestra. Once a self-proclaimed public interest group, the Center for Science in the Public Interest went after olestra with an almost certainly exaggerated and inaccurate campaign that warned of, among other things, anal leakage and increased risks of cancers.

Initially Procter & Gamble was slower to respond than it might have been with the benefit of hindsight. But it quickly recognized that dramatic steps needed to be taken. Steve Donovan the then head of the food division and Billy Cyr, a brand manager, came to us and said basically, "Do for us what you do for your political candidates when they are attacked with negative ads." Our response was an advertising and marketing campaign that calmly and clearly emphasized the product's safety and natural ingredients. By carefully targeting the test markets where CSPI was airing its charges, we were able to neutralize in large part the effect of its guerilla warfare campaign against Olestra. P&G immediately recognized the effectiveness of our approach and embraced the so-called "political model," with great effectiveness. In this case, Olestra's rollout was salvaged, but it never became the huge success Procter & Gamble had hoped for.

In early 1997, soon after AOL had switched to offering its huge customer base unlimited Internet usage for a flat fee, prompting a surge in usage that made getting access to the service very difficult, if not impossible during peak hours. I was brought in by then CEO Bob Pittman to develop a response. Our research quickly identified Steve Case, the founder of AOL, as one of the company's biggest assets. People related to his attractive, awe-shucks persona—many saw him as the friendly "boy-next-door" type. We advised AOL to put Steve Case on the air in a straightforward Jimmy Carter fashion—folksy, honest, wearing a sweater, with no lectures, no evasion—to acknowledge the problem with access and promise that he was doing everything he could to alleviate this problem. He also explained how much AOL was investing to upgrade the quality of their network.

Despite initial skepticism from the media, the approach was an immediate success. AOL customers appreciated the company's efforts—and its forthrightness. The fact that the public responded well to Case's personal appeal underscores the fact that Americans have no problem with great wealth per se. In fact, they admire it. Bill Gates or Warren Buffett, for instance, would be a potent politician.

* * *

Our successes in the corporate world quietly earned us a reputation for fast, creative research and effective tactics. I was only mildly surprised when in the summer of 1993 Mark was contacted by a major New York advertising firm with the following proposition: would we be interested in participating in a bid they were putting together for AT&T's advertising business? We would indeed. At the time, I saw it as a small project. Mark saw it differently—as a chance to do with telephones something similar to what we'd done fifteen years earlier with his personal computer. Our work for AT&T would soon change the character of our firm and the practice of public opinion research. Ultimately, it would change the trajectory of Bill Clinton's fledgling presidency as well.

AT&T's roots ran deep—back to company cofounder Alexander Graham Bell's invention of the telephone in 1875. For the century that followed, "Ma Bell" had provided the United States with the finest telegraph and telephone service in the world. In 1982, however, a federal judge ruled that AT&T was an illegal monopoly. Two years later, the company agreed to break up into eight regional phone companies. These so-called "Baby Bells" became the local phone providers for each region. The long-distance portion of the business and the name went to AT&T.

Most Americans kept their long-distance business with AT&T by default. By the mid-1980s, however, a number of puckish young companies were starting to compete vigorously for AT&T's core business. The most aggressive among them was a little company called MCI.

MCI—originally, Microwave Communications Inc.—was the brainchild of William McGowan, an ingenious, strong-willed consultant who in the mid-1960s took on the quixotic mission of competing with Ma Bell. McGowan originally hoped to compete with AT&T using microwave, not landline technology. However, AT&T's legal efforts to derail the fledgling firm were so ceaseless that in 1974 McGowan sued, initiating a long legal battle that spurred the Justice Department investigation that eventually led to AT&T's breakup. By the mid-1980s, the company McGowan once described (somewhat jokingly) as "a Washington law firm with antennas on top" was a real competitor. But while MCI had been an irritant for years, it never managed to seriously threaten AT&T's hold on the long-distance business. During the mid- and late-1980s, AT&T repelled several attempts by MCI to entice away its customers. In the process, MCI gained a reputation as, in the words of one analyst, "the Michael Dukakis of the telecommunications industry"—a company that often seemed like a hapless competitor. At the end of 1990, MCI was a company in a slump. Its market share was falling; management was in the process of laying off a thousand employees; revenue growth was slowing. AT&T seemed to have a firm lock on its 69 percent share of the $56 billion long-distance industry while MCI and Sprint, with 16 percent and 11 percent respectively, competed for the leftovers. Then came Friends and Family.

In 1991, MCI rolled out a new marketing campaign. The program, called Friends and Family, promised customers who convinced their friends or family members to sign up with MCI 20 percent discounts on their long-distance bills. Friends and Family not only offered lower costs; it also made its customers an integral part of MCI's marketing effort. The young telecom company spent more than $30 million on the campaign—a buy that went virtually unanswered by AT&T. Almost at once, consumers started changing long-distance plans. By the end of the year, AT&T had lost 2 percent of its market share to MCI—a seismic shift in a telecommunications industry where every share point represents about $500 million in revenues. And the defections showed no sign of slowing. By 1992, AT&T was unable to develop an effective counterresponse, throwing the company into a panic. Because AT&T relied on the "Baby Bells" to handle its long-distance billing, it couldn't immediately identify its customers and respond with a similar marketing effort. Instead, it sought to portray MCI as an intrusive "Big Brother," intent on unleashing telemarketers on customers' nearest and dearest. Despite spending $150 million on this campaign, AT&T's counterattack was a failure. People simply did not buy into the idea that a suggestion from a close friend to switch over to MCI was a machination straight out of *1984*.

In response to their failed campaign, AT&T and its ad agencies, led by N.W. Ayer, spent most of 1992 attempting to develop a new strategy. The result was the "iPlan"—a discount program so complex that even marketing professionals were hard-pressed to describe it. By the summer of 1993, AT&T had had enough. It hired Joe Nacchio to run its consumer business and announced that it was looking for a new lead advertising agency. That's when Foot, Cone & Belding (FCB) called us and offered us seventy-five hundred dollars to do a poll for their bid. When you are talking about a multimillion-dollar account, that isn't very much money. Nonetheless, Mark was particularly excited about the prospect of getting involved with such an important campaign. Our work for corporate America had convinced us that businesses in competitive situations didn't just need to gauge public opinion in the same way that politicians did; namely, via polls. They actually needed to *compete* like politicians. Boosted in part by their innovative proposal to work with us, FCB got the account. Our work for AT&T would soon become one of our major corporate campaigns.

As political strategists, Penn, Schoen & Berland had always been guided by a core set of principles. (1) Take the offensive and play to win. (2) Start yesterday: front-runners tend to be winners. (3) Define the issues on your own terms—before others define you. (4) Keep your finger on the pulse and be ready to respond. (5) Let no attack go unanswered.

So far, AT&T had failed to act on any of these principles. Twenty-two months after the attack began, AT&T was still playing defense—and not in a good way. At the time, AT&T was the ultimate conservative corporation. Its brand name represented generations of experiences and investments. As a result, changes were made gradually and only after endless deliberation. This approach might be okay for a monopoly, but in a highly competitive environment like the long-distance phone business had now become, it struck us as badly misguided. No sensible campaign would wait so long to field an effective response. It was past time to hit back. Developing a viable alternative to Friends and Family certainly needed to be part of the response, but it wasn't enough. AT&T had to go negative.

During the mid-1980s, AT&T had responded effectively to MCI's attacks, but the failure of its recent attempts to counterattack the campaign had left some executives spooked. The conventional wisdom supported their cautious approach. Most marketing experts viewed direct attacks on their competitors as strictly anathema. Coke never criticizes Pepsi; McDonald's never disparages Burger King's Whopper—and there's a reason for that. To do so would just give them free publicity. Or so the prevailing wisdom went.

Mark and I thought this was nonsense. In our eyes, Friends and Family had to be stopped, immediately, and the only way to do it, would be to go negative directly against the brand. However, going negative had its drawbacks. While such advertisements certainly worked—on that point, the record was completely clear—a decade of harsh negative political advertisements had turned the public against them. Politicians still used them because campaigns had a clear end (victory) and a clear end-date (election day) after which fences could be mended. For a company like AT&T, the situation was somewhat different, since it had spent a century building and fine-tuning its brand. AT&T executives weren't willing to adopt tactics that would risk spurring a public backlash. A new tool was needed.

By the early 1990s, political consulting was bedeviled by two serious problems. The first had to do with focus groups. For decades, corporations and advertisers had relied on focus groups to test and develop advertising campaigns. Politicians relied on them too. The basic format of both was similar: convene a representative group of people, show them a range of advertisements, pick the one they liked best, and figure out what they liked about it. By the early 1990s, however, political consultants had noticed a problem: focus groups were panning negative ads. Yet campaign after campaign demonstrated that negative advertising still worked. Our own experiences in New Jersey, Alabama, and Indiana verified this. But when focus group participants insisted, implausibly, that they disliked any negative ads, it was impossible to craft effective negative ads in a scientific way. In this environment, how was a consultant to evaluate negative messages?

The second problem had to do with response rates to telephone surveys. In order to gather statistically significant information, a poll has to garner responses from a representative cross sample of the public. One of the astonishing things about polling is how small this sample can be. Give me five hundred people, randomly selected, and I can tell you what the nation thinks about any given issue. But when potential respondents hang up on interviewers, things get more complicated. While it's possible to call a replacement, the possibilities for sampling error increase as response rate declines.

We were no strangers to these problems. Now, backed by AT&T's resources, we set out to solve them. To do so, we went to the mall.

It was Mark's idea, and it was based on a striking fact of life about modern America: seventy percent of Americans go to a mall at least once a week. As long as you chose malls with a modicum of care, mall shoppers bore a startlingly close resemblance to American consumers (and voters for that matter) as a whole. Democrats, Republicans, AT&T customers, MCI customers—whoever we were looking for, we could find at the mall.

Mark's second insight concerned what to do with the shoppers once we had secured their cooperation. Most focus groups present consumers with ads and then attempt to gauge their reaction, all the while searching desperately for the quasi-mythical consumer savant brimming with marketing insight. Our goals were different. Instead of developing a message in a vacuum and then focus grouping it, we began by looking at MCI's ads. Already, we had impressed our partners at FCB and AT&T with our ability to quickly assess visual and verbal slogans: "Let's 'Penn Schoen' it," they'd say. Now, using the mall test, we were able to relentlessly compare every MCI advertisement with AT&T counter-ads before an audience of representative consumers until we found a message that won hands down. In the course of our research, we made an important finding. MCI might not be vulnerable to charges of "Big Brotherism," but its complex rules for Friends and Families did leave it vulnerable to something else—something even more deadly—ridicule. In short order, we had a new game plan for AT&T: make fun of MCI.

Our recommendations for AT&T flowed out of this critique: the way to beat MCI was to offer discount plans that were both attractive and easy to understand and to provide concrete and appealing rewards such as frequent flier miles and long-distance credits. We also worked with AT&T to find discounts that would seem attractive but not cost the company much, such as discounted international long distance. Once FCB shot the ads, we subjected them to the same relentless testing process, refining them again and again until our mall tests showed that AT&T's message consistently and clearly bested MCI's. All the while AT&T looked on nervously. Internal data showed that MCI had secured about 20 percent of the market and was rising fast. If AT&T didn't succeed in stopping the flow of customers leaving its business, the company's viability would soon be called into question.

On December 18, 1993, AT&T unveiled its new "True" plan—and kicked off a $200 million advertising campaign based largely on our research. Negative ads—the "TrueMath" portion of the ad campaign—questioned MCI's claims and ridiculed the complexity of dealing with AT&T's rival. The "TrueSavings" and "TrueRewards" portions of the buy touted the benefits of AT&T's rival discount program. It was an enormous gamble, and, fortunately for us, it worked. More than 15 million people enrolled in the TrueUSA Savings program; more than 13 million registered for TrueRewards. By the end of 1994, AT&T had signed up 14 million new long-distance customers. MCI's hasty response—Friends and Family 2—was a complete dud: in fact, it further underscored our claims that Friends and Family was complex, silly, and not worth the bother. By the end of the year, giddiness had replaced fear in AT&T's corporate suites in Basking Ridge, New Jersey. Its market share was now stable, at around 60 percent. At its next earnings report, AT&T was able to report record earnings of $4.71 billion.

Our work with AT&T and with a host of other businesses gave us a remarkable blueprint for translating our success on the campaign trail to the success in the boardroom. This blueprint was one that brought Penn, Schoen & Berland continued success over the years, as companies who employed us engaged in an increasingly competitive struggle for market share with businesses from around the world.

Over the years, this struggle has begun to take place on a number of fronts, and now, in today's post-Enron era, businesses face greater political and legal scrutiny than ever before. As such, it has become vital that companies understand how market research can determine their overall success and how the basic principles that elect our leaders can be used to insure the fiscal health of corporations. Here are campaign principles for businesses to live by:

- *Start yesterday: front-runners tend to be winners.* Early polling makes a difference. It's important to use polling early to monitor public opinion and to align a large corporation's activities with its brand. When problems do arise, the more closely the organization follows a campaign structure, the easier and faster it is to implement specific initiatives or deal with crisis management. Polls set up the blueprint and strategy for the entire effort.

- *Define the issues on your own terms—before others define you.* In both business and politics, an early effort to inoculate your product is almost always worth the effort. Get a positive story out about your product before the attacks begin; it's often useful to have a message that at least implicitly acknowledges potential problems and then preempts them. An example of a successful inoculation would be our work for Eli Lilly on Prozac. By carefully building up Prozac as a revolutionary product whose benefits far outweighed its costs, Eli Lilly inoculated it against the attacks that were eventually aimed at it. For any highly visible, potentially controversial product, inoculation is essential.

- *Know the competition and be ready to respond.* Make the competition's negatives your positives. Develop a message that is superior to what consumers are hearing from your rival. Onboard course correction is critical as well. Your opponents can (and will) move too.

- *Understand the issues that move people to action and organize the campaign around those issues.* Many products fall prey to baseless but predictable lines of attack. The failure to anticipate such attacks (and to plan a counteroffensive) can be the difference between success and failure.

- *Let no attack—overt or implied—go unanswered.* A good brand deserves a vigorous defense. In today's ultra-competitive marketplace, no brand can rest on its laurels.
- *Take the offensive, play to win, and know what that means.* Who can imagine a political campaign where the whole organization was not focused on electing the candidate? No one. Yet how many corporations exist where the whole organization is not focused on the values and desired actions of the brand? Our experience has taught us that corporations need to keep their teams focused on building and maintaining their brand in the same way politicians cultivate their electability. That means monitoring consumers' opinions on an ongoing basis. It also means respecting the numbers. If the numbers clash with your gut feeling, trust the numbers.
- *Stay nimble.* Constantly monitor the landscape for changes in issues, challenges from competitors, and perceptions among target groups. Remember, competition can come out of nowhere or from places you least expect.

While the use of these techniques and our mall test had played a major role in saving AT&T's most lucrative franchise, ultimately the greatest impact of our work for AT&T came at a different phone number—(202) 456-1414—the number that rings at 1600 Pennsylvania Avenue. The mall test system that we developed to compete with MCI would also allow us to gauge the voting public's complex feelings about President Bill Clinton with unprecedented accuracy. It would also lead us to a new way of viewing the electorate. Instead of seeing only voters, Mark and I began to look at people in a more holistic fashion—as multidimensional individuals whose votes, purchases, and lifestyle decisions all expressed important values. In time, our work for AT&T would lead Mark to an inspired innovation—the neuropersonality poll—that would enable us to segment the public with unprecedented nuance. Together, these innovations would play a critical role in our successful effort to develop a comeback strategy for President Clinton.

In many ways, the Democratic establishment in the mid-1990s was more like the old AT&T than the new one. It had lost touch with important swaths of America. It was allowing the GOP to define it as the "tax-and-spend" party and failing to hit back with a persuasive counterattack. It viewed conservatives with condescension, and it was anything but nimble.

Happily, there were exceptions. One was the exuberant young Democratic governor of Arkansas, Bill Clinton.

CHAPTER EIGHT

THE STEALTH CAMPAIGN AND THE RESURRECTION OF THE CLINTON PRESIDENCY

While Penn, Schoen & Berland had been a successful consulting firm for almost twenty years, we still had yet to work on the ultimate campaign: a U.S. presidential election. It was a call that we had long been awaiting, and oddly enough, when it finally came, it was from my old friend from New York's Upper West Side, Dick Morris.

During my years at Harvard, I had lost touch with Dick. After all, if I wasn't canvassing, I wasn't very useful to him. Yet the shadowy strategist continued to resurface in unpredictable places. In 1990, we made a bid to work on the campaign of former Governor Mark White, who was mounting a race to return to the Governor's mansion in Austin. I was asked by Governor White to send some materials to Morris's home in Redding, Connecticut. When I saw the address, I realized it was a mere five-minute walk from the summer house I had reluctantly rented at the behest of my then-wife,

Peggy. I walked the documents over to Morris to say hello; soon, we were friends again. Of course, Morris's circumstances had changed. The card tables and folding chairs had been replaced by a French country motif. Though Gita was gone, Morris had remarried, and Peggy and I began seeing him and his new wife, Eileen, regularly. Asceticism was out; aestheticism in the form of art on the walls was in. While the furnishing (and the mealtime fare) were much improved, it was clear to me that Dick Morris hadn't changed. For all the new flourishes, there I was again, drinking fresh orange juice, having long, conspiratorial lunches, and talking about bistros on the Left Bank, rather than diners and coffee shops on the West Side. After twenty years of barely exchanging a word, Dick Morris and I were working together again.

Dick had changed in one way. Though Morris had proven to be a talented political consultant, he'd also become something of a political mercenary. Gone were the days when ideology or even political affiliation dominated his thinking; now, the issue for Dick was who would pay the largest fees. (France isn't cheap.) In a field where most political consultants associate exclusively with one political party, Morris had gone to work for both Democrats and Republicans, including some of the most conservative members of the GOP such as Trent Lott and Jesse Helms. Of all of Morris's clients, none were more important, than one of his earliest—a young attorney general from Arkansas whose large ambitions were matched only by his extravagant talents, Bill Clinton.

Morris worked for Clinton in his successful first run for governor in 1978. Clinton soon abandoned him, but when he failed to win reelection in 1980, he returned to Morris. Together, the two men restored Clinton's political fortunes and reclaimed the governor's mansion in 1982. From that point on, the two men had maintained a Svengali-like relationship. One magazine compared them to Elvis Presley and his manager Colonel Tom Parker.[1] As such, it was no surprise to the people who knew Clinton best that, in the wake of the disastrous 1994 congressional election, which took both houses of Congress from the Democrats, the president and the first lady turned to Morris to help revive his floundering presidency.

Ironically, by the fall of 1994, I had spent a good portion of my political career working *against* Bill Clinton. In 1986, and then again in 1990, we were hired by the then-Arkansas attorney general, Steve Clark, who was considering a gubernatorial campaign against Clinton. After ruling out a run in 1986, Clark announced his intention to enter the Democratic primary in 1990. However, news of a scandal involving Clark's use of state funds for personal purposes quickly ended that idea. It would not be our last encounter with the murky world of Arkansas politics. (I was struck that there seemed to be no hard feelings between the two men. Clinton and Clark remained friendly and occasionally jogged together.)

Later, we were enlisted by Jim Guy Tucker to look into his chances of unseating Clinton in the 1990 governor's race. We did some polling of the Arkansas electorate and put together some focus groups with African American voters, but, unfortunately for Tucker, the conclusion was clear: Tucker didn't stand a chance. African Americans in Arkansas loved Clinton—in what would prove to be a harbinger of the black community's extraordinary support for Clinton when he was president. White males were persuadable, but white females were likewise loyal to the charismatic governor. As a result, Tucker instead settled for the lieutenant governor's slot, which he won in a race against a Morris-backed candidate, Little Rock ophthalmologist, Hamp Roy.

As the 1992 campaign got underway, I was surprised that— knowing Morris' history with Clinton—Morris rarely, if ever, mentioned any work he was doing for the campaign. Except for bits of advice he gave the Arkansas governor during the contentious New Hampshire primary, Morris seemed to be uninvolved in 1992. I didn't know if the two were having conversations or not, and I really didn't think it was my place to ask. Then, two years later, Clinton was in trouble, and the dynamic changed.

In the winter of 1994, I got a page from Morris. When I called him back from a pay phone, he got straight to the point.

"Did you want the Clinton campaign?" he asked. "I can get you the whole thing—polling, strategy, media."

The idea that I would be hired in such a manner to be Clinton's main strategic adviser was mind-boggling. It was hard to know if it was even a serious offer—but of course I said yes. Morris later told me that he had to stay in the background because he was working for Republican candidates. His code name would be Charlie, a reference to Charlie Black, a prominent GOP consultant (and incidentally, to Charlie of *Charlie's Angels,* who stayed in the background but ultimately controlled everything).

Even through in many ways Mark's and my political philosophies and backgrounds were a perfect match for Clinton, our experience working against Clinton in the past seemed like an obvious impediment to working for him now. However, by late 1994, Clinton was not concerned about past differences. He was fighting for his political life.

During the 1992 presidential campaign, Bill Clinton had effectively portrayed himself as a moderate Democrat. He had spoken frequently about "ending welfare as we know it," come out in support of free trade, advocated the death penalty, vowed to crack down on crime, and vociferously attacked President George Bush for breaking his "no new taxes" pledge. As a result, Bush was simply unable to tar Clinton with the tax-and-spend, old-fashioned-liberal label. However, once he assumed office, Clinton took the path of least resistance. Because the locus of power in Washington was on the left, particularly among Congressional Democrats, Clinton felt forced to move in that direction, proposing an economic stimulus bill, calling for tax increases, and putting forward an ambitious health care plan whose scope frightened Americans suspicious of big government, instead of focusing on issues like welfare reform, which would have been a surefire way to strengthen his New Democratic credentials. In fairness, if one looks closely at the initiatives Clinton proposed in the first two years of his presidency, they were more "New" than "Old" Democrat, but the perception across America was that Clinton was a liberal. Our first task would be to change that.

At the time, Clinton was in an extraordinarily weak political position. Republicans, led by Newt Gingrich, had just ended forty years of Democratic dominance in Congress and Clinton's poll ratings were in the mid-thirties. The Contract with America was the dominant political manifesto in Washington. The president was even forced in early 1995 to extol his own relevance as commander-in-chief. In short, Bill Clinton was looking more and more like the man he defeated in 1992—a one-term president.

I did our first poll for Clinton in December of 1994 on the issue of targeted tax cuts. Our polling showed clearly that, for Clinton to get reelected, he would need to take dramatic steps to repair his image. He was perceived as a cultural and social liberal and his approval rating was well below 40 percent. Things needed to change, but the working arrangement itself was decidedly odd. The president was clearly listening to me—a national television address on the subject of targeted tax cuts made that clear enough, and yet we hadn't officially been hired. As far as the world knew, Stan Greenberg remained Clinton's pollster. I was breaking out my rusty French to track Morris down in France to talk strategy, and just when I'd begin to suspect that this was not what Morris said it would be, Clinton would go public with one of our ideas. The situation was intriguing but unsatisfying.

Morris assured me that Clinton understood the situation. He wanted to go on television and quickly begin the repair job and, more important, reassert his policy relevance after the electoral debacle of November 1994. In a speech that Morris drafted and sent the president from Paris, Clinton proposed a so-called Middle-Class Bill of Rights, which included a $500-a-child tax cut, a proposal to make college tuition tax deductible, and a plan to allow tax-free IRAs to be used for major events such a medical emergency, purchasing a first home, or paying educational expenses. The pundits derided the speech as "Gingrich-lite," a co-option of GOP themes. Our polls told a different story: among those who watched the speech, Clinton's approval rating jumped nine points.

The apparent buy-in of the president notwithstanding, my partner Mark Penn thought this whole arrangement was a bit fishy and, truth be told, so did I. Then, in January 1995, I got a page from Morris. When I called him back, he said that we had a meeting at the White House on February 6 with Bill Clinton. My immediate reaction (which I blurted out) was that I had a speech in Indianapolis that day, and I didn't know if I could make it.

"You will," Morris coolly replied. Then it sank in, that I was going to meet the president, with Morris, and I was going to be the one who was going to tell him what had gone wrong. Of course I went.

I was shocked by the man I saw there.

The president of the United States is the most powerful man in the world, yet when I first met Bill Clinton, he appeared defeated. He was physically in the room, but mentally he was somewhere else. I knew Clinton to be powerful, brilliant, and famously charismatic, yet the Bill Clinton I found that night seemed disengaged from his own administration. The way he told it, others were making decisions in the White House, and he was one of many voices influencing the process.

I wanted to get up, walk across the room, grab him by the shoulders, start shaking him, and tell him to snap out of it. It was my first meeting with a sitting president, but, far from being awed, I was dismayed. Still, Clinton's analysis of the situation was right on. "I'm way out of position," he complained. He had run as a moderate Democrat in 1992 but was now seen as a cultural and social liberal. In particular, he was bitter about the advice given to him by his pollster, Stan Greenberg, who had counseled him to pursue health care reform but then later criticized the decision and said it was the reason Democrats had lost the 1994 midterm election.

Clinton understood that I took a different view. He was intimately familiar with my work in Arkansas, and he even chided me about some polls I had done four years earlier for Tucker. Nevertheless, he knew I was a committed centrist Democrat, and that I had played a central role in Evan Bayh's successful campaigns.

Morris and I were agreed on our prescription for the president. He needed to first and foremost present himself to the country as a fiscal conservative. Otherwise, he would never be able to escape the tax-and-spend label. My solution was clear: present a balanced budget. In January, Clinton had sent a budget to Capitol Hill that called for massive budget deficits in both the near- and long-term. Not only was it dead on arrival, it was ridiculed by congressional Republicans who were pushing hard for both an end to deficits and a constitutional amendment that would force Congress to balance the budget. They had the American people on their side. According to our polling numbers, 80 percent of Americans wanted Washington to get its fiscal house in order.

Throughout the winter and spring of 1995, I was adamant that Clinton commit to a balanced budget. At one point, Morris called me in California, and I told him in no uncertain terms that, if Clinton failed to act, the electorate would simply tune him out. The president would never be reelected without a plan. Clinton, however, was not yet ready to take that step. We were supported by Treasury Secretary Bob Rubin and Vice President Al Gore, but the liberals in the White House—Chief of Staff Leon Panetta, George Stephanopoulos, and in particular Harold Ickes—disagreed. They claimed that the president would be selling out congressional Democrats. They also quite passionately believed that Clinton's only hope was to maintain a confrontational posture with House Republicans. Passionate disagreements would become a familiar dynamic in the Clinton White House—with political advisers like Morris and myself along with moderates like Gore and Rubin on one side and the liberals on the other.

That was only one of the administration's problems. It was also clear that the operations of the White House staff had become less than ideal. The president clearly did not have direct control over the operations of his staff. At one point, I remember Clinton boasting to me and Morris that after several months of effort he had gotten rid of a press aide who had a reputation for leaking and moved him to the Treasury Department. Nevertheless, as I listened to the story, I couldn't help thinking, "This is the president of the United States and he can't take action against someone on his own staff, a guy who was actually hurting him politically?" It was clear that Clinton felt enormous frustration about his inability to manage—from an administrative standpoint—a seemingly disorganized and divided White House operation.

Having come from the outside into this political maelstrom, I quickly realized that Morris and I had entered a brutal battle for the president's soul. In early 1995, it wasn't clear who was going to win—or even if we were going to be around when that moment arrived, not to mention the next week.

One of the most difficult parts of working for Clinton was the extraordinary uncertainty surrounding what we did. We watched the president drop Stan Greenberg, and, while we were glad to have our own role confirmed, we knew that we could be shown the door just as easily. We had no contract or even the expectation of getting one, and without security, our work was fraught with peril. Clinton inadvertently added to the uncertainty by bringing up the names of people from past campaigns that he was thinking about adding to the team. It was a move Morris and I resisted mightily, knowing that any additions would potentially undermine our influence.

It certainly didn't help my standing that I was working with Dick Morris, who was deeply mistrusted in the Clinton White House, by Harold Ickes in particular. Ickes and Morris had locked horns on the Upper West Side in the late 1960s and early 1970s, and I was amazed to find that Ickes still held a grudge. He couldn't find it within himself to rise above the ancient New York political battles. Initially, Mark and I sought Ickes out and discussed finding ways for us to work together in a cooperative manner, but Ickes gave us an automaton-like response: "I serve at the pleasure of the president. I'll do what he tells me to do." He kept repeating that mantra—no matter what we said. He also kept sticking his pencil in his ear while we talked, as if he could barely be bothered with what we were saying. It was clear that Ickes would work with us only so much as the president wanted him to—no more.

Later, when we become more ensconced in the campaign, Ickes took to throwing roadblocks in front of us, fighting us over expenses, trying to cut our commissions, and refusing to negotiate a long-term contract. It was obvious from day one that Ickes was determined to find a way to get rid of us. He wasn't alone. We understood all too well that Stephanopoulos would do all he could to undermine us—all the while overtly cooperating with Morris in order to assure his own access to campaign deliberations. Some

White House staffers seemed more interested in preventing us from getting access to the president, than they were in fixing his political problems. It was obvious that the White House operation was failing to serve the president's needs. We were committed to furthering the president's agenda, and he wanted us on board; but from the start, many White House insiders saw us as the enemy.

Sometimes the obstacles the White House staff threw in front of us were rather juvenile. For example, Morris had Bill Curry installed on the White House staff. Curry was a moderate-liberal Democrat who had lost the Connecticut governor's race in the 1994 Democratic debacle. We saw him as an important ally. But staffers, not surprisingly, resented his presence. Curry was an affable figure, which made him ripe for hazing. When he showed up at his office, there was no desk. He would be told about staff meetings and then given the wrong time and room. Chief of Staff Leon Panetta, who was working to maintain some sense of coherence in the White House operation, particularly resented Curry's presence. He made every effort to freeze him out of the political operation and kept him cooling his heels across the street from the White House in the Old Executive Office Building, far away from the president's political team. Panetta told Curry that if his help was needed, he would call him. Suffice it to say, Curry's phone rarely rang.

Relations with Panetta were a recurrent problem. After our initial polling effort in December, we had another meeting with Clinton and Panetta. His feelings of rage and anger were clearly palpable. He didn't like answering to Morris, and he liked even less that he was effectively told he had to do Morris's bidding. Making matters worse, Morris seemed to revel in his power, often treating Panetta like an errand boy.

It was a strange situation. Bill Clinton, a man who ran as a New Democrat and was fundamentally a centrist, surrounded himself with liberal Democrats. Vice President Gore excluded, virtually none of Clinton's major advisers embraced the centrist approach we were advocating. Early on, Clinton had me do a presentation for George Stephanopoulos, Ickes, and top aides Rahm Emanuel and Doug Sosnik. I strongly urged them to embrace fiscal conservatism as a way of usurping key Republican themes. Not surprisingly, there was clear resistance to my presence, as virtually everyone in the room had a prior relationship with Stan Greenberg, if not philosophical and political differences with the approach I was advocating.

Luckily, however, the president seemed to be on our side. In truth, Clinton seemed genuinely exasperated by much of his staff. For most of our first year in the White House, many of these well-known staffers were frozen out of the political operation. As it became clearer that I would be playing a central role in Clinton's 1996 reelection effort, I realized that we would also need a new crew of staffers to properly organize and orchestrate the campaign. I feared that, unless we added intellectual heft to the team, we would be vulnerable to a countercoup by those in the White House—and they were legion—who saw Morris (and, by extension, me) as anathema. Strength was in numbers, and I realized that if I could assemble a full-fledged political team, it would be far harder for the Harold Ickes of the world to get rid of us. On the surface, this may have appeared to weaken my hand, since I would now have to share the credit and the access with others—a rare event in Washington, D.C. But if it helped to keep us around to the end, I thought it would be a smart move. Morris, to his credit, agreed.

The first person I wanted to bring on board was Mark Penn. Morris was not so sure, since he'd heard rumors about Mark's inflexibility, but from the outset, I made clear we needed someone with his background and intellectual ability. Morris agreed to what he saw as a compromise: Mark would be allowed to come in for special projects at the beginning of our meetings with Clinton but he would then have to leave. In fact, Morris proved so controlling that he demanded Mark be kept in the White House usher's office until it was time for him to actually do his presentation. Once he was done, Morris would have the usher escort him out. Not surprisingly, Mark was hardly enamored of this idea and he was not pleased about Morris's proclivity for failing to pay his bills on time. Nonetheless, his concerns not withstanding, Mark was chomping at the bit to be more involved. Eventually, I sat Morris down and told him that Mark had to play a key role, and after much discussion, Morris reluctantly backed down, but he tried his best to keep Penn on a tight leash—"like a caged animal" he once told me, with evident satisfaction. It was a cruel way to treat a colleague. I knew Mark's presence was essential for the venture to succeed, and I did everything possible to make this very awkward situation work both personally and politically.

Next, I contacted my colleague Bob Squier, with whom I worked on the Lautenberg campaign and who was close to Vice President Gore. Until his death in 2001, Bob was one of the top media consultants in Washington, but more important, he had worked with Clinton in the past and understood a centrist Democratic philosophy. Above all, he brought a gravitas to the campaign team that was sorely needed. We didn't have the type of DC connections that Squier took for granted, and Bob gave us those connections and more generally added credibility. It was something that Clinton appreciated. It was of little surprise that, during Clinton's famous political meetings in the run-up to the election, Gore sat to his right and Squier sat to his left.

In addition to Squier, Morris brought on Hank Sheinkopf, an old-school New York political consultant. He had wonderful street smarts and was an excellent tactician. Another addition to the team was Marius Penczer, a producer of music videos from Nashville who was quite an interesting character. Though he was very cerebral, his video background gave him an MTV sensibility. He and Clinton shared a great love of country music.

Together, we assembled a group that we knew we could rely on and that also shared our approach about the best ways to move Clinton to the center. Most important, we put together a group that knew how to keep a secret. If there was one defining aspect of Bill Clinton's first two years in the White House, it was that everybody leaked. The place was sievelike with low-level and high-level aides constantly talking to *Washington Post* and *New York Times* reporters. By making every one of his decisions appear calculated and overly political, it was threatening to undermine Clinton's entire presidency. Clinton knew it, and the best way to stay on his good side was to stay quiet. One time, some of the new campaign advisers and I were meeting with the president in his private sitting room when National Security Adviser Tony Lake came in to update him on the fate of Scott O'Grady, an U.S. Air Force pilot who had been shot down over Bosnia during a bombing raid. Lake looked uncomfortable speaking in front of us, but the president reassured him.

"These guys don't leak," he said. In the Clinton White House, it was the ultimate compliment.

Amazingly enough, Morris, Mark, and I worked for Clinton for nearly half the year before news of our employment became common knowledge. While in Washington, that was some kind of a record, in the Clinton White House, it was practically a miracle. I remember being called by Jane Mayer from the *New Yorker* in the spring of 1995. She wanted to confirm that Mark and I—as well as Dick—were working for the president. I was petrified even talking to her. I quickly declined to answer questions and got off the phone. I knew if there was any hint that I was leaking, Ickes would make absolutely sure I was fired. I later found out, much to my dismay, that it was mostly likely Ickes himself who had divulged the information. In fact, after the story broke, Ickes taunted us in a White House meeting, by smugly saying, "I confirm you didn't leak." One could not escape the feeling that no matter how good our work for the president, we would always be looking over our shoulders.

News of our working for Clinton came during the most precipitous moment of the reelection bid. It was in June 1995, that we took one of the most important steps toward ensuring the president's reelection.

For months after our initial February meeting, we continued to push for a balanced budget. Our polls showed that Americans cared about the issue but that they didn't feel that the president shared their concern. In our opinion, there was no way to take this issue off the table without the president committing himself to balancing the budget in a definitive time frame. Two obstacles, however, stood in our way—Congressional Democrats and the White House staff.

From the beginning, whenever we would propose a balanced budget the response from Stephanopoulos, Ickes, or Panetta would usually center on the damage that such a tactic would do the president's position in Congress. "The Dems would never forgive us," was a usual refrain. "Why should we take the pressure off the GOP just when we have them on the ropes?" was another.

While it was unmistakable that we had little support in Congress for a balanced budget, these arguments missed the larger truth. First, the president would never win reelection if he was unable to redefine himself as fiscally responsible. The issue was simply too important to the American people. The continued failure to stake his position on a balanced budget would only help ensure that when Republicans called Clinton a liberal tax-and-spender, the charges would stick.

Second, with Republicans in control of both houses of Congress, it was only a matter of time before a balanced budget made its way to the White House, which Clinton would most likely have to veto. With his own balanced budget on the table—featuring less draconian cuts than what the GOP was offering—he would have far more credibility on the issue. Clinton could (and eventually did) make the argument that the federal government was able to balance the budget and do it in such a way that avoided huge cuts in social and entitlement programs that Americans valued. By agreeing to the idea of a balanced budget, we were shifting the debate to budgetary specifics. When it came to the details, particularly social issues, the Republicans were so out of step with the American people that we believed Clinton could not lose.

Last, the liberals failed to grasp the impact of Ross Perot's candidacy on the electorate. People often forget this, but Perot largely ran on a message of fiscal conservatism and balancing the budget—and in doing so, managed to get close to 20 percent of the vote. Despite a highly erratic personal performance, Perot's relatively high percentage of the vote demonstrated that people cared about fiscal discipline. The sooner we could get ahead of the issue, the better for the president's political prospects.

From a tactical perspective, it was an easy decision, but the liberals in the White House refused to go along. Even after the president signed on to the idea, they still groused. Dick and I would shake our heads, wondering why the president couldn't just tell his staff that this was his decision and that they needed to get on board—or get out. I wonder if, much like Ehud Barak, Clinton enjoyed the give and take among his staff and liked having clear alternatives from which to choose. Or perhaps Clinton was simply afraid of upsetting the liberal wing of the party. If he jettisoned Stephanopoulos or Ickes, the outcry from the Left might lead to a primary challenge—which, in June 1995, still seemed like a real possibility. In particular, the thought of a Jesse Jackson run in the primaries worried Clinton to no end, and the consequence of this was often political paralysis in the White House.

Still, in our view, a balanced budget was inevitable. If he wanted to avoid being marginalized in the balanced budget debate, the president would need to put his plan on the table. By the late spring of 1995, it was clear that the president was moving in our direction. In an all-important meeting in May, we again made our case and the president seemed swayed. Then again, with all the competing pressures, we could never be sure Clinton would not be swayed in the opposite direction.

A few days later, Clinton did an interview with a New Hampshire public radio station and said that he was considering a plan for a balanced budget. The liberals in the White House were aghast. We were delighted. It was clear that Clinton was tired of getting attacked in the press and watching the most important political debate in the country take place without his involvement. Clinton was back in the game. We pressed our advantage by urging him to eschew a detailed budget proposal and instead deliver a speech to the nation that committed the White House to working with Republicans in achieving a balanced budget. More important than specifics was a commitment, by the president, to push for a balanced budget that protected his key priorities—Medicare, Medicaid, education, and the environment.

On June 13, 1995, the deal was sealed. The president delivered an address to the nation. Pledging to balance the budget in ten years (three years sooner than the Republican plan), he outlined significant spending cuts—including cuts in Medicare—that were far less draconian than the reductions Republicans were calling for on Capitol Hill. The president presented his plan as a common sense and fairer way to end government deficits. It was a brilliant gambit and it worked even better then we had have expected.

The outcry from Democrats on the Hill only made the moment that much better. For example, David Obey, a liberal congressman from Wisconsin, blasted the president, saying "if you don't like the president's position on a particular issue, you simply need to wait a few weeks." Pat Schroeder from Colorado said, the Republicans "are just playing him like a kitten with a string." Schroeder and Obey's responses were indicative of the way Democrats on Capitol Hill were thinking. In a way, their criticism worked in our favor. The president had not only positioned himself squarely in the middle but also drawn a clear and unmistakable distinction between himself and the liberal wing of the Democratic Party. As John Breaux, the moderate Democratic Senator from Louisiana said, "The President has been reborn."[2]

Dick liked to call it triangulation—taking the best from both parties. To him, it was the most effective way not only to redefine the Democratic Party but also to respond to the more centrist mood of the electorate. In fact, I never truly bought into the triangulation theory. In my view, the reality was more complicated. The country was trending to the Right. For Clinton to survive, he had to move closer to conservative political positions—because quite simply, they better reflected the mood of the electorate. But he had to talk in terms of values—effectively speaking a new language that many Democrats did not understand and appreciate. By late June, Clinton's poll numbers were already inching upward. Slowly but surely, things were coming together.

O ur call for embracing a balanced budget was only the first step in our effort to make over the president's image. If there was one issue that Republicans consistently used to attack Democratic candidates, it was crime. I had watched how in the 1970s Ed Koch had played on New Yorkers' fears of rising crime. Like Koch, Bill Clinton had consistently gone against traditional Democratic positions on the issue. Clinton supported the death penalty, pushed for stricter sentencing laws, and worked to put more cops on the street.

In the run-up to the 1994 midterm elections, Republicans ridiculed the president's support for social programs, such as "midnight basketball," and took him to task for his support of the Brady bill and the assault weapons ban. (It was the latter two that played a role in costing a number of good Democratic congressional repre-sentatives and senators their seats in the 1994 midterm elections. This was particularly true in the South and Plain States, where De-mocrats' gun control policies have long been a source of political vulnerability.) The result was that the president was getting little public support for his tough stance on crime—even though his posi-tions were supported by a majority of Americans. Most Americans support gun control measures, but the ones who care the most—and vote on the issue—tend to be the most fervently anti-gun control.

We decided that highlighting Clinton's crime agenda would be a critical first step in rehabilitating his image with voters. We recom-mended a $2 million ad buy—run through the Democratic National Committee (DNC)—that would focus on the president's support of the assault weapons ban and his efforts to put a hundred thousand po-lice officers on the street. We aimed the ad campaign at swing voters in swing states—places like Washington, Missouri, Wisconsin, and Penn-sylvania where we believed the election of 1996 would be determined. What we didn't do was run the ads in media centers such as New York, Washington, and Los Angeles. Our goal was to fly under the media radar screen and get our message out directly to the American people, unfiltered by an increasingly hostile and cynical media.

Before running the ads, we went back to virtually the same malls that we visited to test the AT&T ads, making sure that they moved the numbers in the president's direction. This type of ad campaign was unprecedented. No one had ever done this kind of massive adver-tising more than a year before an election. With most major political reporters located in New York, Washington, and Los Angeles, we felt fairly confident that our reinvention of the president would raise few eyebrows. We were right. It took months before the national media caught on and, by that point, Clinton's numbers were showing marked improvement—exactly as our mall tests had predicted.

Not surprisingly, Ickes fought us tooth and nail, claiming the ex-penditure was a waste of money and that, by Election Day, the ads would be forgotten. In fact, the ads worked brilliantly. The presi-dent's favorabilities immediately rose among viewers who had seen the ads, and the GOP did nothing to negate our gains. We were re-casting the president's image before the Republicans had the chance to weigh in. In short, we were taking an issue that for years had weighed on Democrat candidates and turned it into a positive. This was exactly what we had preached years earlier as the key way for Democrats to tilt the political battlefield in their favor.[3]

The crime ads were only the first step in a massive advertising campaign. From the summer of 1995 until Election Day, in November 1996, we bombarded the nation's airwaves with ad after ad. In crafting these ads, we were careful to always start on a positive note, first emphasizing that the president was a positive campaigner who shared and understood voters' concerns and values, and only then responding to Republican attacks with counterattacks that, say, excoriated congressional Republicans for their "heartless" budgetary plans. The swing in the president's poll numbers was immediate and extraordinary. Within a short period, Clinton's favorabilities were cresting.

All of these attack lines were made possible without the president's embrace of a balanced budget in June. When it came to our advertising plans, we had an issue where the president was in complete agreement with us. In our first meeting, Clinton complained bitterly about the way that the GOP, and in particular the health insurance industry (with the infamous Harry and Louise ads), had spent millions on paid advertising to sink his health care plans. It was still a bitter pill and Clinton vowed that he wouldn't let it happen again. Of course, there was a price to pay—an almost constant fundraising effort by the president, which would lead to greater political problems down the road.

Those problems notwithstanding, the early advertising blitz was probably the single most important factor in turning public sentiment in the president's favor. Without the ads, I'm convinced Clinton would not have been easily reelected. They not only were an invaluable tool in reshaping the president's image but also in laying the groundwork for the biggest political battle of the president's first term—the government shutdown.

Meanwhile the Republican's budget, based on their "Contract with America," made its way through Congress, and Clinton made it clear that the bill would never become law. In press conferences and through White House statements, he warned the GOP that he would veto its budget if it contained the cuts in Medicare, Medicaid, education, and the environment it was calling for, but the

Republicans remained unconvinced. They were so convinced of Clinton's vulnerability that they believed, in the end, the White House would bow to pressure and sign the bill. A veto could prompt a government shutdown, and GOP leaders thought that Clinton would never let that happen. They were wrong. The president was determined to do the right thing. Moreover, we understood that if there was a shutdown the political benefit would go to President Clinton and the Democrats.

Again, it was the ad strategy that made a tough stand by the White House possible. Through the summer and fall, the DNC ran ad after ad attacking the Republican budget bill—in particular its large Medicare and education cuts—and presenting the president as the defender of these popular programs. Public sentiment was clearly turning in our favor. In August 1995, at Clinton's request, we ran a number of polls based on various scenarios involving the budget and a possible government shutdown. Every one of them led to the same conclusion: stand and fight.

I remember sitting with Clinton in the Map Room in August 1995, and saying we couldn't lose on the budget. "Poll it again," Clinton said. "I just don't believe it. Test five or six more hypotheses." Clinton knew he was betting his presidency on this approach; if it didn't work, he could lose the White House. He wanted to be absolutely sure it would work. He knew we were driving the strategy and, without the support of the rest of his staff, Clinton wanted to be sure we weren't advocating a move that could be his political suicide. But the numbers always came back the same. In one poll, by 58 percent to 25 percent, Americans said they would blame the Republicans if the government was shut down. We viewed the budget battle as the single best opportunity to repair the president's image and show him as a strong and principled leader. We had set the trap for the Republicans, and amazingly, they walked right into it.

The first government shutdown lasted only a few days, but the results were extraordinary. In its aftermath, Americans now saw Clinton as a strong leader who was willing to fight for what he believed in. His approval ratings (often the hardest political numbers to move) rocketed up. It was as if overnight people had begun to view Clinton in an entirely new light. We couldn't have written a better script, but of course, it was due in large measure to the fact that we had so effectively laid the groundwork with our massive ad campaign. President Clinton's behavior in the first government shutdown only served to reinforce the ad campaign that we were running, pledging to defend Medicare, Medicaid, education, and the environment. In fact, we ran so many ads trumpeting Clinton's stance on these four issues that we used to simply refer to them as MMEE.

Remarkably, the Republicans felt that committing political suicide once was not enough. After weeks of fruitless negotiations, they shut the government down again: this time for more than a month. Again, the White House stood firm. And again, the Republicans paid a heavy political price—none more so than Senator Bob Dole, the presumptive Republican nominee for president. Instead of spending much needed time campaigning in New Hampshire, he was tied down in budget negotiations that went nowhere and only further tarnished his image by connecting him to the increasingly toxic Newt Gingrich. It would have made sense for Dole to strike a deal immediately and distance himself from Gingrich. His failure to do so ultimately doomed his candidacy. It was a point not lost on anybody in the White House. Throughout the general election, we never let an opportunity slip to refer to the Republican ticket as "Dole/Gingrich."

If the immediate benefits of a shutdown were clear, the answer to the question of how long we should allow it to continue was not. The White House was already talking with Congress; Dick Morris soon became adamant that Clinton push for a deal. He thought it would ensure reelection. Unsurprisingly, Ickes and Stephanopoulos fought him at every juncture. As I watched Gingrich and the GOP be blamed for the shutdown, I figured we could weather the storm, but our poll numbers also showed that people were getting frustrated at the lack of compromise. At some point, I realized that the numbers could start to slip and the voters were likely to hold the president at least partly responsible. So, from my perspective, if we could get a deal on favorable terms, it was a win-win for us. Otherwise, we continued to occupy the high ground in the mind of the public and thus we should stay and continue the fight.

Of course, it was the GOP that really needed to end the impasse, but it never seemed to realize that. It wanted all or nothing from the president. The more the White House moved closer to their point of view, the more the Republicans would dig in their heels. Clinton moved quite far in their direction (many in the White House thought it was too far), and yet the Republicans couldn't find it in them to move just a little bit toward him in order to make a deal. They simply didn't understand the idea of compromise. They expected the president to surrender completely to their demands. In the process, they paid a terrible political price for their pointless intransigence.

As important as polling was in guiding our tactics, Mark and I soon realized that it was more important that we not let short-term successes distract us from developing a longer-range winning strategy. After all, the GOP would shoot itself in the heart only so often. In order to win in the future, the Democratic Party desperately needed to reconnect with middle-class voters. I was eager to find a tool of analysis that would help us better understand voters' concerns and how they might react to various Democratic messages. It was Mark who had the inspired idea of looking at the electorate in a more three-dimensional manner, much as we had with AT&T.

Mark's great insight was that it wasn't enough to understand the electorate demographically, by such parameters as age, race, gender, or even by political affiliation. Instead, he argued, we needed to look at the electorate in an entirely new way, in order to figure out what issues were of greatest concern. Lifestyle characteristics, attitudes, values, and even personality traits—all of these things were relevant. In short, he sought to create a psychological profile of the electorate. I was intrigued by Mark's approach; together we thought that with the proper funding and a willing client, we could move beyond the traditional static approach to politics and instead look at issues from a multiple set of perspectives.

Our approach came straight from the playbook we had developed for AT&T and other corporate clients. In the corporate world, clients are focused both on their current customers and also their potential customers. As a result, we had devised entire strategies around helping corporations broaden their base and reach out to new customers. In our view, we could do the same thing with politics. Roughly 40 percent of Americans were aligned with Clinton and the Democrats, 40 percent with the GOP, and 20 percent were in the middle. The key, in our view, was to develop a message that appealed to that middle 20 percent. In elections, particularly presidential races, it is often those "potential customers" or swing voters that make all the difference. In order to get those swing voters, we realized that we would need to segment and divide the electorate in ways that had never before been attempted. Mark called it a neuropersonality poll, and it would eventually become the defining document of Clinton's reelection effort.

We began by examining the electorate from three distinctive perspectives—issues, personality, and lifestyle. The issues part was relatively easy. Through our polling, we were able to divide the electorate into four distinct parts—(1) the Clinton base (28 percent of the electorate); (2) the Dole base (18 percent); (3) Swing I voters (29 percent), who were leaning toward Clinton but still undecided; and (4) Swing II voters (25 percent), who were more conservative and less likely to support Clinton. From our research, we deduced that Clinton needed the support of 60 percent of Swing I voters and 30 percent of Swing II voters to win.[4]

B y directly asking voters what issues were most important to them—and what issues were most likely to effect their votes— we were able to get a snapshot into people's thinking. We couched our surveys in such a way as to determine the intensity that voters felt about the issues. Instead of asking people whether they supported, say, cutting taxes, we read them a statement like "The federal government is running a budget deficit and it would be irresponsible to make any tax cuts now given the size of that deficit." Then we would ask them, "If a presidential candidate made such a statement, were they much more likely to vote for the candidate, somewhat more likely, not affect your vote, somewhat less likely, or much less likely to vote for that candidate." The point wasn't to figure out what issues were most important but rather to understand which ignited the greatest passion—in other words, what issues were most likely to impact their vote. Equally as important, we could test the messaging and see what formulations were most resonant. Unlike most traditional pollsters, we were looking less at issues and more at testing attitudes.

Our polling indicated that Swing I voters were far less partisan than other groups and far more concerned with family-oriented, middle-class programs. They were liberal-moderate politically, and while they supported some progressive causes, they focused primarily on the type of government programs that made them feel safer, both personally and financially. They were worried about health care insurance, and while they supported crime prevention and gun control, they also wanted to see tougher sentences for violent crime. And they were concerned about the costs and burdens of taking care of their elderly parents.

Swing II voters were a harder nut to crack. They were fed up with Washington and were very supportive of congressional term limits. They were more likely to scapegoat illegal immigrants, not so much for racial reasons but because they were angry at the idea of having to compete with them for jobs and scarce government resources. However, they shared much of the Swing I voters concerns about economic fairness, affordable health insurance, and health care for seniors. But unlike Swing I voters, they had little patience for a progressive agenda and identified more closely with conservative, middle-class values.

The results of the poll surprised us. It's often believed that income level and age are the great defining variables in U.S. politics. Not so, we discovered. In 1996, the most important factor was marital status, particularly among swing voters. Moreover, having and raising children was the crucial dividing line in the electorate. Families were far more inclined to support the GOP—by a ten- to fifteen-point margin. To close the family gap, we needed to devise a set of issues that focused on family-oriented concerns like tobacco advertising aimed at children, crime prevention programs, elder care, raising the minimum wage, and preventing health care companies from denying coverage. These issues appealed to both

Swing I and Swing II voters as well as nearly all Clinton voters and even some Dole supporters. A second-tier of issues particularly resonated with Swing II voters. These included a tough trade policy with Japan, ending welfare assistance after two years, and a wider application of the death penalty.[5]

Based on our results, we counseled a two-tier strategy, one aimed at communicating family-oriented programs, the other at emphasizing crime, balancing the budget, and espousing a more muscular foreign and trade policy. Another issue that played well—even among the GOP base—was Medicare cuts. Voters wanted to see health care costs cut, but not at the expense of benefits to seniors. We also discovered that certain other traditional issues wouldn't do much to help us. For example standing up for abortion rights, which appealed to Swing I voters, did little to attract Swing II voters. Many foreign policy issues were not helpful, including supporting Russian President Boris Yeltsin or bailing out the Mexican economy (although to Clinton's credit, this didn't stop him from acting proactively and responsibly on both issues). Finally, among every single group there were strong majorities in favor of school prayer—something the president instinctively understood. In August 1995, he gave a speech talking about the important role of religion in schools—although from a noncompulsory perspective and consistent with the Constitution. While many on the Left were aghast, Clinton understood the electorate's concern on prayer. I well remember him telling us that he knew from his experience in the South that the prayer issue would be of significant benefit. "There's a way around the constitutional problems," he told us. "You'll see how effective it is for me after I talk about it."

To help us further segment the electorate, we organized voters into what we called issues-based clusters. This allowed us to determine and measure the intensity of the electorate's views on issues. We broke these voters into nine segments: Economic Liberals, Social Liberals, International Liberals, and 1992 Clinton Voters, all of whom were part of the president's base of supporters; Balanced Budget Swing voters, Crime Stoppers, and, surprisingly, Young Social Conservatives were also swing voters; while the Dole base consisted of Senior Citizen Conservatives and Rich Conservatives.

Now that we understood the issues we needed to focus on, the next question became, "How do we craft the messaging technique for the campaign?" In modern politics, how you get the message across can often be as important as the message itself. To do this, we stepped away from public opinion research and into the field of psychology. Using a modified Myers-Briggs personality test, we sought to classify the electorate by personality traits: extroversion versus introversion, sensing versus intuition, thinking versus feeling, and judging versus perceiving. We asked a series of questions that rarely show up on the questionnaires of political pollsters:

- Are you the life of the party or a wallflower?
- Do you value common sense or a vision?
- If you were a teacher would you rather teach fact courses or theory courses?
- Which do you value more, logic or sentiment?
- Are you spontaneous or do you plan ahead?

What we discovered was fascinating but troubling. Voters who demonstrated intuition, feeling, and perceiving were the most likely to support Clinton. When you consider President Clinton's own personality traits, these results were not altogether shocking. The problem was that the swing voters we needed to reach were more likely to be sensing, thinking, and judging. They were more drawn to centrist, action-oriented measures like stiffer sentences, ending welfare benefits after two years, raising the minimum wage, and limiting affirmative action. They responded to logic- and fact-based presentations, as opposed to visionary, emotional plans. These were voters who made practical, pragmatic decisions based on detailed information. In other words, President Clinton had to make a conscious effort to modify the way he presented arguments in support of the policies he was advancing depending on which audience he was trying to reach.

Now that we understood the personality types in the electorate, we wanted to take our research a step further—by understanding lifestyle choices. By gaining insight into how voters lived their lives and what traits they most valued, it helped us to further hone the president's messaging.

Our poll focused on the types of questions that pollsters rarely ask: "What are your favorite television shows?" (Clinton voters liked HBO, MTV, and *Oprah;* Dole voters preferred *Home Improvement* and *Larry King;* swing voters liked *Seinfeld* and *Friends.*) "What are your favorite types of music?" (Clinton supporters listened to classical, rap, and top 40; Dole voters were more into talk radio, '70s music, and classic rock; swing voters liked news/sports and '50s music).

Then we dug even deeper. We discovered that Clinton voters jogged; Dole voters hunted; and swing voters watched sports and enjoyed the outdoors. (Swing I voters liked soccer, while Swing II voters preferred hockey.) Clinton voters were most likely to speed but were often afraid to go out at night. While they lived the most active and health conscious lifestyles, they also liked to overeat and shop in department stores. Dole voters were more likely to be found in a religious setting, and they were less tolerant of gays. Swing voters generally worked in a technical setting, were more focused on their family, and were worried about retirement and caring for an aged parent. Even among the Swing I and Swing II voters we detected important distinctions—Swing I voters preferred PBS, Swing II voters were *Home Improvement* fans; Swing Is were more accepting of homosexuals and were less likely to own a gun then Swing II voters.

From these lifestyle questions, we were able to develop an extraordinarily sophisticated view of typical swing voters. They were around the age of forty, had an income of between thirty thousand dollars and sixty thousand dollars, were married, well-educated, moderate, and nonunion. They enjoyed music, sports, the outdoors, videos, talk shows, *60 Minutes* (a favorite with both Swing I's and Swing II's), and soap operas. More specifically, Swing I and II voters were afraid. They were worried about crime, retirement, technology, pressure at work, and illness. How they reacted politically to those fears was what made our work so challenging.

Using this research, we developed strategies to better appeal to each group. For example, Swing II voters were more inclined to watch *60 Minutes* and *20/20.* So if we ran pro-Clinton ads during those times, we would show the president in less formal settings, which Swing II voters preferred. Swing II voters also preferred country or rock rather than pop, and the issues that resonated with them involved elder care as opposed to the Family Leave Act, which worked better with Swing I voters. Never before had a political candidate so carefully dissected the electorate.

The final step in our recommendations involved the language used to sell the president's ideas. We knew that swing voters were interested in results and not more promises or grandiose proposals. This result prompted Bill Clinton's almost mantralike description of having created ten million new jobs, half of them high-wage jobs. But more than results, they wanted a sense from the president that he understood their concerns and could offer solutions that they could understand and relate to.

Once again it was Mark who came up with the breakthrough answer. In the early weeks of our work for him, Clinton had asked me an important question: why had he lost the centrist political positioning he worked so hard to achieve in 1992? I had answered that the problems were based almost entirely on tax-and-spend issues, but Clinton was not so sure. He kept saying there was more to the problem than the old tax-and-spend label. One day, as we were meeting with the president to discuss results of a recent poll, Clinton posed the question once more. This time, Mark responded with an inspiring answer.

"Mr. President," he said. "It's not about economics, it's about values."

It was a brilliant insight and one that played a critical role in saving Clinton's presidency. For years, the Republican Party had taken Democrats to task on the issue of family values. However, the GOP's definition was insufficiently and absurdly narrow. The Republicans' view on family values revolved around religious issues like abortion, school prayer, or personal morality. While the GOP base loved the family values debate, many Americans found it unresponsive to their day to day concerns.

Mark was able to deduce that the way to appeal to this new breed of swing voters was to justify Clinton's policy decisions in terms of all-American values such as "opportunity, responsibility, and community." By doing so, Clinton struck a responsive chord with many swing voters who had begun to tune him out. We realized that just as Clinton needed to propose a balanced budget in order to restore credibility with the American people, so too he needed to talk about religious and cultural values as well as issues like prayer in schools and welfare reform, to reverse the sense of alienation that many Americans felt from their president. Once Clinton had credibility on cultural values, he could refocus the values debate on what we called public values—things like protecting children, caring for elderly parents, safeguarding the environment, looking after the neediest, expanding economic opportunities—instead of the Right's narrow focus on family values. Our polling suggested that these public values were far more resonant in the electorate's minds than the more narrow "family values" agenda that the GOP trumpeted. In short, we found a way for Democrats to win the values argument.

The constant partisan battling in Washington had caused most voters to simply tune out when it came to specifics. Believing that both sides were manipulating facts, voters were far more likely to take a position based on their internal barometers of right and wrong. When asked to judged programs by how they affected them directly compared to the values they represent—voters were overwhelmingly more concerned with values.

Take, for example, the typical pocket-book defense of the Head Start program: "We spent $20 billion on Head Start so that one million children will have a job in twenty years. We will save billions in future welfare costs through the success of the program." Sounds good, but our research found that this was a far more effective approach: "We recommend spending $20 billion on Head Start in the next ten years because every child deserves an opportunity to have an education and go to college in this country. This is what America is all about—having opportunity regardless of where you were born or your economic circumstances." Values trump economics.

From that point forward, the president's rhetoric justified his priorities in values terms—our duty to our families and communities, and our responsibility to protect seniors and the environment. It was a dramatic break with the past, and it secured Mark's presence in the campaign. In fact, Morris was actually quite jealous that Mark, whom he had long feared and distrusted, had had such an important and useful insight.

With the help of our neuropersonality poll, we were able to cast Clinton as a defender of the public values that Americans hold dear, as opposed to the Republicans overly preachy and moralistic focus on values. Of course, all this focus on values begs the question, "Which values are most resonant?" According to our research, four main points stood out.

1. *Standing up for America or showing leadership against a difficult enemy.* Value was key whether we were talking about unfair trade practices by Japan or even homegrown terrorism such as that found in Oklahoma City.
2. *Providing opportunities for all Americans.* So, for example, when President Clinton outlined his defense of affirmative action, he did so not in terms of redressing past wrongs but instead providing equal opportunity to African Americans.
3. *Doing what's right, even when it's unpopular.* Standing up to special interests, being independent and fearless, were all critical to ensuring the American people understood Clinton was on their side.
4. *Preserving and promoting families.* This one was the key. In every poll we took, it was always the top-scoring value, and, considering the president's vulnerability to the "family gap," it was clearly the most important. This value manifested itself in several ways, including the president's support for the V-chip and antismoking initiatives as well as family leave, educational opportunities, and elder care.

Bringing all these points together was the notion that the country needed to find common ground. After years of partisan fighting, voters wanted leaders who sought consensus. The public, quite unlike many politicians and pundits, believed that compromise and conciliation were the single best ways to get the government and the country moving in the right direction.

In 1992, the issue had famously been "the economy, stupid" because the economy was in tatters. Our values approach worked in 1995 and 1996 because Americans were optimistic about the economy and were focused on softer, social issues. While Bob Dole was talking about the need for tax cuts (and the liberals in the White House were fretting about economic uncertainty), we were focusing on issues that mattered to people directly: keeping their kids away from cigarettes, clamping down on violence on television, improving the nation's schools, expanding opportunities for a college education, and so forth. Although pundits ignored many of these initiatives (save to occasionally scoff at us for thinking small), these were the issues Americans cared about. Some may wonder how those voters could be brought to Bill Clinton's side when his personal behavior had already been questioned by many. What we discovered from our polling was that public values trumped private character. Questionable conduct would be set aside so long as public character was seen to be at the highest levels. It was a thesis that would be severely tested in the near future.

The president's focus on values through the fall and winter became the precursor for his State of the Union address in late January 1996. The State of the Union is one of those unique opportunities for the president to have the nation's political stage all too himself. We weren't going to squander the opportunity.

Working closely with the White House staff, Mark and I put together a list of more than a hundred policy ideas. We then polled each of them. Again, the results matched our neuropersonality poll. People were concerned about teen smoking, crime, television violence, and the environment. They were optimistic about the future and in particular the economy. Luckily, these were all issues on which Clinton and the Democrats had a natural advantage. We made them the central focus of the president's State of the Union address.

By this point, Mark was far more intimately involved in day-to-day White House planning. He was working closely with the president's speechwriting staff to turn our poll-tested ideas into reality. For both of us, however, it was an incredibly exciting time. All our years of trying to move the Democratic Party to the center were finally coming to fruition. From our first campaign together, Mark and I had come to understand that elections are won in the middle—by those candidates who best understand how to neutralize their negatives and shift the debate to their terms. We had long believed that Democrats could be the dominant party in American politics, but only after they began to better understand the electorate and how to more effectively get their message across. Now we were actually doing it.

For too long, Democrats played national politics from the same class warfare script, without realizing that their base had changed and developed a whole new set of concerns and opinions about politics. Pundits always complained that President Clinton ran to the Right in his second term, but the reality was far more nuanced. He was still a Democrat. He still believed in the role government plays in people lives. He simply realized that he needed to tailor that message to voters' concerns. Frankly, at a time when antigovernment antipathy continued to run strong, he developed a set of policies that showed government could be a force for good and fiscally prudent.

This delicate balancing act was characterized best in the president's 1996 State of the Union address. When he declared to great acclaim and some derision that "The era of big government is over," it was the culmination of fifteen years of effort by Mark and me. There were those who will argue that when Bill Clinton said the era of big government was over, he destroyed activist government as a force in American society. I would argue quite the opposite. When he made that statement, Bill Clinton saved activist government. By his words and actions, he had redefined the role of government so that it could once again play an affirmative role in people's lives, which he outlined with a comprehensive set of government proposals, from school uniforms and higher educational standards to curtailing tobacco advertising geared toward kids and combating television violence. He was again making government relevant in the lives of Americans. His words speak volumes about the new direction Bill Clinton's presidency had taken:

> We know big government does not have all the answers. We know there's not a program for every problem. We have worked to give the American people a smaller, less bureaucratic government in Washington. And we have to give the American people one that lives within its means.
>
> The era of big government is over. But we cannot go back to the time when our citizens were left to fend for themselves. In-

stead, we must go forward as one America, one nation working together to meet the challenges we face together. Self-reliance and
teamwork are not opposing virtues; we must have both.

It was an amazing performance, and it embodied all the lessons we
had learned from our neuropersonality polls: a focus on opportunity, standing up for what's right, promoting families, and seeking
consensus. Watching Bill Clinton deliver that speech on a chilly
night in January 1996, I knew that barring an unforeseen event, the
race was largely over. In just under a year, we had achieved all of
our strategic goals. We had recast the president as a fiscal and cultural conservative, as a defender of public values, and had positioned him to become the first two-term Democratic president
since Franklin D. Roosevelt.

The president's transformation was about more than simply
political repositioning. The president was a changed man. The
sullen and diminished figure we had met a year earlier had been replaced by a vibrant and commanding figure. Clinton exuded a
level of confidence that gave him a true presidential air. I remember that when we first began working for Clinton not a single
member of his staff stood up when he entered the room. A year
later, it would be unthinkable to do otherwise. Clinton had become a larger-than-life figure.

He was also a delight to work with. He was and remains the best
reader of polls I've ever seen—always aware of what questions to ask
in order to challenge us and force Mark and me to give him better results. He went over many of the pieces of direct mail that we sent
out, pored over television commercial scripts, and was a tireless
fundraiser of campaign dollars. Based on the results of our neuropersonality poll, Clinton would introduce and promote a series of
small programs—like tuition tax credits, V-chips for televisions in
order to keep inappropriate content from children, school uniforms,
and curfews—that responded to people's daily concerns. These were
often derided in the press, but for swing voters suspicious of big-
government solutions they were perfect—and they resonated.

By early 1996, I felt confident about Clinton's prospects for reelection. However, as is so often the case in politics—and in particular

with Bill Clinton—there were surprises lurking around the corner. Three in particular would appear to make 1996 a more challenging year than we had ever anticipated—welfare reform, Ken Starr, and Dick Morris.

Even more than the taxes and crime, welfare was the Democrats ultimate challenge. Rightly or wrongly (usually wrongly), Democrats had been portrayed as supporters of providing generous payments to undeserving indigents who refused to work. No Democratic politician had yet figured out an effective means for blunting Republican attacks that support for the current welfare system was tantamount to encouraging a culture of dependency. The welfare debate in America has rarely been grounded in reality; both sides have used scare tactics and anecdote to justify support for or opposition to welfare policies. However, one thing was clear: a majority of Americans believed that the welfare system was broken and desperately needed to be fixed.

Like the issue of taxes and crime, we felt strongly that unless Democrats took the welfare debate off the table, they would continue to be vulnerable to Republican attacks. In 1992, President Clinton made a name for himself as a New Democrat by pledging to "end welfare as we know it." It was a brilliant political gesture, which unfortunately was not followed up by legislative action. I remain quite convinced today that had Clinton proposed welfare reform before health care reform in 1993, Newt Gingrich would never have been Speaker of the House (and I might have never had the chance to work for a U.S. president).

For Clinton, however, welfare reform was a very personal issue. He understood that welfare was more than a political issue: breaking the cycle of poverty and despair risked inflicting further suffering on the poorest Americans. The president knew that change would have to have at least some Republican inflection, but he worried about the damage that might be done in the process. We in turn knew that until there were stronger work requirements in the legislature, the Republicans would continue to use the issue as a weapon to attack Democrats. Half measures would not do; Democrats would have to agree to fundamentally change the system. We argued that taking welfare off the table would not just help Democrats politically but also reenergize activist government. No longer would Republicans be able to make the claim that Democrats were wedded to an ineffective and counterproductive welfare system. Clinton would put them on the spot to prove if they were serious about providing poor people with a social safety net.

That did not mean we thought the president should sign just any bill. Many people forget that Clinton vetoed the Republicans' first attempt at welfare reform in January 1996, for its overly punitive measures against legal immigrants as well as the replacement of free health care benefits for poor children and nursing home care for senior citizens with discretionary block grants. In both cases, the president did the right thing and took the political hit. (In fact, the GOP almost certainly made the bill harsher than necessary in order to ensure a White House veto that it could use against the president.)

As the year progressed, Republicans wanted something to take home to their constituents, and so they began to move closer to the president's position. From our perspective, there was no way the president could avoid signing a welfare bill. Poll after poll showed that a White House veto could undermine all of our efforts to recast the president as a New Democrat. According to one poll, a veto of welfare reform could actually have led to a possible defeat in November. That's how serious the issue was. Nonetheless, the White House staff pushed hard for a veto. They argued that anything the Republicans came up with would be too harsh, too punitive, and would harm too many children.

Morris, Mark, and I argued that not signing the bill might well lead to a GOP victory in November, which would result in an even harsher bill becoming law. We felt that by ending welfare as an entitlement, we could begin the process of remaking a failed system during the president's second term. The president went through an exhaustive deliberative process—innumerable staff meetings, late-night calls to friends, poll after poll testing different scenarios. But I think the result would have always been the same. Clinton was not going to give Republicans the sort of ammunition they wanted from a welfare veto. I believe that while deep down he wouldn't have minded vetoing the bill, the proximity of congressional passage (August 1996) pretty much ensured that he was always going to put his signature on it. He would find a bill to sign and take another issue off the table.

While working for the president took up most of our time, I still worked for a host of other clients, one of whom was Clinton's successor in Arkansas, Governor Jim Guy Tucker. This was hardly an easy task. Tucker was on trial for fraud and conspiracy, which grew out of the then-ongoing Whitewater investigation, and I had been hired to help with jury selection. It was an unnerving experience. From the first, it was clear that this was more than a legal trial. There was a sense that Arkansas was under siege by Special Prosecutor Ken Starr and his staff. I felt as though I was in a police state and was one of the hunted because I worked for Democrats like Tucker and Clinton. I remember sitting behind the defense table and at one point rising to go outside to the bathroom. When I came back, an FBI official was peering down at the papers I had left on my seat. There was a sense that anything was acceptable. Later, I saw Starr seated in the courtroom, serenely watching the day's proceedings, holding nothing more than a Bible.

Finally, a member of Tucker's security detail pulled me aside and asked me to speak with him alone. He had previously worked for Clinton, and he told me that the FBI was spending hundreds of hours threatening and cajoling former members of Clinton's security detail into discussing the president's personal life. He said everything that was happening in Little Rock, from the Tucker trial to the investigations into Jim and Susan McDougal, was really about Clinton and finding someway to get him. It was frightening stuff. When I got back to Washington, I went to Clinton and warned him about what was happening. By that point, he needed no warning; he understood where Starr was headed. For an hour we talked, just the two of us. He told me he'd been hearing many of the same details from his friends in Little Rock—tales of intrusive FBI interviews and heavy-handed tactics. Tucker and Clinton were never close, but he told me he felt terrible for Tucker since he knew that he'd done nothing wrong and that Tucker was just the latest casualty in the war against Bill Clinton. Considering Starr's desire to get him it was hardly surprising that Tucker was convicted—but on only two of seven charges—and based largely on the dubious testimony of a former Arkansas judge and convicted perjurer by the name of David Hale.

The more we talked, the more I got a sense of a man under siege. Clinton was clearly angered by the never-ending investigations, which he likened to a jihad by Starr and the right wing. "We didn't do anything wrong," he would repeat like a mantra. Clinton saw himself as an upright, decent person who was trying to do the right thing. He recognized that he was far from perfect, but he felt that his own behavior was no different from that of other presidents. The biggest difference was the level of scrutiny he was being subjected to. He believed—accurately, it would turn out—that the Far Right would do anything and everything they could to get rid of him. While I appreciated the opportunity to speak with him so frankly, it was hard not to feel sorry for the him. Here was the most powerful man in the world, seemingly at the very mercy of political opponents who would stop at nothing in their effort to undermine him. It was an eye-opening insight to the limits of political power.

As the campaign entered the final stretch, our strategy sessions settled down into a routine, save for one increasingly erratic exception, Dick Morris. Dick had always been a bit of a contradiction. He demonstrated extraordinary political insight, but at the same time he had a mercurial, often bombastic personality, marked by insecurity and self-absorption. My sense of Morris has always been that he desperately wanted to be liked. He may have gone from a spartan Upper West Side apartment to a tastefully decorated home in Redding, Connecticut, but some things never changed. I had long ago learned to shrug off his excesses, but Mark and the other members of the campaign team were not so understanding. As time went on, even I had to admit that Morris's behavior was becoming increasingly bizarre.

The rise in Clinton's poll numbers led to a significant rise in Dick Morris's already inflated view of himself. He would think nothing of calling high-level administration officials and pushing his pet ideas. Reporters would line up outside his suite at the Jefferson Hotel for the latest political tidbits and leaked poll results, while Mark and I would be busy at work. In the lead-up to the Democratic convention, Morris pushed our client, Indiana governor Evan Bayh, and Vice President Gore to set aside their own ideas and speechwriters and instead deliver speeches he had written. Both resisted, but Morris refused to back down. Bayh came to me complaining that Morris was increasingly difficult. Morris's ideas were becoming stranger, and yet Morris himself was becoming more and more intolerant, refusing to compromise on even the most minor issues.

The relationship between Morris and Mark was a particular sore point. After Mark moved to Washington full-time, Morris became increasingly obsessed with Mark and his activities. Things went so far that Morris actually issued an edict saying Mark could not have any meetings with Clinton without his prior knowledge and approval. I tried to stay out of the mounting feud, but of course I couldn't. Morris insisted that I monitor Penn's activity on an hour-to-hour basis; Mark insisted that I take a stronger stance against Morris. Neither course of action seemed practical, to the frustration of both.

In late January 1996, Penn joined a group of staffers in the Oval Office to discuss an important upcoming speech. I was in Las Vegas meeting with casino industry clients when I got an urgent call from Morris, who had somehow tracked me down.

"Get him out of there," Morris screamed, claiming that Penn had broken the rules and gone into an unauthorized presidential meeting. I was at a loss. I wasn't about to call the president's secretary, Betty Currie, and tell her to walk into the Oval Office and ask Penn to leave a meeting with the president of the United States. Morris saved me the trouble. He called the White House personally and asked that the staff remove Mark. Luckily, the meeting was close to breaking up, though it didn't stop Morris from telling everyone in attendance that he never wanted something like this to happen again.

Behind the scenes, I was caught in the middle between my two feuding colleagues. While Morris wanted me to drop Mark and work with him exclusively, I tried to explain to Morris that the name of the company was Penn, Schoen & Berland—I couldn't exactly drop my partner. Conversely, Mark would pester me with questions about my relationship with Morris. "How can you be so solicitous of someone like Morris?" he would ask. But regardless of the infighting, by the spring of 1996, it was increasingly clear that Penn and I were basically doing much of the polling and consulting work and thus facilitating day-to-day decision making for the campaign. In theory, Morris was now charged with developing general strategic goals and initiatives. In practice, it sometimes seemed that he was now most focused on burnishing his image with the public.

The changes in Morris went beyond mere bureaucratic politics. When Mark and I first started working on the campaign, we frequently ventured off to dinner at Washington's Old Ebbitt Grill. Morris would usually join us. As time went on, he seemed distracted and usually begged off. We had assembled a fairly close-knit team, so his absence seemed odd. We had no idea what he was doing, and it wasn't until right in the middle of the Democratic convention that the tabloid newspaper, the *Star,* broke the story that explained his absences: Morris had been enjoying the services of a DC call girl. There were even photos of the two of them kissing on the balcony of Morris's hotel suite. The story ignited a news storm and threatened to seriously harm the president on the eve of what was supposed to be his moment of triumph.

It was clear that Morris had to go.

After receiving a call from Morris that he was in trouble, I went to his hotel room at the Democratic National Convention in Chicago. Dick was in no mood to go quietly. He blamed a GOP conspiracy for his downfall. He claimed that he hadn't done anything worse than the president. He railed against "yellow journalism." I sat with him in his hotel room, reminding him of the irreparable harm that he could do to the president and his own career if he didn't resign. I told him, "Go with dignity." Finally, Morris read the writing on the wall and resigned, but only after complaining vociferously to White House Chief of Staff Erskine Bowles, who was sent to solicit his withdrawal from the campaign. Yet he would not disappear completely.

About a week after the convention, Clinton convened a meeting at the White House with me, Vice President Gore, and Mark. His basic message was clear: "I have complete confidence in you and I want you to take the lead between now and the election." While we were always thankful to Morris for his role in bringing us into the White House, we also understood that he had become an albatross. I told Clinton, in no uncertain times, that for political reasons he needed to cut his ties with Morris. I was greeted by silence. I repeated it. Silence again. One thing I learned about Clinton was that when he was silent, he definitely doesn't agree with you. I never saw Clinton be anything but respectful of Morris: about the only thing he ever said to me about him was his constant amazement at Morris's zeal in trying to get his bills paid as soon as possible. Yet knowing Dick, I feared that he would turn on Clinton when the moment was ripe. Those fears would be realized when Morris leaked sensitive polling he had done privately and secretly for the president during his impeachment crisis. First, however, I had my own Morris crisis to deal with.

Not long after he was officially thrown off the campaign, Morris called me with what he said was an offer I couldn't refuse: He wanted me to serve as his secret conduit to Clinton. I politely declined. Then he said he was considering writing a book about his experience advising Clinton. "Did I want to be the secret coauthor?" Again, I demurred, but I sensed that Morris would make it a decision I would regret, and I feared that he would begin leaking disparaging information to reporters about Mark and me.

However, with Election Day around the corner, I could no longer worry about Dick Morris. We had a campaign to run. In the short term, Clinton took our advice and sailed through the campaign, easily fending off Dole's attacks and staking out his centrist positions. While Mark took up Morris's role on the campaign, I was responsible for organizing state polls for the reelection effort. One of my key responsibilities became the Tennessee race. The state was a swing state and as Vice President Gore's home base, we thought it was essential to win big. Unfortunately, neither Clinton or Gore was terribly popular in Tennessee. We were forced to run a very conservative race—focusing almost exclusively on Swing II voters and emphasizing the issues of gang crime, welfare, and law and order generally. We also worked hard to find Christian conservative ministers who would endorse Clinton in television commercials. We joked that we out-"Republican-ed" the Republicans. And still, we barely won. Even then it was clear to me that Vice President Gore was going to face some serious hurdles in his home state of Tennessee when he ran for president. Gore never was able to change his image in Tennessee, contributing significantly to his defeat in 2000. Had he been able to replicate what we had done in Tennessee in 1996, he might well have carried the state and won the presidency.

I didn't see the president during the Inaugural in 1997. In fact, we didn't talk again until the first strategy meeting of his second term. At the end of our meeting, the president pulled me aside. "Doug, thanks for all your hard work," he told me. "I couldn't have done it without you."

I don't care how many races I work on or how jaded I become, that was a memory that will last me a lifetime.

CHAPTER NINE

JON CORZINE, COMPETITIVE LIBERALISM, AND THE IMPORTANCE OF MESSAGING

There's a widespread belief that money is the key to success in politics—that with enough money, almost anyone can buy the elected office of his or her choice. There's no denying that money is important. It allows candidates to introduce themselves to voters (one thing money can buy is name recognition), to respond forcefully to attacks, and, when necessary, to go negative.

But money alone isn't enough. Modern political campaigns require a particular set of attitudes and skills that not everyone has. Competition for a party's nomination is fierce; the pace of campaigns is brutal; and the skills they require are not easily mastered. Learning to talk about yourself and your values, staying on message, speaking fluently on the stump, fielding questions from single-issue activists and a probing press corps, projecting leadership on television, and most of all doing it day after day after day—these are skills that even the most gifted politicians have trouble mastering. They're also skills that money can't easily buy. If it could, former Californian Congressman Michael Huffington would be a member of the U.S. Senate; Tom Golisano would have been a serious candidate for governor of New York; and Ross Perot or perhaps Steve Forbes would have been the forty-second president of the United States. In fact, while it's true that there are more self-financing candidates running for office today than ever before, very few of them actually win. Of the thirty top spenders in the 2004 House and Senate elections, only one—Texas Republican Michael McCaul—won.

In spite of these poor odds, by the late 1990s, Penn, Schoen & Berland had amassed an unusual record of success with wealthy candidates. The trend began with Frank Lautenberg in 1982, and during this campaign, Lautenberg spent what at the time was seen as a staggering sum—$5 million. Two years later, we replicated our success with Jay Rockefeller in West Virginia, combining poll-based information about the voter positions on issues with an innovative direct-mail operation that allowed Rockefeller to speak directly to voters with unusual attention and precision. Because of these and other successes with wealthy candidates, it was only natural that when Lautenberg decided that he was going to retire in 1999, I found myself drawn into discussions about who might replace him.

New Jersey was (and is) an unusual state. Though geographically small, it is the tenth most populous state in the country. For many Americans, New Jersey's image is the Jersey Turnpike, the Sopranos, Bruce Springsteen, industrial sprawl, retro diners, and big malls. Some of those stereotypes are based in truth: New Jersey does have the highest population density and the highest ratio of highways per area of any state in the nation. However, it also has spacious horse farms, affluent New York suburbs, a sparsely populated agricultural south, and vast pine forests along the coast.

New Jersey is also a curious state politically. Although the state was clearly trending Democratic by the late 1990s, particularly on the federal level, it had a Republican governor and a Republican legislature. This anomalous situation was attributable to one man— former Democratic governor Jim Florio. As a pugnacious lawyer, a former Golden Gloves boxer, a staunch proponent of gun control, and a favorite son of South Jersey, Florio had swept into the governor's office in 1989 on a platform that promised not to raise taxes. Then, a few months after being sworn in, the new governor rammed a 2.6 billion dollar tax increase through the legislature. The result was an outpouring of public anger—followed by the kind of electoral bloodletting rarely seen in American politics. Republicans went from a minority to a veto-proof majority in the state legislature. New Jersey's brightest political star, former Princeton University Rhodes Scholar and New York Knicks player Bill Bradley, narrowly escaped defeat at the hands of an obscure horse country Republican, Christie Todd Whitman in the 1990 Senate election. In 1993, Florio was voted out, and Whitman became New Jersey's first female governor.

Florio's blunders had cost the Democratic Party control of New Jersey for most of the 1990s, and most Democratic county organizations disliked him accordingly. Nevertheless, the former governor retained the sympathy of many liberals in New Jersey. When Lautenberg stepped down, Florio saw an opportunity for political vindication and decided to through his hat into the ring, horrifying Democratic leaders. Whitman was widely expected to be the Republican candidate, and a Whitman-Florio rematch looked like a race that Florio was sure to lose—unless, that is, someone stepped in to stop him. That was exactly what some of the most powerful Democrats in New Jersey, including the state's junior Democratic Senator, Bob Torricelli, were determined to do.

Florio's opponents had several ways to thwart his comeback at their disposal. New Jersey county party leaders determine the order in which candidates' names appear on the primary ballots. Since poor placement often translates into fewer votes, securing county level support is a critical part of any New Jersey elections. Florio's opponents in populous northern New Jersey would ensure that Florio got bad billing in the northern part of the state. This would hurt—as long as they had a strong opponent, and that was precisely what Torricelli set out to find.

Known as "the Torch" for his fiery temperament, Torricelli was a ferocious competitor and a prodigious fundraiser. He had taken control of the Democratic Senate Campaign Committee at a time when most other Democrats were demoralized by Republican gains and brazenly announced plans to raise $104 million. Amazingly, he did it. He had also launched a major effort to recruit wealthy individuals to run as Democratic candidates, so I wasn't entirely surprised when one evening in March 1999 at a Torricelli fundraiser in Newark headlined by President Bill Clinton, Orin Kramer, a prominent New Jersey fundraiser, turned to me and said, "What about Jon Corzine?"

It was an intriguing suggestion. I knew that Corzine was the chairman and CEO of the investment-banking firm Goldman, Sachs & Company, and I had heard that he was preparing to step down, taking with him an estimated $300 million fortune. I also knew that he was a Robert Rubin protégé and a major Democratic donor at the national level. At Torricelli's behest, I offered to call and sound him out on whether he might be interested in becoming a U.S. Senator. Several days and several phone calls later, I found myself on the phone with a genial but decidedly uncommitted man. "I'm about one in twenty in terms of doing this," he told me. Still, we decided to meet in person to talk further; he suggested an early breakfast at the Carlyle Hotel—6:45 AM sharp.

This first meeting was a lot like our first conversation. Corzine in person was a friendly midwesterner; a tall man with a salt-and-pepper beard whose vigorous bearing still reflected the walk-on college basketball player he had once been at the University of Illinois. The Carlyle hadn't yet opened, so instead we ended up talking at the Three Guys coffee shop on Madison Avenue. I had expected to find a Robert Rubin Democrat—a fiscally conservative, socially liberal financier. Instead, I found a man who saw politics primarily as a form of public service. He was low-key, genuine, and candid. In the course of our conversation, he revealed that he thought that part of the Social Security trust fund should be invested in the stock market—not exactly a position that would appeal to Democratic primary voters. What did interest him was philanthropy and public service. In truth, we didn't really talk about domestic politics very much. Mainly, we discussed Corzine's upcoming trip to South Korea, where he was scheduled to meet with President Kim Dae Jung.

Despite his apparent lack of interest in politics as it was normally practiced, Corzine had many attributes that might make him an appealing candidate. He had grown up on a farm in Illinois, gone to public schools and universities, served in the Marine Reserves, and by dint of unstinting hard work made a fortune on Wall Street. Clearly no one becomes chairman of Goldman Sachs without being smart and tough. In addition, he was a generous philanthropist who made a point of sharing his vast wealth with a wide range of organizations in New York and New Jersey. It wasn't a war hero, John Kerry kind of biography (or even a John Edwards, poor-boy-made-good biography), but his story did reflect the American Dream, and I believed it could be parlayed into a positive public image. In the general election, his moderate instincts and fiscal prudence might play well.

The prospect of a Corzine candidacy presented an unusual challenge—and an intriguing opportunity. Clearly, Corzine had the resources to fund his own campaign. In that regard, his candidacy did offer a chance to see just what money could do. From my perspective, however, the real challenge of a Corzine campaign was one of messaging. How could we find a message that would win in the primary against a liberal favorite like Florio and propel Corzine to victory in a general election against a moderate Republican? It was a challenge that would draw on all my skills as a political strategist.

The other question concerned Corzine himself. Would a man who had scaled the pinnacle of success in the corporate world have what it took to win in the humbling and ruthless world of politics? Could someone who really just wanted to serve find a place in contemporary American politics?

The search for a message began with what political consultants call a benchmark poll. In late April, we queried 1,038 New Jerseyites to find out how the investment banker from Wall Street might play in the Garden State.[1] On May 5, we presented our results to Corzine and a handful of close advisers at the Harmonie Club, a well-known midtown spot. On first impression, we found that voters were suspicious of a rich Wall Street Democrat without political experience, but when presented with a brief (and flattering) account of his life that emphasized his mainstream values and his standing as a self-made man, attitudes became much more favorable. We found that this Jon Corzine matched up well against Governor Christine Todd Whitman in the general election. After respondents heard all of the arguments for and against both candidates, the financier was within seven points of the governor. For a total unknown, it was a remarkable showing. It was a clear sign that Whitman's position was weak—very weak. That was the good news.

The bad news was the Democratic primary. Primaries tend to attract the party core—for Republicans, the Far Right; for Democrats, liberals. Among this group, there was simply no way Corzine could win as a moderate New Democrat. Even if Corzine spent lavishly in presenting a positive self-image to New Jersey voters and went negative against his opponent, Florio still won, decisively, by a margin of roughly 25 percent. No matter how negatively we described Florio, his support never fell below 50 percent. The logic of our findings seemed inescapable: among the committed Democrats who make up the majority of primary voters, Florio's liberal positions were still quite popular and even a negative campaign was unlikely to change things. Corzine was highly unlikely to beat Florio in the primary. As a result, I recommended preliminarily that Corzine not enter the race, given the unlikelihood of success.

Corzine's reaction to my recommendation was interesting. Far from being distressed or upset, he seemed to appreciate my candor and my willingness to advise him against embarking on a campaign that would clearly benefit my business. But he didn't accept my advice. Instead he said, "Go back to the drawing board." In doing so, I turned to a brilliant and often unfairly maligned political strategist, Bob Shrum.

Despite our philosophical differences, I had a good relationship with Shrum, who had established himself as the wunderkind of liberal Democratic politics by drafting Senator Ted Kennedy's famous "the dream shall never die" speech at the Democratic Convention in 1980. Shrum was famous for wanting to run populist campaigns that addressed "big issues." A Corzine campaign presented him with just that opportunity. It helped that Corzine trusted Shrum, who had joined our brainstorming sessions with a strong endorsement from Treasury Secretary Robert Rubin. Not surprisingly, it was Shrum who had the inspired idea of running Corzine as an unabashed liberal.

When I asked Corzine if he'd be willing to let me do another poll, one that would test the appeal of a liberal candidate, he readily agreed.

This time we focused the polling on one question: Was there a way to beat Florio in the primaries? A close reading of the data convinced us that the answer was yes; even among liberal primary voters, Florio had vulnerabilities we could exploit. To win, Corzine would have to be positioned as an out-and-out liberal—someone who believed in the need to significantly increase funding for key social programs like health, education, and diversity policies focusing on the needs of minorities and New Jersey cities were critical to victory. It was time for another meeting in midtown.

Corzine listened to our presentation and accepted the proposition at once. The problem at hand was how to beat Jim Florio and win the nomination, and Corzine seemed pleased that we had found a way to do that. In the course of a long discussion about a liberal message, Corzine made it clear that he was comfortable with a decidedly populist, left-wing message—save for social security reform, which he simply would not compromise on. (The other item Corzine preemptively took off the table was his beard, which he announced was staying come what may.) Corzine then turned to another subject—money. "I hate to talk about money," he began, "but what would I need to spend?"

No one seemed to want to answer this question so I spoke up. "Seven, eight million dollars in the primary, and another seven million dollars in the general," I replied. To me, it seemed like a reasonable amount. New Jersey is situated between the country's first and forth most expensive media markets (New York City and Philadelphia) and has no statewide station of its own. As a result, campaigns in the Garden State tend to be very expensive. Corzine, however, turned absolutely white. It was clear that he was taken aback by the figure.

"I can't even imagine spending $15 million," Corzine said. There was an uncomfortable silence before the conversation shifted back to the potential campaign.

By early summer, we had assembled the nucleus of a team. It was now time to work on a strategy. In July, we gathered at his beach house in East Hampton (New York) to finalize the strategy. For two days, Corzine and the group I had assembled plotted out his campaign. I had the misfortune of arriving late and ended up being assigned a bunk bed in Corzine's attic, which goes to prove that political consulting is not all glamour, even when you're working for a multimillionaire.

To develop Corzine's message, we fielded another poll that measured how voter attitudes changed toward our (potential) candidate in response to various policy proposals. The goal was to select the set of policy proposals that most boosted the candidate's numbers. Our research soon led us to a strategy that I called "universal everything"—universal health insurance, universal long-term care, universal college education, universal gun registration. For good measure, we also decided to emphasize higher salaries for teachers, lower class sizes, and "values" education. It all tested extremely well. Among New Jersey Democrats, at least, it seemed that being an ultra-liberal was the path to popularity. Of course, our strategy did pose problems down the road. If it worked, Corzine would enter the general election as a liberal running against a moderate Republican, but the alternative was not making it past the primary at all.

Our polling also showed that voters responded best to Corzine when he presented himself as a passionate outsider who was committed to changing the status quo. Shaping an appealing story about a candidate's background was critically important. After potential voters heard about Corzine's positions and biography, the horse race looked a lot closer—Florio at 46 percent, Corzine at 43 percent.

We also tested ways to parry potential Florio attacks on our candidate and his controversial position on Social Security. We determined that the best way to respond was simply to dismiss the attacks as "old-style politics" and emphasize that he (Corzine) was interested in the future and would never put seniors at risk. Our polling also assured us that our positions on the issue weren't too liberal. While "the L word" may be the kiss of death in some parts of the country, New Jersey wasn't one of them. In short, we believed that the potential for a winning campaign was there, however, it was still just potential; the actual horse-race numbers hadn't changed much. Throughout the early summer, Florio had enjoyed a 62 percent to 14 percent lead over Corzine. Most voters were still unfamiliar with Corzine. However, we assured Corzine that the numbers would change once our media blitz started.

The second challenge would be communicating the fact that even though he had been the head of Goldman Sachs, Jon Corzine was a committed liberal. This was not a straightforward task. Already, newspapers such as the *Philadelphia Inquirer* had described him as a candidate [who] "would appeal to the middle." Now we needed to make it clear that Jon planned to embrace an agenda that would make him one of the most liberal candidates in the country. Was that something that we could do with a multimillionaire fixture of Wall Street? I was about to find out.

T he combination of vast resources, political talent, and arm-twisting soon did its work. In late June at a hastily called press conference in Trenton, Representative Frank Pallone, the other leading contender for the anti-Florio mantle, announced that the time had come to face up to "the reality of what's going on statewide" and bow out. "I still believe in my campaign and my ability to run statewide," Pallone said with more than a touch of bitterness. "However, I don't have a multimillion-dollar personal bank account."

Despite Pallone's sentiments, money couldn't substitute an engaged candidate, and Corzine still had to make the sale to the state's many important players. In a series of private meetings with county leaders that summer, Corzine did just that. His greatest selling point was his personality. Many county leaders no doubt expected the former chairman of Goldman Sachs to be a haughty financier or a shark. Instead, they found themselves talking with an earnest, hardworking, down-to-earth man who was solicitous of their opinions and concerns. The fact that he was willing to contribute generously to their political organizations didn't hurt either. By midsummer, we had lined up the support of some of the state's most powerful people, including key figures from Hudson and Union counties. We were also making excellent headway with the African American community, securing the support of powerbrokers like Calvin West, the "confidential aide" to Newark Mayor Sharpe James, and Assemblyman Joseph Charles, chairman of the Democratic Party's black and Latino legislative caucus.

These triumphs did not come easily. Florio recognized our efforts and worked hard to head them off. He quickly lined up the support of eleven of the state's twenty-one Democratic county organizations. Florio even attempted to counter Corzine's financial advantage. After Corzine announced that he was devoting five hundred thousand dollars to an exploratory committee, Florio announced that he had raised six hundred seventy-five thousand dollars—and that he had brought in James Carville, the political consultant who had helped shape President Clinton's first campaign, to offer advice and to help him raise even more. All the while, Florio directed a flow of hostile comments toward our candidate, accusing him of mounting "the equivalent of a hostile Wall Street takeover of the Democratic Party." Despite these attacks, by the end of the summer it was clear that we had the support of the greater part of the party establishment. It was also clear that the transformation of Jon Corzine was well underway. By late summer, our effort to position Corzine as a liberal—someone who supported universal health care, heavy federal investment in public schools, mandatory gun licensing and registration, strong labor unions, and increasing the minimum wage, had gone so well that reporters analyzing the race were predicting that, as a committed liberal, Corzine might pose difficulties in the general campaign:

> But Mr. Corzine's efforts to stake out strong liberal positions now could pose problems for him if he wins the primary and goes on to face the likely Republican nominee, Gov. Christine Todd Whitman. In a general election, he will have to appeal to more centrist and even conservative voters, many of whom may strongly object to his positions on tax cuts, abortion and gun control.[2]

It was a fair point. Fortunately, our most formidable opponent was about to exit the scene.

On September 7, four days after the *Times* published its introductory story on Corzine, Governor Christine Todd Whitman stunned the political world by announcing that she would not run for the U.S. Senate. Most pundits concluded that Whitman, who was struggling with her low popularity ratings, didn't have the appetite for a campaign against a man with unlimited money. As a result, the Democratic nominee would be running against a less well-known figure. The road to the Senate appeared to be a lot less bumpy—if we could just get past the Democratic primary.

Two weeks later on September 23, Corzine formally announced his candidacy, first from his home in Summit, then in suburban Union County, and finally in Camden, the epicenter of Florio support. It was our way of taking the fight to Florio. We wanted to let him know that despite his ties to South Jersey there would be no respite from our campaign. However, Corzine was careful not to criticize the former governor directly, opting instead to pledge his allegiance to classic Democratic principles and lay out his support for universal health care, comprehensive long-term care for the elderly, more federal spending on education, stricter gun control laws, and a higher minimum wage. We also began running radio ads as a way of gradually introducing our candidate to key segments of voters. Our goal was to position him as a passionate outsider with big ideas—in other words, a big-hearted liberal who was not Jim Florio.

The campaign got underway in earnest in the first weeks of 2000. On January 18, Corzine delivered his first major speech—an address to the Robert Wood Johnson Medical School calling for universal health care. Three days later, Corzine got another boost when Essex County endorsed his candidacy, giving Corzine the nearly complete support of party organizations in the populous northern part of the state. As the emphasis shifted away from private meetings and toward the public arena, we had started to mall test the campaign ads. To fine-tune them, Penn, Schoen & Berland sent researchers to fifteen malls across New Jersey, from the ritzy Short Hills mall to the more modest Woodbridge Commons Mall. Once again we used our AT&T mall techniques, and our selection of malls was designed to produce a representative cross-section of the electorate. By charting voters' attitudes before and after they were shown potential campaign ads, we were able to develop the most effective ads possible and determine exactly how they would play with different segments of voters.

The candidates' first joint appearance came before the Princeton County Democratic Club. By all accounts, it was a difficult night for Corzine. Looking confident and knowledgeable, former Governor Florio rattled off detailed policy proposals while Corzine spoke haltingly and vaguely about his campaign. Clearly, Corzine needed to improve his speaking style. As we predicted, Florio also attacked Corzine for his plan to invest 15 percent of the Social Security fund in the stock market.

Despite these difficulties, we knew voters wouldn't cast their ballots for the best debater, and that ultimately the primary forum for the campaign would be the state's television screens. In early March, we stepped up the tempo of the campaign by running weekly television ads, an expensive proposition, and calling likely voters. The ads cost roughly a million dollars a week, but fortunately, we had the resources necessary to implement this strategy. We suspected that much of Florio's lead rested simply on his higher name recognition, and that three months of advertising would make Corzine an equally familiar name.

While Florio's campaign could do mailings, they didn't have the funds to put advertisements on the air. As a result, we had a clear opportunity to define our candidate's image in a positive fashion. We even decided to attempt to inoculate Corzine against the attack we feared most by running ads that touted the fact that Corzine wanted to invest a small portion of the Social Security trust in the stock market. The voters wouldn't like it, but at least they would hear it from us first.

By April, the Corzine campaign had gone through more than $10 million. In contrast, Florio had raised only $1.38 million—not enough to go on the air. Despite the distractions, our ads were having an effect. In just a month's time, Corzine had closed to within striking distance of Florio. However, no matter how skillfully we sold Corzine, we couldn't take the lead without presenting contrasts. By mid-April, the tracking polls showed that our upward momentum had stalled out. It was time to go negative.

We honed in on Florio's greatest vulnerability—taxes. In late April, we started to air ads that reminded voters of Florio's 1990 decision to increase taxes sharply. It was a devastatingly effective tactic. Polls soon showed us with a sixteen-point-lead over Florio, who had yet to air an ad.

Finally, on May 24, the Florio campaign put its first ads on the air, a trio of commercials that accused Corzine and Goldman Sachs of a somewhat confusing variety of misdeeds. Ads sponsored by the Camden County Democratic Committee that portrayed Corzine as a Wall Street raider who preyed on the elderly soon followed and were more effective (though still somewhat incoherent). The former governor also benefited from the revelation that Corzine's personal lawyer had hired private investigators to do opposition research. While hiring opposition researchers is a standard part of an important political campaign, hiring private investigators is not. Though the story passed quickly, when it first broke, it was unclear what impact it might have on the campaign.

While the media shifted its scrutiny from Florio's floundering candidacy to the Corzine campaign, another important campaign was playing out behind the scenes—the campaign to turn out the vote on Election Day. To ensure the support of the Democratic base, Corzine spent lavishly, doling out approximately eight hundred eighty-five thousand dollars to local organizations and important state politicians. By late spring, virtually every political operative in New Jersey seemed to be on his payroll.

In addition, there was money for get-out-the-vote operations in cities such as Newark and Trenton. The weekend before the primary, I met with Corey Booker, a talented young African American lawyer who at the time was planning a run against incumbent mayor Sharpe James (and who today is the mayor of Newark). We turned Booker down as a client—we already had a relationship with Woodbridge Mayor and future Democratic Governor Jim McGreevey, who was a James ally—but at the end of the meeting I asked Booker and his advisers what he thought was going to happen in the primary. Corzine was going to win big, they said. Booker predicted that the former financier would win more than 90 percent of the African American vote. I was surprised by their confidence and pushed back.

"Florio was an incumbent governor," I pointed out, "and our polls show that he has at least some support."

"Just watch," Booker responded. "He has no support."

I hoped Booker was right. The numbers we were seeing showed that the race had tightened considerably. We were now polling daily, and our polls had Corzine hanging on to a narrow four- to five-point lead. In the final week of the campaign, we spent $2.5 million on ads. We were doing everything we could, and it certainly looked like he was going to win, but I wasn't breathing easy just yet. Nor was Corzine.

By mid-May, Corzine had spent nearly $25 million on his campaign. At first, the spending had been relatively easy. He wrote big checks and watched his numbers rise. When his lead started to decline in the late spring, Corzine became more anxious—understandably so. Ultimately, he would spend $36.7 million in the primary election; to lose would be an expensive and embarrassing disappointment. I tried to allay these concerns. Races naturally tighten toward the end, I told him. But as his pollster, I had to be honest. There was no guarantee he would win.

The fact of the matter was it had been a brutal primary season. Running for office is a grueling experience. It was clear that Corzine did not take naturally to the campaign trail the way that some politicians did, but he had proven to be a disciplined campaigner and one of the hardest working candidates I'd ever seen. He was incredibly tenacious, and he was a fighter. He simply would never quit. He learned the skills he needed to learn in order to become a successful politician and made the tough calls. He spent what needed to be spent, even when the sums greatly exceeded early estimates. By the end of the primary, I understood why he had risen so high on Wall Street: in his dogged, unpretentious way, the man was simply determined to win. It was amazing how well a man who had spent his career in the gilded world of finance had adapted to retail politics. I hoped his mood would lift when the nomination was his.

Tuesday, June 6, was Election Day. That morning a voter turnout operation like nothing New Jersey had ever experienced before swung into action. As the day began, Penn, Schoen & Berland phone banks began calling voters our phone banks had identified as favorably inclined toward Corzine across the state. But the real action was in urban neighborhoods in cities like Newark and Trenton. There the roughly six hundred fifty thousand dollars that campaign operatives had distributed to everyone who could get a voter to a polling station paid off handsomely. African American voters turned out in disproportionate numbers and voted overwhelmingly for Corzine. By the end of the day, it was clear that Corzine was headed for a comfortable win—buttressed by his overwhelming margin with minorities. The county party endorsements Corzine had gotten also paid off in the northern part of the state. It was these two factors—monolithic minority support and his party line backing that stretched his single-digit lead in the polls into a very comfortable double-digit victory on primary day. He ended up trouncing Florio, 58 percent to 42 percent.

I was delighted with the result. We had taken a political novice, a Wall Street tycoon, transformed him into a passionate liberal, developed a compelling message and platform, and steered him to victory in a Democratic primary over a former governor who was a favorite with New Jersey liberals. That evening at campaign headquarters, I congratulated Corzine on his win and asked him to preview his speech for me. He looked at me shrewdly and silently handed me the speech—a speech that passionately laid out his liberal agenda. It was clear that he planned to do the next stage of the election his way.

Alarmed and, to be honest, somewhat surprised by this unexpected turn of events, I said, "Look, Jon, you need to tack back to the center on fiscal prudence for the reasons we discussed."

I reminded him that independent voters outnumbered Democrats and Republicans and tended to be more fiscally conservative than the Corzine base, which was more liberal, lower-income, and more likely to belong to a labor union or a minority group than the electorate as a whole. While our polling showed that Corzine led independents at the moment by a 45 percent to 30 percent margin, 25 percent of these voters were unwilling to commit. The election would ultimately turn on these voters.

Jon just shook his head. That's when I realized something important: While I had been thinking strategically about liberalism as a strategy to win the primary, Jon had been thinking about it philosophically. In the process, he had become a truly committed liberal. I realized with a start that he had no intention of altering the thrust of his campaign to accommodate the desires of the first political consultant he had ever hired. This is who he was, and this was the campaign I was now going to run. That evening, Corzine gave an impassioned, resolutely liberal speech. His only nod in the direction I was recommending was a passing reference to fiscal prudence. Later that evening, I realized that I was now in the position of trying to elect a man to higher office who was now becoming one of the most liberal politicians in the United States.

So much for the all-powerful political consultant.

Corzine's admirably principled positions presented me with a daunting challenge. Corzine's opponent, Republican Congressman Bob Franks, was a moderate, pro-choice Republican who had the potential to appeal to centrist Democrats. Corzine needed to secure those voters' support by reassuring them that he would be fiscally prudent, but many members of the Corzine campaign looked at Jon's fourteen-point lead over Franks and concluded that the election was effectively over. I was focused on a different number—the 20 percent of the electorate that was still undecided, despite having heard an enormous amount about Jon Corzine. We needed to convince New Jersey voters that Franks was really a right-wing Republican. In my conversations with Jon, I advised him to hammer away at substantive issues while the campaign ran ads that portrayed Franks as an acolyte of former House Speaker Newt Gingrich and called attention to his efforts to cut spending on health care, education, the environment, and the elderly. Initially, the Corzine campaign held off acting on my advice. It declined to go negative and instead cut back its early media spending. That's when the trouble began.

Our problem was not really the Franks campaign, it was the press, particularly the *New York Times*. The *Times* ran article after critical article on Corzine's campaign spending. These stories didn't particularly hurt us with the electorate, at least at first. At one point, we presented voters with the statement that Corzine spent almost $37 million against Jim Florio and was now spending millions against Bob Franks and asked how that knowledge affected their vote. Forty-six percent of people responded in the way the media would have expected and said it made them less likely to support him. However, 40 percent of respondents said it made them *more* likely to vote for him. Nonetheless, the unceasing criticism from the media gradually undermined Corzine's image with the electorate. During the primary campaign, he had been the underdog. Now he risked being seen as becoming a free-spending financier trying to buy his way into the U.S. Senate.

The attacks also took a personal toll. Corzine was not a battle-hardened politician; he did not yet see the press as a tough, sometimes capricious player in campaigns who likes nothing more than to topple the leading candidate. He thought of himself as big-hearted, well-intentioned person—a philanthropist who had given millions of dollars to charities in the metropolitan area and who was now running for office with the best of intentions. As a Democrat who had spent most of his professional life working in New York, Corzine naturally wanted to be thought well of by the city's paper of record. The fact that it failed to recognize his good intentions and instead portrayed him as a plutocrat bent on buying his way into the Senate was deeply hurtful. However, by late August, Corzine had a more serious problem: his lead was shrinking. By the end of the month, our fourteen-point lead had shrunk to eight points.

Then things got worse. For months, Corzine had been dogged by demands from the press that he release his tax returns. At first, Corzine resisted, citing a confidentiality agreement with Goldman Sachs. In September, however, he finally decided to release information on his charitable giving. It showed that between 1996 and 1999, he had earned $145 million, paid $43 million in taxes, and given away $25 million to charities—hardly a damaging allegation. But when reporters found that a considerable portion of these donations had gone to charities in urban northern New Jersey, the bastion of Corzine support, and to black churches and organizations like Operation Rainbow/PUSH, they insisted once again on depicting them as an attempt to "buy" political support. Corzine suffered as a result.

I had always expected Corzine's lead to narrow as the election drew nearer: most races do, and the fact that Corzine was now determined to run as a liberal made the race more challenging than it might have otherwise been. Still, I respected his convictions and felt confident that we would still win—if we went negative. The rest of the campaign team finally agreed, and in late September, we rolled out grainy, black-and-white ads that tied Franks to the most unpopular Republican in America, Newt Gingrich. The effect was immediate: Franks's approval ratings fell sharply.

Two weeks before the general election, Franks finally took to the airwaves. He had $2.5 million in the bank. Now he started to spend it on ads that pointedly attacked Corzine's vast fortune with the tag line, "There are some things you can't buy" and touting his own candidacy with the line "Experience, Money Can't Buy." The national Republican Senatorial Campaign Committee also entered the race, opening their coffers to fund the Franks campaign. The race narrowed further.

The Thursday before the election, I had dinner with Corzine, his wife, and his campaign finance chair at an Italian restaurant in Summit. Over dinner, he asked me where I thought he stood. I told him that I thought he had a narrow but durable three- to four-point lead. "New Jersey is basically a Democratic state," I said. "Vice President Gore has opened a wide lead over Governor Bush, and you'll benefit from Gore's coattails." I also assured him that we were running the best voter identification and get-out-the-vote operation in the state's history. I told him that I was confident he was going to win.

Two days before the election, our polls showed Corzine with a five-point lead—with a three- to four-point margin of error. We were now polling daily and running frequent mall tests to determine how well the messages of the two candidates were playing. That weekend the mall tests picked up something interesting: we found that pro-choice women were concerned about whether Franks, who professed to be pro-choice, was really reliable on the issue. Corzine quickly grasped the importance of the finding and rushed ads onto the air highlighting the issue. We also launched a massive buy that attacked Franks for being weak on Medicare and Social Security.

The day before the election I spoke to Nancy Dunlop, Corzine's personal lawyer who was handling budgeting and payments for the campaign.

"Quinnipiac has just reported that Bob Franks is a point or two ahead," Dunlap said, sharply. The pause that followed made it clear that I was now supposed to justify myself.

"I think you will see Jon win by a few points tomorrow," I told Nancy in an even tone of voice. It was no secret the race had tightened, but with Al Gore holding a sizeable double-digit lead over George W. Bush in New Jersey and our numbers continuing to show Corzine a couple of points ahead of Franks, I felt confident in my prediction.

"We'll see," she said sharply. It was clear that I would be held accountable if I was wrong.

I wasn't. On Election Day, Corzine eked out a 50 percent to 47 percent win, thanks to a huge showing in New Jersey's cities and a high level of support from female voters. It was a narrow win for a Democratic candidate in a state that Al Gore carried by sixteen points. But in the end, Corzine had stayed true to his principles and won. We had fused the candidate's principled liberal stance with an effective message and pitch-perfect tactics and won two tough elections. We had faced an extraordinary media onslaught and come up with a strategy to overcome it. We had done what only a small percentage of self-financed candidates had done; we had won.

Our work for Corzine during his bid for the U.S. Senate and then again in 2005 when he ran for governor was a perfect example of how a resource-rich, well-thought-out campaign could help an candidate without much political experience win office. It was also an example of how principled politics and smart messaging can and should co-exist. During his time in political office, Corzine has become one of our most intriguing political figures. As a member of the U.S. Senate, Corzine was a passionate and effective liberal. As governor of New Jersey since 2005, he's been an even more interesting figure, a chief executive willing to take considerable risks (most notably shutting down the state government) in order to realize important goals like shoring up state finances and reforming the state's broken system of property taxes.

While these positions have sometimes stirred opposition, they've also contributed to the widespread recognition that Corzine is precisely the kind of leader that too many Americans believe no longer exist—a politician of principle.

Our success in New Jersey working for Wall Street CEO-turned-politician soon attracted the attention of another potential new client across the Hudson River—media mogul Mike Bloomberg.

CHAPTER TEN

SEPTEMBER 11, MIKE BLOOMBERG, AND THE RISE OF THE OUTSIDER

In the fall of 1998, Kevin Sheekey, previously one of the late Senator Daniel Patrick Moynihan's closest aides, dropped by our office for a visit. He spoke admiringly of the work we had done for President Clinton and noted playfully that our successes were not at all apparent in the look of our rather spartan offices. He then suggested that we might be able to redecorate if we went to work for his boss, financial media mogul Mike Bloomberg.

I was enthusiastic about the prospect—and not because I wanted to redecorate. Bloomberg was one of the most successful entrepreneurs in New York City, a true media and financial services information visionary. With Bloomberg L.P., he had revolutionized the way Wall Street processed information and then gone on to build one of the world's fastest growing media companies. The prospect of working with a company that valued innovation, insisted on cutting-edge technology, and put the highest premium on high quality information was an exciting one, so when I received a call about meeting with Bloomberg in person, I readily agreed.

Our first encounter occurred over lunch at the Paper Moon restaurant on East Fifty-eighth Street where we were joined by one of Bloomberg's most trusted confidants, Patti Harris. I started in on an overview of services that my firm could provide to Bloomberg's media empire, but Bloomberg didn't seem that interested. After listening to my pitch, he started peppering me with detailed questions about New York politics. I quickly realized that it was not Bloomberg L.P. we were there to talk about: Mike Bloomberg was considering a run for mayor.

As enticing as the prospect was, I wanted Bloomberg to understand the magnitude of what he was considering. Walking back toward Bloomberg's office, I made clear to him that running for office would result in extraordinary media scrutiny; a lot of people would be eager to take him down.

"That's never stopped me before and it's not going to stop me now," he said forcefully. He went on to say that he was concerned about New York City's growing racial and social divide. Listening to Bloomberg talk took me back to my earliest days as a political operative. Back then, the New Deal coalition had been collapsing; the key to success had been recognizing that and dealing with the fallout. Now, as I listened to a potential candidate explaining his vision of a city that worked for everyone and expounding on plans for unifying the city, I couldn't help but see an opportunity to put an updated version of that coalition back together again—that is, if he ran for mayor and if he won. At the time, both of these seemed like a long shot, but then Ed Koch had been a long shot in 1977 too.

In short, I was intrigued but not at all certain what Mike Bloomberg would do. We parted amiably, promising to keep in touch. Little did I know that within a year's time, I would embark on a series of campaigns that would change the fate of New York City and chart a new course for the practice of politics in America. Through the use of "microtargeting" databases and other outreach tools and the novel deployment of volunteers, the 2001 and 2005 Bloomberg campaigns would transform the practice of politics just as surely as the Bloomberg terminal had transformed the trading floors of Wall Street.

In October 2000 during John Corzine's campaign for Senate, I ran into Bloomberg after appearing on the *Charlie Rose* show, which at the time was filmed at the old Bloomberg headquarters on Park Avenue. "How is Corzine doing?" Bloomberg asked. It was a natural question. In some respects, a Bloomberg campaign would present challenges similar to those encountered by Corzine. Both were wealthy, self-made men who were new to electoral politics but who were driven by their desires to serve. In other ways, however, their situations were quite dissimilar. Corzine was a liberal who had defeated a flawed opponent to become the nominee of the Democratic Party. Bloomberg was an independent-minded centrist who had no real chance at winning the hotly contested nomination of the Democratic Party in New York.

"He's going to win," I replied confidently.

Several weeks later, on Election Night, I saw him again at a party. It was a classic New York scene: in one room the celebrities were mingling; in the other the politicos and business types were nervously watching the returns. Early on, the major television networks had called Florida for Vice President Al Gore. If true, it seemed Gore would almost certainly become the forty-third president of the United States. Several hours later, however, the call was reversed, and the state was put back in the undecided column. At that point, I figured it would go for Bush in the end. Bloomberg sidled up to me and asked me who was going to win. I told him, "George Bush will be the next president of the United States."

Later that evening, as the uncertainty deepened, he came back again, "So what do you think?" he asked. "Bush is going to win," I repeated. Even with a lead of only a few hundred votes, I knew it would be hard to dislodge the presumptive winner with the Presidency on the line. Over the course of the next month, Bloomberg would call, half-playfully, to ask if my opinion had changed. It never did. Bloomberg apparently was impressed, if not by my judgment, then at the very least by my steadfastness. Soon thereafter, Patti Harris called to ask me to help find someone to start briefing Bloomberg in more depth on the issues facing the city. By January 2001, it was clear the Michael Bloomberg mayoral campaign was about to begin and that I would be playing a major role in crafting the campaign.

As I was starting to work on a benchmark poll for the Bloomberg campaign in the spring of 2001, my partner Mike Berland approached me with a request: could he add ten questions to the poll about voter attitudes, values, and beliefs? In 1996 for President Clinton's reelection campaign, we had segmented the electorate in terms of their lifestyle choices using information from Claritas, a marketing research firm that divided consumers into fifty segments worldwide. Berland thought we could do something even more sophisticated in New York City by creating our own proprietary model.

It was a simple idea, but one that would ultimately have a profound impact on the Bloomberg campaign—and on the practice of politics. What we wanted to do was to unify voters across the city based on attitudes, beliefs, and values, hence our name for this endeavor, "the unification model." Traditional political segmentation divides voters into groups based on party affiliation, race, or other socioeconomic factors. In contrast, we would be developing a new blueprint for unifying, appealing to, and persuading voters—*individual* voters. By combining information gleaned from our extensive polling data with voting lists and the most advanced marketing data available, we were able to build a database with detailed information on *every* voter in New York City. The result was nothing less than a revolution in political outreach and communications. For more than thirty years, political consultants such as myself had based our careers on the accuracy of a random sample and projecting to large universes of voters, which allowed us to use a small number of people to determine what a large number were thinking. Our work for Bloomberg in 2001 and 2005 set a new standard—instead of relying on a sample, we were using technology to create a virtual census—a database that would allow us to "microtarget" individual voters with custom-tailored messages in a way that had never been done before. We were using technical sophistication to get back to digital, one-on-one retail politics.

In the decade since I had last been involved in a mayor's race, New York City had changed dramatically. While Ed Koch's 1977 campaign for mayor had succeeded through identifying Jewish voters as the swing voters of city politics, by 1999, the number of outer-borough Jewish voters in the city had declined, and Jewish voters were no longer New York's swing vote. The new swing voters were the ones Ed Koch had clashed with twenty years earlier—African Americans and Hispanics. African Americans now made up roughly a quarter of the electorate, and Hispanics made up another 27 percent. Together, the two groups accounted for more than half the voters in New York City, which is not to say that these two groups (if you can describe these diverse constellations of ethnicities as two groups) had the same political instincts. They did not. The failure of the other candidates to recognize this distinction would prove quite important in the race to come.

New York City's diversity was one of the main reasons Michael Bloomberg had gotten into the race. In addition to believing in his ability to serve as the city's chief executive, Bloomberg was also deeply concerned about New York City's divisions. What most appealed to him about the position of mayor was the opportunity he believed it would afford him to unite the city. (In a sign of his commitment, he began taking Spanish lessons so he could better communicate with Hispanic voters—lessons he continues to this day.) While many New Yorkers believed that Mayor Rudolph Giuliani had done a good job in combating crime and getting the city back on its feet, the incumbent mayor was seen as a divisive figure in many minority communities. One of our earliest polls found that people were seeking a "Giuliani without the divisiveness."

The stage could hardly have been better set for Michael Bloomberg, who married centrist ideas and managerial know-how with a passion for unifying the city. First, however, we had to get him onto the ballot. Bloomberg was a registered Democratic. However, he realized early on that his best shot at getting on the ballot and winning involved running as a Republican. In order to maintain the possibility of running for mayor as a Republican in 2001, Bloomberg had decided to reregister as a Republican before the 2000 presidential elections. It was a step he felt comfortable taking, since by nature, Bloomberg was not an ideologue or partisan. Rather, he looked at issues through the prism of problem solving, and as the legendary New York mayor Fiorello LaGuardia once noted, there is no Democrat or Republican way to clean the streets.

Bloomberg saw himself as a leader whose sole interest was getting things done. If running as a Republican was the best way for him to get things done, he had no problem running as a Republican.

Somewhat surprisingly, when I learned of this decision, I felt quite comfortable with it too. I knew that Bloomberg was not a Republican ideologue and that he would have few connections with the national Republican Party. Also, my growing familiarity with my client had bred real admiration. In my heart, I thought he was the best choice for the city and, frankly, his political leanings were probably closer to mine than were most of the other Democrats who were running. The sheer challenge of crafting a winning campaign no doubt appealed as well.

Still, Bloomberg's chief political counselor, Kevin Sheekey, wanted to be certain Bloomberg couldn't run as a Democrat. As a result, he asked me to undertake a benchmark poll to test Bloomberg's electability as either a Democrat or a Republican, and when the results came in, they were clear: There was no way Bloomberg could win the Democratic primary. He simply didn't have the name recognition and base of support among Democrats to beat the four announced candidates for the Democratic nomination. Unlike Jon Corzine's opponent in New Jersey, the terribly flawed Jim Florio, New York City's Democratic contenders were all experienced politicians with healthy ties to local party organizations and primary voters. There was no way Bloomberg was going to outflank them on the Left.

As a Republican, however, he had a chance. He would most likely face only token opposition in the primary, making it likely that if we ran a professional and well-financed campaign, Bloomberg would win a spot on the November ballot. After a private conversation with Republican Governor George Pataki, Bloomberg let it be known that he had made his decision. Now we had to ensure that Bloomberg won the Republican nomination while simultaneously presenting him to New York City's overwhelmingly Democratic electorate in a favorable way.

Despite the media empire that bore his name, few voters knew much about Bloomberg personally. It was imperative that we quickly put Bloomberg in a favorable light. Once people have a positive view of the candidate, they are far more receptive and open to his or her campaign themes. Taking a similar approach to the one we had with Corzine a year earlier, we found that playing up

Bloomberg's background as a self-made man and a lifelong philanthropist resonated with voters. New Yorkers liked the idea of someone who could bring business skills to the halls of government. They also liked the fact that Bloomberg had come from a working-class background. But most of all they wanted someone who would make sure New York stayed the course when it came to crime and economic development. Our mission was to position Bloomberg as the man who would keep New York moving in the right direction.

To do that, we had to take a number of issues off the table. First, was his lack of political experience: more than double the number of the respondents to our poll saw a political background as more important than a successful business background when considering who to support for Mayor. We had to show the connection between Bloomberg's business experience and what he would do for the city. We also needed to inoculate Bloomberg from the broad range of attacks that were sure to come. Republicans are such a minority in New York City that if Bloomberg won that nomination (as we expected he would), he would not enjoy the kind of committed supporters than normally come with the nomination of a major political party. We would have to build that from scratch.

We started with an effort to expand Bloomberg's name recognition. This is one task that can be accomplished with money alone. In June, we mailed glossy brochures titled "The Bloomberg for Mayor Report," which highlighted his independence from the political fray, to approximately a hundred thousand voters, followed quickly by a mailer to Republicans (whose support Bloomberg of course needed to secure the Republican nomination). We also played up his role in running a successful company and creating thousands of new jobs. Our initial polling uncovered that those who received the material immediately developed a more positive view of Bloomberg. It was the beginning of a direct-mailing campaign unprecedented in American politics, one that would ultimately blanket the city with some forty-one *million* pieces of mail.

The next challenge was to preempt criticism that the billionaire was buying the election. Our initial mentor David Garth had joined the campaign, and it was Garth who hit on a brilliant strategy of turning this potential liability into an asset. At Garth's behest, the Bloomberg campaign ran an early commercial that *highlighted* his wealth with a pledge to refuse outside money from either private or public sources. Instead of sounding defensive about his fortune, we were trumpeting the fact that "Mike" (as we now began to call him) would be his own man, without ties or obligations to outside donors. We also ran a series of ads that sought to humanize Bloomberg by playing up his humble roots (he grew up in the blue-collar Boston suburb of Medford, went on to Johns Hopkins and then to Harvard Business School, became a partner at Salomon Brothers, and then left to found Bloomberg L.P.) and his commitment to public service. (A particular favorite was an ad showing Mike painting schools.) These ads sought to portray our candidate as an independent outsider, and they worked spectacularly well. (After the campaign, I ran into Corzine and he asked me why the issue of money hadn't been a problem for Bloomberg. I told him that we used this particular ad to inoculate the candidate.) Despite sniping from the press, Bloomberg's favorability rating continued to rise throughout the summer. It was now time to find his voters. That's where our effort to segment the voters came into play.

In June 2001, Mike Berland and I met with Bloomberg to present our preliminary findings—and our unification strategy—to Bloomberg and his top advisers. I began the presentation.

"New York City has 3.6 million voters, of whom 2.4 million are Democrats," I began. "We know who those people are, and we know how they vote. Unfortunately, the way they vote is overwhelmingly Democratic, so voting history doesn't tell us very much. That's why we've set out to segment the electorate—to place every voter in New York City—into one of six distinct groups—Liberals, Traditional Democrats, Former Archie Bunkers, Traditionalist Republicans, Middle-Middle Democrats, or Successful and Happy Moderates."

"Successful and Happy Moderates," who made up 19 percent of the electorate, were Bloomberg's base. They tended to be white, Jewish voters with very high incomes who lived primarily in Manhattan (and Queens). Like Bloomberg himself, these voters were social liberals and fiscal conservatives. They shared his goal of continuing Mayor Giuliani's reforms in a more inclusive fashion; they liked his support for culture and the arts; and they respected him for his success.

"Your swing voters," I continued, "are the Middle-Middle Democrats. These voters—low- to middle-income Democrats, mainly

Jews and Catholics, with mid- to low-levels of education, who live primarily in the Bronx and Brooklyn—will determine whether you win or lose the election. They're moderate politically, appreciate some of Mayor Giuliani's accomplishments, but don't like him personally.

"These voters don't know who you are yet, but when we told them your story, they loved it," I told Bloomberg.

In fact, Middle-Middle Democrats were the group who responded best to Bloomberg's story and message: before listening to Bloomberg's most effective message, only 11 percent of these voters had said they would consider voting for Mike as mayor. After hearing his message, however, that number rose to a heartening 46 percent. The single largest voting bloc in New York, making up about 24 percent of voters, this was very simply a group we had to win.

One of our first surprises was that non-Puerto Rican Hispanics made up a sizable portion of this group. In fact, our research showed that these voters had attitudes that were very similar to long-established white ethnic groups like Italian and Irish New Yorkers. They thought New York was the greatest city in the world and were passionate about living here; they often owned their homes and relied on the public schools. These voters wanted a mayor who would protect and enhance these vital public services. Much of our subsequent advertising would target members of this Middle-Middle Democrat group.

Other segments would be easier to reach. The Former Archie Bunkers and Traditional Republicans (who made up 11 percent and 13 percent of the electorate, respectively) were decidedly on the right. Former Archie Bunkers were hard-core conservatives—older, with slightly higher incomes, and better educated than the Traditional Republicans. Both had hostile views of politicians and were opposed to the taxpayer funding of campaigns. They were particularly concerned that, if elected, the presumptive Democratic nominee, Public Advocate Mark Green, would raise taxes. They also wanted a mayor who would work to get New York's fair share from the federal government. Traditionalist Republicans were attracted by a message of education reform that held teachers and schools accountable. In addition, they wanted a mayor who would be fiscally responsible and proactive in dealing with city problems. These voters were Bloomberg's strongest potential supporters. The Former Archie

Bunkers responded less well to Bloomberg's messages (presumably because Bloomberg himself wasn't particularly conservative); however, their support would undoubtedly go to the Republican nominee in the actual election. Together, these two groups made up roughly a quarter of the New York electorate.

On the Left were the Liberals and Traditional Democrats, representing approximately one-third of the electorate. While members of these groups voted in a nearly identical fashion, they had little in common personally. Liberals were generally white, secular, middle income but highly educated, and lived mainly in Manhattan. Traditional Democrats were younger, predominately black, more female, more Baptist, lower income, and much more likely to live in the Bronx, Brooklyn, or Queens. These were the segments of the electorate least likely to cast a ballot for Bloomberg.

In order to better understand where these voters were located, we developed a segmentation algorithm that allowed us to classify every one of New York City's five thousand election districts. The resulting map of the city looked like this:

All Areas
- ■ FAB
- ■ Traditional Republicans
- ■ Middle-Middle Democrats
- ☐ Successful Happy Democrats
- ■ Liberals
- ■ Traditional Democrats

Our original goal in analyzing the electorate was to better understand voters' attitudes so that we could develop effective messages for our candidate. However, as soon as we geo-coded election districts, we noticed a startling fact: parts of the city that were very different demographically actually had the same attitudes and political orientation. We found that many upper-middle-class Jewish voters in Manhattan and Queens thought about the challenges facing the city in ways strikingly similar to, say, Italian Catholic homeowners in Brooklyn or middle-class African American homeowners in Queens. This claim raised some eyebrows with the Bloomberg campaign. However, I was convinced that we were right and argued strongly for targeting voters with such attitudes with similar messages. The data were clear: core beliefs were more important than the old demographic demarcations. For a candidate committed to unifying the city, it was a welcome message.

Up until this point, our work had been highly sophisticated but not unprecedented. However, our next step proved even more radical. Past targeting campaigns had been crude: campaigns had bought voter lists, which contained names and addresses and then made informed guesses about voters' genders, ethnic backgrounds (based on names), and their ages (based on when they had first registered to vote). But the effort that was taking shape under Mike Berland and our associate, Bradley Honan, was something new. We were now proposing to move beyond sampling and extrapolating toward something new—a census of the city, a database with information on every voter, not a sample.

"This election is going to involve a little over a million voters," I said. "We don't have to sample, we can do it all." By purchasing consumer information and using phone banks, we could build a profile of *every* swing voter that combined demographic, voting history, and consumer data. We could tag every voter as a member of one of the groups and develop a dialog with them that emphasized the issues they cared about most. Every piece of mail, every phone call, would be targeted precisely on them.

Bloomberg's reaction was simple and direct. "Let's do it," he said.

So with Bloomberg's blessing, we set out to build something that had never been constructed before—a detailed database of every voter in New York City. The theory was simple and elegant but assembling the actual database was an enormously complex endeavor: after merging voting, demographic, and attitudinal information, it was not uncommon to have as many as two hundred and fifty variables per voter. Multiply that by four million adult voters, and you start to get a very large database indeed. Nonetheless, by midsummer our database and targeting effort, spearheaded by Bradley Honan, was well under way. Now it was time to use that database to start talking with these voters. For help with that, I turned to Duane Baughman, the San Francisco-based direct-mail wizard I had brought into the campaign.

The best way to begin to reach these various segments was through direct mail. Our first mailing—some 396,000 letters—went out to registered Republicans. After all, Bloomberg first had to win the Republican mayoral nomination. Several weeks later, we send out 1.2 million pieces of mail to the Bloomberg base—Successful and Happy Moderates, Traditionalist Republicans, and Former Archie Bunkers. Thus, the Former Archie Bunkers and Traditionalist Republicans received a classic Republican message emphasizing fiscal responsibility and Bloomberg's commitment to fighting crime. The Middle-Middle Democrats and the Successful and Happy Moderates heard about Bloomberg's economic development plans. Traditional Democrats learned about Bloomberg's commitment to inclusiveness. Our goal here was not so much to win them over—it was clear that Liberals and Traditional Democrats were not going to go for Bloomberg in large numbers (at least not in this campaign)—so much as it was to defang them. In other words, we sought to convince them that Bloomberg was a decent candidate—not someone they should mobilize against.

Everything worked together. Bill Knapp's television advertisements laid out the primary themes of the campaign while a highly targeted direct-mail effort, directed in part by Emma Bloomberg and supplemented by prerecorded phone calls, reached the most important groups with the most effective message possible. Each aspect of the outreach campaign reinforced the effectiveness of the other. Our direct mail made the air campaign—Bloomberg's television and radio ads—more believable. People listened more closely to

advertisements because of the mail they received. We soon supplemented these efforts with something that at the time was quite
new—prerecorded phone calls. Once again, our database guided
our efforts. Traditionalist Republicans heard from George W. Bush.
(That was a new experience for me.) Former Mayor Ed Koch called
voters in four different segments, with different messages for each.
Black Entertainment Television cofounder Robert Johnson and Nor-
man Seabrook, head of the powerful city corrections officers union
and a major figure in the black and Hispanic community, called minority voters in the segments whose lifestyle characteristics suggested they might be open to a Bloomberg appeal. Former New
York Governor Hugh Carey called outer-borough voters who were
among the Middle-Middle Democrats and Successful and Happy
Moderates. It was a classic "under the radar" operation. Later, after
the campaign, the Democratic nominee, Mark Green, would rue-
fully congratulate me on what we'd done; at the time, however, our
efforts went virtually undetected by the city's media.

By late August, we had moved Bloomberg to the position we had
hoped to have him in. His positives were high, and we had largely min-
imized his negatives. His favorable ratings were at 53 percent, and his
unfavorable rating was just 22 percent. With only weeks before the
September 11 Republican primary, Bloomberg was on track to crush
his token opponent and secure the nomination of the Republican Party.

However, the general election was another matter entirely. Despite having spent $20 million, polls showed that Bloomberg still
trailed the Democrat's likely pick, Mark Green, by a double-digit
margin. Mike was understandably concerned about his position vis-
à-vis Green. Toward the end of August, during one of our regular
meetings, he pulled me aside and plaintively said, "I'm already in
this race for $20 million; when do I start to move?"

It was an eminently fair question, one that I did my best to answer. I assured him that we were headed in the right direction. The
race was far closer among those who had seen our advertisements, I
told him, which suggested that our strategy was working. The polls
were moving; our strategy was working.

Yet I had to concede that not all the signs were positive. Our
polling showed that most New Yorkers saw Bloomberg as a better

chief executive than Green. However, voters didn't see a pressing need for those skills. However, we had not yet convinced voters that they needed someone with Bloomberg's business skills in office. Times were almost too good for our candidate. Convincing voters that New York City needed someone with Michael Bloomberg's skills would be one of the major challenges of the general campaign. What I did not know was that the world was about to change.

The day of the Democratic and Republican primaries began uneventfully. Polls showed Bloomberg with a sizable lead over his sole GOP opponent, former Congressman Herman Badillo. All the intrigue was on the Democrat side, where an increasingly heated campaign pitted several long-time Democratic office holders. I awoke early and headed down to the polling station on Eighty-Eighth Street and Park Avenue where I ran into Bloomberg's campaign manager, Patti Harris who was handing out campaign literature with big smiles. She asked for reassurance that Bloomberg would prevail. I said I was confident, but that she shouldn't abandon her post.

Later that morning, I met my friend, Mike Kramer, who at the time was a *Daily News* reporter, for breakfast. As we were walking on the sidewalk afterward, we noticed that neither one of our cell phones seemed to be working. We wandered into a phone store on Seventy-third and Third Avenue to fix the problem. Suddenly, a woman came in screaming, "They've hit the Trade Center."

The day was September 11 and the Republican mayoral nomination was about to get a lot less important.

At first, the woman's outburst didn't fully register with either of us. We hurried over to Kramer's apartment to figure out what was happening. Nothing prepared me for the shock of turning on the television that morning and seeing the twin towers of the World Trade Center in flames. Acrid, black smoke was billowing out of two jagged gashes in the side of the buildings. The horror was just beginning. A few moments later, the South Tower collapsed in a ghastly cloud of smoke and debris. Kramer and I watched in stunned disbelief as the North Tower fell to the ground soon thereafter. As a lifelong New Yorker, it was one of the most painful moments of my life. Though I had spent most of my career immersed in the world of politics, on September 11 politics didn't enter my thoughts. Like many New Yorkers, the next few days were a blur for me. The city was bombarded with the images of lost souls and the omnipresent odor of the smoldering rubble of Ground Zero.

For Bloomberg, it was an especially traumatic event. He lost several employees and friends in the attack. He refused to even talk about politics or have us do any polling until the primary was rescheduled. He felt very strongly that it would be inappropriate to begin campaigning again, and it was a view we all shared. In the days after 9/11, we were all numb. No one really had the heart to hit the campaign trail. For more than two weeks, we were at a standstill as we waited to see what would happen.

Eventually, both party primaries were rescheduled for September 25. As expected, Bloomberg won easily. However, no Democrat emerged from the crowded field with more than 50 percent of the vote. Public Advocate Mark Green and Bronx borough president Fernando Ferrer would face each other in a runoff two weeks later.

For all the obvious uncertainty, one thing quickly became clear: it was now a very different race. The issues on the minds of the electorate were radically changed. Rebuilding devastated Lower Manhattan and spurring a shattered economy were now the two most important issues. Both questions played to Bloomberg's experience as a business leader, which we had highlighted in the weeks and months leading up to 9/11. We had even tested ads before 9/11 depicting a looming financial crisis in the city so we knew that voters would be receptive to our candidate.

One other equally important dynamic was emerging—the embrace of New York Mayor Rudolph Giuliani as one of the country's great heroes. Rudy's behavior on 9/11 and in the days afterward had made the man who before 9/11 had been seen as a divisive figure into a national hero. There was even talk of postponing the election or extending Giuliani's term into the spring—a plan Bloomberg had virtually no choice but to support. Voters were now looking for a candidate who would support the policies of the Giuliani administration. They wanted an economic recovery plan, and they were worried about the city's fiscal condition in the wake of the terrorist attack and the subsequent slow-down on Wall Street. There was a widespread sense that an economic downturn was coming. In short, a horrible tragedy had made Bloomberg into a truly compelling candidate—if we could effectively position Bloomberg as the best candidate for rebuilding and reviving the city.

Bloomberg embraced this brief enthusiastically. Throughout the course of October, he focused much of his energy on enhancing an ambitious five-borough economic plan that would remind New Yorkers of his vast experience in creating jobs and spurring economic growth as a businessman. As I watched him toil away on the plan with real anguish on his face, it was clear that this was not simply a political effort on his part. He was determined to come up with a plan that would work. Eventually, we would mail a copy of the plan (tailored to each distinct subgroup of voters) to virtually every New York City voter who we thought we had a chance to win. It was clear to me that after the traumatic events of 9/11 and the political uncertainty around the campaign, Bloomberg was determined to do what he believed to be the right thing.

While Bloomberg concentrated on devising a plan to shore up New York, the battle for the Democratic mayoral nomination was turning into a bloodbath. The relationship between the two top Democratic vote getters, Mark Green and Fernando Ferrer, already badly frayed, had become downright homicidal. Ferrer ran an especially divisive campaign, constantly evoking "the other New York"—a euphemism for the black and Hispanic voters who felt left out of the city's revival. Green in turn attacked Ferrer as being unprepared to run the city and guilty of underestimating the magnitude of the crisis following 9/11. It was a fair criticism: even in the wake of tragedy, Ferrer did not alter his "other New York" campaign message. Things got even uglier when shadowy groups sought to tie Ferrer to one of his most prominent supporters, the Reverend Al Sharpton. There were serious questions raised about the Green campaign's complicity in these attacks—questions that remain to this day. Even after Green narrowly won the runoff, Ferrer initially refused to endorse him, only relenting under massive pressure from fellow Democratic lawmakers. Hispanic voters were particularly annoyed at Green. That anger would become a recurring liability for Green and a boon to us.

We capitalized on the situation by running a devastating ad that featured the comments of his fellow Democrats saying Green had divided the party and the city. We also played on Green's reputation for arrogance by running an ad showing Green telling an interviewer he could have done at least as good a job as Giuliani after 9/11, if not better. The confluence of internal Democratic fighting and our attack ads only enhanced the perception of Green as that of a politician who would do or say anything to get elected.

Making matters worse for Green was the shortened election schedule. Green had already spent millions to win the Democratic primary. Now he was facing an opponent who was spending a million dollars a week and who had escaped from the Republican primary with strong support in his own party and a relatively positive image among most New Yorkers. Not surprisingly, Green quickly went on the offensive, attacking Bloomberg's substantial resources.

"Message beats money," Green chanted at his victory speech, making clear that he would attack Bloomberg for "trying to buy the election." In response, we immediately labeled Green a "professional politician" who had issued "more than twelve thousand press releases" and yet had little to show for his efforts. Moreover, we tried to play up the public's embrace of Bloomberg as an independent and outsider, who would get things done at City Hall. With only a few weeks until Election Day, the contours of the race were set.

Nonetheless, we weren't exactly brimming with confidence. Green continued to hold a double-digit lead, even after the hard-fought runoff race. New York was still a five-to-one Democratic city, and we were running against a seasoned—and increasingly confident—candidate. Indeed, Green was so confident that he was attempting to duck debates with Bloomberg. Then, just when we most needed to focus, the *New York Times* delivered a shot to the solar plexus of our campaign. An October 25 article by Adam Nagourney claimed that Bloomberg's campaign was falling apart. I had spoken with Nagourney, who had tried hard to get me to confirm that the campaign was tanking. Of course, it wasn't. I pointed out that recent polls showed that Bloomberg was quickly narrowing the gap. Nagourney dealt with this objection by simply ignoring it. Nagourny's piece was potentially devastating, not because it was right but because the article itself threatened to sow dissension in the campaign.

With ten days to go, we were still eight points behind. However, the numbers were actually moving in our direction and, with one in five voters still undecided, we were much more hopeful than Nagourney or many other veteran observers of New York City politics, most of whom expected us to lose. Thanks to our analysis of the electorate, our perspective was rather different. We understood that the issue for the all-important Middle-Middle Democrats was that they wanted to live in New York more than any other place. That in turn meant that in a post-9/11 world, they wanted a leader who could manage the task of rebuilding and strengthening New York City. They also knew that these were extraordinary times that called for a new type of leader. We played to that sentiment, running ads of real New Yorkers talking about how we needed a businessman in times like these, not a politician. However, Mark Green seemed not to appreciate this sentiment. Our polling showed that Mark Green's assaults on Ferrer merely reinforced Middle-Middle Democrats' image of Green as an old-school politician and made them more open to supporting Bloomberg.

Luckily for us the fallout from the *New York Times* article did not last long. Just two days after that article appeared, Mayor Rudy Giuliani agreed to endorse Bloomberg. At a time when our polls showed overwhelming support for Mayor Giuliani and a desire among many New Yorkers to continue his policies, the Giuliani endorsement was a huge boost to the campaign. Suddenly Mike Bloomberg, like him or not, was a credible mayor to everyone in New York.

Much of the credit for this endorsement belongs to David Garth. Giuliani and Bloomberg had not been particularly close. No one on the Bloomberg campaign had known when—or even if—a Giuliani endorsement was forthcoming. By early October, there was even a sense among some Bloomberg advisers that an endorsement, should it come, would be too late to do much good. Garth, however, continued to pursue Giuliani's blessing, and eventually after weeks, if not months, of concerted effort, he arranged for the two men to meet for a private dinner at the Palm West restaurant in midtown

Manhattan. The conversation went well, and Giuliani ended the meal by agreeing to endorse Bloomberg.

The next challenge was to get an ad on the air, fast. City law forbade us from filming in the mayor's office so the campaign rented a suit at the Waldorf Astoria Hotel and re-created the Mayor's office in intricate detail. The ad was made—as official-looking an endorsement as the best minds in the advertising business could devise—and then hustled onto the air. The Bloomberg campaign made certain the endorsement was everywhere—seen on television screens, trumpeted in direct mail, and relayed via recorded phone calls to every household in New York City that was politically in play. Almost immediately, Bloomberg's poll numbers started to move. The timing of Giuliani's endorsement proved to be just about perfect. Mayor Giuliani's embrace of Bloomberg dominated the news, and Green's lead continued to fall. By October 31, Bloomberg was within four points of Green. From our data, it was clear that the Ferrer/Green primary battle was finally having some impact on the Democratic nominee. Bloomberg was gaining traction in the key Middle-Middle Democrat and Hispanic groups. In addition, independents were moving in our direction: Bloomberg now enjoyed a nearly two-to-one advantage over Green. Outer-borough Jews and Catholics were also moving away from Green. One week out, the race was tied, and it appeared like all of our work was beginning to pay off.

On the Sunday before the election, the *New York Daily News* endorsed Bloomberg, which gave a huge boost to the campaign. After enduring their taunts and barbs for months in the lead-up to the Republican primary, Bloomberg now had the support of the city's largest tabloid. While many outside of New York think of the *New York Times* as the major city paper, it is in fact the *Daily News* (and the *New York Post*) that are most widely read by New Yorkers outside of Manhattan, particularly among the voter segments we were targeting. On November 4, our polls showed Bloomberg in the lead for the first time, and with these results I started to feel confident.

In a race this tight, every vote mattered and so the campaign intensified what was already the most sophisticated messaging and get-out-the-vote operation in the history of urban politics. A week before voters went to the polls, we began to inundate them with a series of automated calls from Bloomberg supporters. Once again, our database ensured that voters heard from a figure they admired who spoke from a script that addressed their concerns. Thus, African American voters picked up the phone to find a message from former Manhattan Borough President Percy Sutton; lifelong Republicans found a message from President Bush on their answering machines; outer-borough white voters heard from Mayor Giuliani and former Governor Hugh Carey; Jewish voters got an earful from Ed Koch.

We also mounted one of the most sophisticated direct-mail campaigns in American political history. Bloomberg's daughter Emma played a key role in this effort, editing every piece of mail and ensuring that everything was rigorously fact-checked. Kevin Sheekey, Mike Berland, and Duane Baughman also stepped up the intensity and precision of the effort. In the final week of the campaign, each voter segment got five to six targeted mailings, each of which spoke to the subject that concerned them most. All of these operations took place under the media radar screen. Neither the Green campaign nor the media had any idea just how sophisticated our outreach efforts had become.

The effort continued on Election Day. In order to monitor voter turnout, we deployed thousands of poll watchers to make sure that the voting went smoothly. We knew that we had enough supporters in the database to win; the challenge now was to actually get them to go to the polls and vote. When media exit polls showed Bloomberg trailing by two points at 2:00 PM, we realized that not enough voters had come out and that they would need an enhanced turnout effort. I immediately got on a conference call with Bloomberg, Kevin Sheekey, Patti Harris, and Mike Berland. The future Mayor got right to the point.

"I don't need detailed explanations of what is happening, I want to know what we do to change things," he said. Mike Berland and I both recommended that we send recorded calls to our supporters and to undecided voters in our strongest areas, one from Bloomberg, the other from Giuliani, candidly informing voters how close the contest was and asking them to go vote and make a difference. Sheekey likewise redoubled the efforts of the field organization he painstakingly had developed, pouring thousands of volunteers into the streets in the neighborhoods we believed were Bloomberg strongholds and could provide the margin of victory.

It was an amazing effort. In the final seven hours of the campaign, almost 1 million recorded phone calls got sent; tens of thousands of doors got knocked on; and thousands of leaflets were handed out. As Sheekey said later, the effort was virtually a microcosm of the campaign itself, and it may well have meant the difference between victory and defeat.

Election Night was a somewhat surreal experience. There I was at the Hilton Hotel in Times Square watching the election returns with Governor George Pataki and Mayor Giuliani, Republicans both, as they chuckled with delight over what seemed like an impending Bloomberg win. These were men that I had little in common with politically, and yet we were all on the same team. As delighted as I was by our probably victory, it was an uncomfortable feeling for a lifelong Democrat.

Bloomberg himself seemed a bit stunned. Only days earlier, the reality had hit him that he might actually win. Since our very first meeting at the Paper Moon, Bloomberg had held firm to his desire to bring the city together. His message and reason for running never wavered. He had campaigned from the heart, driven by his commitment to New York, and now it appeared that he would have a chance to make good on his hopes for the city.

In the end, Bloomberg won the mayor's race by three points. We had done precisely what we had set out to do. We had identified the Middle-Middle Democrats as the swing vote, targeted them relentlessly, and won their support. Three subgroups had been particularly important—Staten Islanders, Hispanics (particularly non-Puerto Rican Hispanics), and older women who we thought would respond to Bloomberg. We likewise won a considerable portion of those voters. Most remarkable was Bloomberg's performance among Hispanic voters, where he had essentially split the vote with Green, a 21 percent improvement over Giuliani's performance eight years earlier. We won Staten Island, a must-win borough for any Republican candidate, by almost sixty thousand votes. In many ways, we had done on a citywide scale precisely what Dick Morris, Jerry Nadler, and I had done thirty years ago on the Upper West Side for the Dick Gottfried campaign: we had canvassed the city (or rather, purchased and assembled information from marketers who had canvassed the city) and crafted messages that would resonate with individual voters. It was an amazing victory. Looking back at the poll numbers, we had been behind for virtually every day of the campaign. Not until the final forty-eight hours of the campaign did Bloomberg take the lead.

A stonishing as Bloomberg's victory was, the way in which he was elected also presented the mayor with some difficult challenges when it came to governing. Mayor Bloomberg would not be indebted to any special interest groups, but he would have no committed supporters either. In the wake of 9/11, the city's economy had stalled; revenues were plummeting. Smoke from Ground Zero was still blowing toward City Hall, but rather than working with the Mayor, the city's powerful Democratic establishment was eager to see him fail. Challenges like overseeing the reconstruction of the site, implementing the five-borough economic plan, and reforming the city's schools would be, to a considerable extent, his alone.

Faced with a $7 billion budget deficit, Bloomberg made the politically difficult but necessary decision to slash programs and raise property taxes. His outer-borough supporters were outraged. Polling immediately after the tax hike showed that the mayor's approval dropped precipitously from 57 percent in March 2002, to 41 percent in the Quinnipiac poll in November 2002. When the mayor pressed ahead with an antismoking initiative that required all bars and restaurants to ban the use of all tobacco products, the press went haywire, arguing that this somehow proved that Bloomberg was nothing more than an "out of touch billionaire" who wanted to deny ordinary New Yorkers the pleasures of tobacco—never mind the thousands of lives prolonged and ultimately saved by the mayor's action.

By 2004, another issue had the potential to derail Bloomberg's reelection plan. The five-borough economic plan included one element that threatened to overwhelm the entire initiative in the public's mind: the construction of a new football stadium for the New York Jets on the West Side of Manhattan. Previously, the Giuliani administration had proposed building a new stadium for the Yankees on the West Side. Bloomberg's plan linked constructing the new stadium for the Jets with expanding the Javits Convention Center and using both as a means of trying to win the 2012 Olympics. Many New Yorkers saw this as nothing more than a taxpayer-subsidized initiative for another billionaire, Jets owner Woody Johnson, rather than as a component of a broader economic development strategy. Cablevision, which owns Madison Square Garden, feared that the new West Side stadium would provide unneeded competition and launched a multimillion-dollar television advertising campaign against the plan—and against Mayor Bloomberg—which further hurt his standing with the public.

Topping off the mayor's list of troubles was the fact that President George W. Bush and the Republicans were coming to town: the GOP was having its nominating convention at Madison Square Garden. A less timely reminder that Bloomberg was a Republican would have been hard to devise. His approval rating had dropped to a mere 28 percent, the lowest a mayor had ever tested in a *New York Times* poll, and he trailed his likely Democratic opponent, Fernando Ferrer, 52 percent to 38 percent. In short, as the 2005 Mayoral elections approached, Mike Bloomberg appeared to be in deep trouble. Most observers of New York City politics had already written him off as a one-term mayor. The mayor himself was surprisingly unconcerned by these numbers. He was focused on doing the right thing and seemed confident that the public would eventually come to appreciate what he was doing. However, everyone on Team Bloomberg was understandably concerned. September 11 had given rise to a unique atmosphere that had favored our candidate. However, New York City politics was rapidly reverting to normal. Reelecting Bloomberg would be a major challenge.

The turnaround began with Kevin Sheekey. Sheekey had cut his political teeth as a top aide to former New York Senator Daniel Patrick Moynihan. As Bloomberg's top lobbyist in Washington,

D.C., he had long been the mayor's right-hand man. He never flaunted his power; he never had to. His status as Bloomberg's political alter ego was never in doubt. During the first Bloomberg campaign, he had demonstrated a striking and unusual set of political skills—friendly, low-key, even deferential at times, he had done a masterful job of orchestrating the clashing egos involved in any large-scale political campaigns. He also had a knack for getting to the heart of complex political questions with a few well-chosen questions. Sheekey had one other arresting attribute: no idea was too big. In November 2004, as he prepared to take control of the Bloomberg reelection campaign, he came up with a very big one indeed.

Sheekey had managed the Republican convention for Bloomberg and seen the Republicans' political organization firsthand. He had been particularly struck by their plans for an ambitious turn-out-the-vote operation for the upcoming 2004 presidential race, one that relied on a database strikingly similar to the one we had developed for Bloomberg's 2001 race. At a meeting that fall in my office with Patti Harris, Sheekey proposed using our database to mount the same kind of operation that Republicans had run in 2004.

"Why don't we build the same kind of turnout operation the GOP is building?" Sheekey suggested. He went on to elaborate on his vision: fifty thousand volunteers, chosen and guided by our database, who could fan out across New York and urge voters to reelect Mike as part of a multifaceted approach that would utilize phone banks, direct mail, and the Internet.

It was a brilliant idea, one that built on the work we had done in 2001.

"I think it's a great way to give people a stake in Mike," I told Kevin. "If he's just a rich guy writing checks to a television station, there's no personal connection." Finding people in the outer boroughs—places like Astoria or Bay Ridge—who liked Mike and who would take that message to friends and family would give our candidate a tremendous advantage. We quickly agreed to pursue Kevin Sheekey's idea. In 2001, the challenge had been unifying the city and identifying supporters for Bloomberg. Now we were going to try to mobilize those supporters as Bloomberg volunteers.

Given the expansive atmosphere of the meeting, I decided to suggest some other ideas as well. As long as we were thinking outside the box, why not begin a massive telephone campaign? As in 2001, instead of simply sampling the electorate, we could interact with them, perfecting our database and recruiting volunteers. It would also allow us to learn what virtually every voter in the city thought about the mayor's job performance and what their primary concerns were. We could then target every voter with communications aimed at their concerns. I recommended that we start with independents, a key swing group that we had to win overwhelmingly if we were to win reelection. Sheekey and Harris both agreed. Very quietly toward the end of November, we began a systematic effort to reach independent voters by calling every voter in New York outside of Manhattan. At Sheekey's behest, we also set up an interactive website to supplement this effort. We even purchased lists of email addresses in order to allow us to communicate with voters through their inboxes. Unbelievably, this was done under the radar screen, attracting virtually no press attention.

The results of this effort were striking. Roughly 3 percent to 5 percent of the people we reached agreed to serve as Bloomberg volunteers, while over half of the people we called said they could ultimately be persuaded to support the Mayor. These findings played an enormous role in boosting our confidence. Fifty thousand volunteers now seemed like a reasonable goal, and we could all see a way clear to persuading at least a narrow majority of New Yorkers to support the Mayor's reelection. The campaign would ultimately spend almost $9 million on these field operations. It was an operation so massive and sophisticated that the media never did get its mind around what Sheekey had accomplished—a further testament to his ability to operate quietly.

The power of our approach became evident at our first big event, a campaign rally at B.B. King's nightclub in early February 2005. Drawn by our phone and Web outreach, over twenty-five hundred people showed up at the event. That spring, the Bloomberg campaign opened eight offices across the city and began to build the volunteer apparatus that Kevin Sheekey had so boldly imagined. By late March, the number of volunteers who had signed up reached twenty-five thousand. It was now clear that we would have the ground troops we needed to run a very effective campaign in what we expected would be a very close race.

It was also clear that Kevin Sheekey, Patrick Brennan, and the other members of the Bloomberg campaign were building an on-the-ground outreach operation of unprecedented proportions. In the first five months of the campaign, volunteers identified by our database-driven phone banks, along with union workers and paid canvassers, would knock on seven hundred thousand doors and speak directly with three hundred fifty thousand voters. All willing voters spoken to by campaign workers received a phone call two days later asking them if they in turn wanted to volunteer. Then, six days after that, they got a letter from Mike that spoke to the particular issue they had indicated was most important to them at the door. When our research suggested that seniors responded well to Mike's message, the campaign also organized free senior breakfasts, where seniors watched a video about Mike and heard in person from a local leader who supported Mike (the campaign also carefully recruited twenty-five hundred local "validators" whose endorsements were then used in areas where they carried weight), from past mayors like Ed Koch, or from family members. In the course of the campaign, we would invite seven hundred fifty thousand seniors to such breakfasts; some seventy-five thousand seniors accepted our offer.

This organizing that Kevin Sheekey was undertaking went light-years beyond what the GOP had done in 2004, and he was only getting started. Over the course of the summer, Sheekey doubled the number of campaign offices, hosted another one hundred fifty senior breakfasts, and directed canvassers to another three hundred thousand residences, bringing the total number of voters contacted to just under a million. The campaign also blanketed the city with Bloomberg signs and literature. As someone who got his start in politics plastering subway cars with campaign literature, I particularly appreciated the campaign's approach to the subway system. Twice a week canvassers targeted select locations in every borough for canvassing. The pinnacle of this effort would come in mid-October when the campaign recruited enough volunteers to cover every subway stop in New York City. It largely maintained this blanket coverage of the subway system during the final two weeks of the campaign.

I had felt confident from the first that Sheekey's nontraditional approach was going to work. Sheekey, however, wanted to leave nothing to chance. In order to get an outside perspective on how well we were doing, he set up a meeting with Alex Gage, founder of the marketing firm TargetPoint, the company which had helped the GOP develop its vaunted "Voter Vault" database on the American electorate. Later that spring, Berland, Sheekey, and I gathered in our offices to present our strategy to Gage. Needless to say, I was curious about what Gage would say about our operations.

At Sheekey's behest, Berland began his presentation. He explained how in 2001 we had built a unifying database with information on every voter in New York City, using polling, phone banks, and consumer marketing information. He then went on to discuss how in 2005, we had raised our game, using our database and various outreach strategies to create an immense volunteer field operation that supplemented our high-tech campaign with the oldest and most effective outreach strategy in politics—neighbors calling neighbors.

Gage was clearly impressed.

"That's more sophisticated than anything we're doing," he said. Sheekey looked pleased. It was gratifying to know that we had developed a system for identifying and mobilizing voters that was more sophisticated even than the unprecedented Republican turnout operation that snatched victory away from John Kerry.

As the Bloomberg campaign was gearing up, our early polling showed Bloomberg continuing to trail Ferrer by about five points; it also showed a slight drop in his approval rating, no doubt due to the amount of public attention his advocacy of the West Side stadium was receiving. Both Bill Knapp and Patti Harris, who functioned both as the campaign overseer and a sophisticated and nuanced strategist in her own right, felt strongly that we needed to begin our television campaign as soon as possible so that voters would get information from some place other than the free media, which was obsessively focused on the mayor's stadium woes.

We needed a new narrative, one that told the story our way. Ultimately, we decided to go back to 9/11—to emphasize the accomplishments the city and its mayor had achieved since then. The story went something like this: Mike Bloomberg took over a city still reeling from 9/11, facing an economic downturn, ethnic and racial division, as well as on-going problems with the schools. Faced with these challenges, he developed and implemented a five-borough economic plan, took over the schools and began to improve test scores, and worked tirelessly to bring people together. This message happened to have the very great merit of being completely true. The commercials that delivered this message, which began with a shot of the mayor being sworn in with smoke from the World Trade Center evident overhead, sought to visually demonstrate the progress the city and the mayor had achieved. Almost immediately the mayor began to pick up support in the polls. By mid-June, he had moved to almost a double-digit lead over Ferrer.

We also told the story of Mike Bloomberg personally. Refuting the argument that he was nothing more than an out-of-touch billionaire, we showed him with New Yorkers of all races and ethnic backgrounds and had him talk passionately about his commitment to the city. Later in the summer, we laid out his future plans so as to demonstrate that he had a real agenda for the future that could give people confidence that their quality of life would be arguably even better in a second term.

By the end of the summer, the polling we saw internally showed the mayor with a fifteen-point lead. Public polls showed him doing even better, sometimes registering a lead as high as 30 percent. I warned the candidate and his team that this was an illusion. In a city where Republicans made up less than 20 percent of voters and where George W. Bush was extremely unpopular, the Democratic challenger would rally.

Fortunately for us, the Democrats ended up with a primary process that turned out to be almost as divisive as the one they had faced four years earlier. On primary night in September, it appeared that Ferrer had been drawn into a runoff with Congressman Anthony Weiner, who represents parts of Brooklyn and Queens. However, shortly before a recount was going to determine whether Ferrer had been able to avoid the runoff contest, Weiner withdrew from the race. While the Democrats had avoided a runoff, the fact remained that almost 60 percent of Democratic voters had voted for someone other than the ultimate nominee Freddy Ferrer.

The media, seeing disarray and disorganization in the Ferrer campaign, effectively pronounced the race all but decided. We weren't so sure. In order to better understand the mind of the electorate, we decided to revisit the question of how New York City's electorate should be understood. In addition to the six distinct groups, we had devised in 2001, we now added two new segments, Fearful and Anxious New Yorkers and Cultural Liberals. The so-called Fearful and Anxious New Yorkers—we called them FANS—were downscale city residents sensitive to the potential threat of terrorism and crime. This was a mixed group of lower-middle-class whites, Hispanics, and blacks, who felt vulnerable to terrorists and criminals. A traditional, demographic analysis would have pegged most of these voters as Ferrer supporters. However, our unification model showed that their attitudes made them open to Bloomberg's message. Through both direct mail and phone calls, we emphasized to this group Mayor Bloomberg's record in reducing crime and the comprehensive antiterrorism program the police department had developed under his guidance.

The second new group, Cultural Liberals, was similar to the Traditional Liberals we had identified four years ago—with an important difference: they were primarily *cultural* liberals. Many were affluent whites who had moved back to the city to be close to the cultural opportunities New York offered. While Bloomberg's stance in support of a West Side Stadium had hurt him with this group, we knew that most other aspects of his record played well with these groups. Once we moved past the stadium issue, these voters responded particularly positively to Mayor Bloomberg's support for diversity, the arts, and social programs. Indeed, we found that by talking about poverty and social justice, we were able to win significant support among a group that traditionally gives the bulk of its support to Democratic candidates. It was not really typical to talk about issues like mammograms and free concerts in mayoral campaigns, but our research and analysis showed that these were precisely the kinds of things Cultural Liberals responded well to.

We already knew how to communicate with the other six segments of the electorate. However, Bloomberg was determined to reach out to one of these groups in particular, the so-called Traditional Democrats. Largely black and lower income Hispanics these voters formed Ferrer's base of support. It was also a group Bloomberg wanted to win. He was proud of the support he enjoyed in the Hispanic community, particularly among voters from the Dominican Republic, and was eager to break through to African American New Yorkers in the same way. Indeed, so determined was Bloomberg to speak to these voters that he repeatedly rejected my suggestions to focus on sympathetic segments of the black community, such as middle-class black homeowners and people from Caribbean nations.

"Doug," he'd say, "I don't want to win a segment of the African American community, I want to win the community."

All the while, we were building our database—and discovering interesting and unexpected linkages among the voters of New York City. Time and time again, we discovered surprising convergences in the political attitudes of, say, happy, successful Italian moderates living in Bay Ridge, Brooklyn, and middle-class African American Democrats living in St. Albans, a thriving neighborhood in southeastern Queens. Such seemingly disparate voters might in fact both be well suited for an outreach effort that focused on Bloomberg's efforts to improve the public school system. A Jewish voter on the West Side might well get a similar message on homeland security to a lower-middle-class Hispanic voter in Queens. Our in-depth understanding of the electorate meant that we no longer focused on white voters, black voters, and Hispanic voters or on Democrats, Republicans, and Independents. In fact, such gross categorizations seemed hopelessly crude. For instance, among Traditional Democrats, we found that many Caribbean blacks in Queens as well as many non-Puerto Rican Hispanics who were home- or business-owners were attitudinally similar to Bloomberg's committed voters. It was a hopeful discovery in a city that is often seen as being deeply divided.

In the last eleven days of the campaign, we unleashed our final campaign blitz, which we dubbed "GOTV." By this point in the campaign, the greatest challenge we faced was not the Ferrar campaign but an apathetic electorate. One year earlier, many supposedly savvy observers of New York City politics believed that there was no way Mike would be reelected. Now, virtually everyone expected that he would be. Our polling certainly supported such an optimistic appraisal; however, it also showed a pronounced drop in the number of Bloomberg supporters who intended to vote. Direct mail, phone banks, prerecorded phone calls, canvassing—everything was now targeted on turning out Bloomberg supporters.

With all of these operations in motion, it was difficult to imagine how Kevin Sheekey and the Bloomberg campaign staff could do more; nevertheless, more is precisely what they did. Another nine campaign offices were opened across the city. Approximately twenty thousand bodies were out on the street every day. By the end of the period, Bloomberg workers had knocked on another five hundred fifty thousand doors, bringing the total number of doors knocked on to over one million. By Election Day, Bloomberg volunteers had knocked on roughly a million doors citywide.

On Tuesday, November 8, Election Day, with roughly forty thousand Bloomberg workers fanned out across the city. As in 2001, early indications were that turnout was light in areas of strongest support for the Mayor. Once again, Kevin Sheekey moved decisively. This time he was able to deploy an additional twenty thousand voters/volunteers onto the streets to urge Bloomberg supporters to go vote. He and Mike Berland also did another two recorded calls to supporters and undecideds from the mayor, urging people to cast a ballot and not stay home. Once again, these efforts worked, contributing to Mayor Bloomberg's sweeping 59 percent to 39 percent win over Ferrer.

Against a Hispanic opponent, Mike won a third of the city's Hispanic vote. He also won 46 percent of the African American vote—a striking validation of his efforts to unify the city. Bloomberg's victory was so complete (92 percent of Republicans, 79 percent of independents, 48 percent of Democrats) that what had struck many as impossible one year early now seemed inevitable. It was a stunning example of how a strong candidate with a targeted, moderate campaign could sweep even the most heavily Democratic of cities. In New Jersey, we had demonstrated that under certain circumstances (a heavily Democratic state, a lavishly funded campaign), an old-fashioned Democratic campaign could still win. In New York City, however, we'd established that an independent, moderate candidate could win even bigger. Moreover, with Bloomberg's willingness to invest in technology,[1] we had been able to develop a novel and, I believe, revolutionary approach to politics. We had moved beyond sampling toward a census approach—toward interacting with every voter as an individual. New York City has 3.4 million voters. The United States as a whole has about 110 million—just thirty times more. Can there be any doubt that the techniques we pioneered in New York will one day be applied to the country as a whole?

I think not. Sampling and polling, focus groups, and mall tests will continue to be an important part of American politics for a long time to come. However, I have no doubt that a generation hence, the future of political communications and campaigns will look a lot like Michael Bloomberg's 2005 campaign in New York.

I also feel strongly that this is a good thing. For more than two decades, politicians and, yes, political strategists have won elections by inflaming political divisions—by pushing voters into liberal or conservative camps and by running negative air campaigns that seem irrelevant to their lives. What Clinton started to do in 1996—and what Bloomberg actually did do in 2005—was find a more honest and hopeful way of communicating with the public. Instead of targeting Democrats, Republicans, or Independents, we spoke to people—to individuals. Thanks to our database and microtargeting strategy, Bloomberg was able to speak to voters about the issues they cared about, and the voters noticed and welcomed that. That's why more than fifty thousand of them signed on to volunteer with the Bloomberg campaign.

Of course, from the viewpoint of our clients, the ability to simulate microlevel voting behavior and then shape our message and communications strategy accordingly was a huge leap forward in political tactics. However, I genuinely believe that the movement toward interacting with voters as individuals represents a hopeful development for our democracy. Vicious political partisanship is the way that contemporary American politics is currently played, but it's not the way most Americans see the world. By looking beyond political orientation to the sum total of citizens, voters, and concerns, I believe that techniques like those we developed for Mike Bloomberg can play an important role in moving beyond current political divisions and reminding people—and their elected leaders—of how much Americans have in common. Far from trivializing politics, the type of voter segmentation and message targeting that we used in New York actually reconnected voters to politics—and politicians to voters—in ways that were profoundly beneficial.

PART FOUR

AN UNCERTAIN FUTURE

CHAPTER ELEVEN

THE POWER AND POSSIBILITIES OF EXIT POLLS

In between Bloomberg's campaigns during the spring of 2004, Penn, Schoen & Berland was contacted by a group of business-men from the Dominican Republic who were becoming con-cerned about the fairness of their country's upcoming presidential election. The campaign marked a comeback for former President Leonel Fernandez, a lawyer and former university professor who from 1996 to 2000 had engineered an economic boom that greatly bene-fited one of the hemisphere's poorest countries. His successor, the in-cumbent president, Hipolito Mejia, had been a disaster, presiding over a period of high inflation and growing corruption. Fernandez should have been a strong favorite to retake the presidency, and in-deed polls suggested that he was in the lead. However, the campaign had become so heated that many Dominicans, including the business-men who had approached us, feared that the Mejia government might actually tamper with the process. The fact that the Bush administra-tion was close to Mejia's government made them even more nervous:

would Washington really prevent fraud by such a close ally? They wanted us to conduct a series of polls to document Fernandez's lead as well as a final exit poll to keep the process honest.

It was the kind of assignment we excelled at. By 2004, Penn, Schoen & Berland had amassed a record of success in Central and South America, thanks largely to Mark's long history of work in the region. Since 1979, we had helped elect more than half-a-dozen heads of state in Latin America. At the same time, we'd developed a considerable expertise in fielding exit polls in tense, high-stakes elections. In 2000, not long after our exit polls had prevented Milošević from stealing the election in Serbia, I conducted an exit poll that confirmed the defeat of Mexico's ruling party, the PRI, ending seventy years of one-party rule. Though intense pressure had been brought in an effort to dissuade us from performing this referee role, we had disregarded that pressure—and in the process, helped change the political destiny of these nations.

Our dramatic success with exit polling in Serbia and Mexico marked the emergence of an important new deterrent against government fraud. For years, exit polls had been a crucial tool in U.S. political campaigns. In the 2001 New York mayor's race, for example, early exit poll results had Mike Bloomberg running two points behind Mark Green with low voter turnout. We responded with recorded messages to swing voters from Mayor Rudolph Giuliani and Bloomberg that urged voters to go out and cast their votes for the Republican candidate. In a close race, those calls may very well have tipped the balance. Now, in the Dominican Republic, we were using exit polls as we had in Serbia and Mexico, making them a crucial check against electoral fraud and a key tool for promoting the spread of democracy across the globe.

The businessmen who approached us about monitoring the election in the Dominican Republic believed that our presence in the Dominican Republic would deter the Mejia government from trying to manipulate the results. Since Mark was busy with other projects, I took on the responsibility of directing the polling operations in Santo Domingo. Marcela Berland (the wife of my partner, Mike Berland, and an excellent analyst of Latin American politics in her own right) and Raj Kumar were our representatives on the ground.

The moment the polls closed, we were able to predict with confidence that Fernandez had won an overwhelming twenty-point victory.

It was a prediction that clearly unnerved the incumbent government. For three hours after the voting had ended, even as CNN en Español reported our numbers, official Santo Domingo was quiet. It was clear that Fernandez's triumph was in danger. Marcela and Raj had personally presented our results to Leonel Fernandez; now Raj told me that the candidate had privately confided that rumors were circulating that Mejia's henchmen were preparing to send thugs to steal the ballot boxes. I was dismayed by this report and wondered how Washington would react. Would the Bush administration really allow Mejia to tamper with the election results? If it did, what would that mean for our employees in the Dominican Republic? I worried that they might be in danger. After all, at this point, our exit poll was the Fernando camp's only claim to office.

As evening stretched into night, I began to receive more reassuring news. Washington accepted our poll's legitimacy; the Organization of American States (OAS) was actively working to persuade the government to accept the results of the election, which coincided with the results our exit poll had recorded. Later that night, and after hours of tension, the central government finally posted numbers showing Fernandez ahead, but only slightly. Instead of the twenty-point win we were predicting, the central election commission showed a mere seven- or eight-point lead. It was a worrisome sign: by acknowledging that it was behind—but not by much—the Mejia government was keeping the option of fraud on the table.

Then, just as I was starting to get truly concerned, Mejia conceded the election. A government that might once have been tempted to cheat had been forced to play fair. A few days later, OAS Secretary-General Cesar Gaviria would confirm in the press the important role our exit poll had played. It was a dramatic example of the power of polling to protect democracy.

Just a month after our work in the Dominican Republic, I was approached by one of our first clients, Diego Arria, Venezuela's former ambassador to the United Nations and a special adviser to

Secretary-General Kofi Annan, about a far more explosive situation—the upcoming Venezuela referendum that would decide the fate of Latin America's most divisive elected official, Hugo Chavez.

Chavez is a sort of Latin American Milošević—someone who abides by the trappings of democracy while undermining its substance. He is a leader whose politics represent a bizarre blend of cult-of-personality authoritarianism and retro socialism. As a young lieutenant colonel in 1992, he had attempted to seize power through a military coup. The effort failed and Chavez was arrested, but instead of being thrown into prison, Chavez was allowed to go on national television to urge his co-conspirators to abort the coup. His brief address made him a hero to Venezuela's most disaffected citizens, particularly the rural poor who for generations felt neglected by Venezuela's moderate, two-party political system. Even though the young lieutenant colonel acknowledged only that the coup had failed "for now," Chavez was sentenced to a light, two-year prison sentence and later released from prison with an honorable discharge.

In 1998, Chavez decided to take a more legitimate route to political power and announced that he would be a candidate for the presidency. Utilizing an aggressive anticorruption, antigovernment platform, Chavez won with one of the largest victory margins in the country's recent history. Nevertheless, Chavez wasn't content with a big victory: he was also determined to upend the country's political system, insisting on a series of "reforms" that helped to strengthen the office of the president and solidify his already growing power. Chavez's increasingly authoritarian policies, his open admiration for Cuban dictator Fidel Castro, his delusional association with the great liberator Simon Bolivar, and his strange penchant for spontaneously bursting into poetry and song—often on national television—disturbed a broad swath of Venezuelan society. Middle-class, urban Venezuelans were particularly alarmed by his machinations.

In early 2002, Chavez finally went so far as to attempt to install loyalists in the corporate suites of the state oil corporation, Petroeos de Venezuela, SA (PVSA), the single largest source of hard currency for the country. The takeover triggered massive counter-demonstrations from the country's unions. When the demonstrations took a violent turn, the army sided with the protestors and took Chavez into custody. Once again, Chavez emerged unscathed. The provisional government that replaced Chavez soon collapsed, and by the end of April, Chavez was back in office and stronger than ever. He soon purged the armed forces and took complete control of PVSA, leaving him in a position of unchallenged dominance in Venezuela. By 2004, Venezuela was firmly under Chavez's control.

Chavez was clearly a formidable—and unscrupulous—character. However, I was confident that we could still operate effectively in Venezuela, given our history there. In 1979, David Garth had asked Mark if he'd be interesting in doing polling for the presidential campaign of Luis Herrera Campins. At first glance, it had seemed a daunting prospect that elicited a variety of logistical questions. Could he overcome the language barrier? Would it be possible to accurately frame questions in a country with a political culture so different from the United States? What model should we use to best understand Venezuelan politics? Mark's first trip to Caracas was a leap into the unknown, a venture whose success was by no means assured.

As it turned out, the 1979 Venezuelan presidential election was a precursor to the type of international political consulting that has since become *de rigueur* for many of us and many other American consulting firms. We were one of three groups of American pollsters operating in the country at the time. The ruling Democratic Action party had hired Joseph Napolitan, the man who coined the phrase "political consultant" during John F. Kennedy's 1960 presidential campaign. A smaller, independent party had brought our then rival Pat Caddell aboard as well as John Deardourff, a media consultant we had worked against during the Carey campaign. Despite this august competition, Mark quickly came up with a winning strategy. Since Venezuela did not have universal phone service, he decided to go door-to-door and work closely with some of the embryonic polling organizations that were already on the ground. Within only a few weeks, Mark was conducting comprehensive polls and churning out presentations not terribly unlike those we had done for the

Koch campaign two years earlier. He helped the campaign develop a slogan of "*Ya Basta*" or "Enough," which reflected popular anger with the ruling party's free-spending ways and its lack of attention to solving the country's problems. In the end, Herrera won by a healthy margin.

The presence of so many American consultants was unusual in 1979, but Venezuela in 1979 was an unusual country. Venezuela was a vibrant democracy, and with ten different candidates and a collection of political philosophies that ranged across the political spectrum, it was one of the few true democracies in a region full of authoritarian regimes. Moreover, the campaign had been going on for more than two years, with millions of dollars being spent on huge street rallies and constant political advertisements on radio, television, and in print. Unlike other Latin American countries, most Venezuelans had televisions and radios, and political ads ran constantly during campaign season.

Now, a quarter century later, we had another chance to work in Venezuela. Anti-Chavez protestors had succeeded in gathering enough petitions to force a nationwide vote on a recall referendum in August 2004. Over a lunch near the United Nations, Arria was candid about his fears. Chavez's history suggested that he wouldn't hesitate to steal the election. What, he asked me, could be done to prevent that?

The answer was obvious: an exit poll.

"If there isn't an exit poll," I said, "you *will* lose."

We got the job. Little did we suspect how difficult and dangerous the work would be.

The first sign of trouble came when one of my associates, Raj Kumar, began telephoning potential partners in Venezuela to conduct our exit poll. Raj called twelve vendors. Ten said no at once, without even bothering to ask the standard questions pollsters normally would (such as, "What would this entail? How much would it pay?") Two other firms agreed, only to back out a few days later. In short, it was clear that people in Caracas, once one of the most open cities in Latin America, were living under a cloud of fear.

I confess to thinking at the time that local firms were suffering from a bit of paranoia. We would soon learn otherwise.

For the moment, however, I was more focused on solving the technical problem before us. Clearly, without Venezuelans to ask the questions, it would be impossible for us to proceed. They didn't have to be professionals—we could train other people to ask the necessary questions, as long as they had certain minimum skills—but we had to find someone. Only one organization in Venezuela was willing to take part in a project that might put them in conflict with Chavez, an organization called Sumate (which means "Join Up"). This was the group that had been largely responsible for organizing the recall election, and as such they were already planning to conduct an exit poll of their own. Raj suggested that we piggyback on Sumate's efforts—basically using their volunteers to conduct our poll. I was intrigued. At a meeting with María Corina Machado, one of Sumate founders, I laid out the guarantees we would need to proceed with such an arrangement: we would choose the methodology; we would develop the guidelines and procedures; in short, we would have the final say-so on all decisions related to the poll. The goal, I emphasized, was an exit poll that was beyond reproach. There was no need to argue the point. María accepted my conditions almost instantly. They were as eager to run a professional operation as we were and indeed would soon hire several other consultants to ensure that field operations were conducted properly and smoothly. The exit poll was on.

Of course, I knew that by using Sumate we were opening ourselves up to enormous scrutiny. Though we made clear to the Sumate leadership that we were going to operate our exit poll fairly and objectively, we were concerned that Chavez and others would use the relationship to undermine the legitimacy of the poll. In truth, we had little choice. The only way we were going to be able to do our job in Venezuela would be with Sumate's assistance.

Once we had set up our connection to Sumate, we were ready to begin organizing, but while I expected that our people on the ground, Marcela Berland and Raj Kumar, would be operating in a tense environment, I didn't expect that their lives would be in danger.

Just a few days before the election came the first indication that I had underestimated the threats posed by working in Venezuela: Chavez went on television to accuse Sumate's leadership, our partners in the exit-polling process, of treason. He then gave out the names and home addresses of key Sumate members and warned of potential "action" that might be taken against them. Although no mention was made of Penn, Schoen & Berland, it was a threat that left me seriously concerned about my people on the ground. About a week before the election, Marcela and Raj had flown into Caracas. I was in daily phone contact with both of them, and I didn't like what I was hearing. During a meeting at his hotel, Raj reported that beefy men in wraparound sunglasses and visible earpieces sat at the table next to theirs and tried to eavesdrop on their conversation. Raj and Marcela found that their movements through the city were frequently shadowed. I hoped this was just an attempt at intimidation, but neither I nor anyone else knew what Chavez might do when faced with the imminent loss of power.

Sunday, August 15, Election Day, was an extraordinary scene: turnout surpassed all expectations. Across the country, Venezuelans lined up for hours to vote; some people waited for more than twelve hours to cast a ballot. Although the huge turnout created gridlock at many polling stations, the mood in the country seemed to be generally upbeat. In many areas, residents who lived near the polling stations supplied drinks and sandwiches to those waiting. As a professional pollster, I knew how to interpret this kind of behavior: in my nearly thirty years of experience, turnout of this sort is almost always inspired by dissatisfaction with the incumbent. Our early exit poll results confirmed my intuition: they showed that Chavez was in serious trouble. This wasn't going to be a nail-biter; Chavez was going to lose, and lose big. Our results showed the motion to recall Chavez passing by an extraordinary eighteen-point margin, or 59 percent to 41 percent.

As soon as I had the early results, I quickly called Diego Arria in Caracas to consult on the next steps. We both agreed that the earlier we released our findings, the more effective they were likely to be.

As had been the case in Dominican Republican, we hoped that our results would put irresistible pressure on the Chavez government to respect the will of the public. That afternoon, I directed Raj to draw up a press release spelling out the results. At 7:30 PM, we faxed the release to Western media outlets. The document was clearly marked, "Embargoed till 8:00 PM."—the time when polls actually closed. It was critically important that the poll results not appear early. In Venezuela, as in many countries, it is illegal to publish exit poll results before the polls actually close. Nonetheless, it's absolutely standard to send out embargoed results in this manner.

However, due to higher then expected turnout, the election commission in Venezuela decided to keep the polls open until midnight. By then, however, it was too late to retract or delay our press release. Our results were out and circulating fast. Strategically, we thought this was a good thing. After all, part of the purpose of our poll was to keep the government honest. However, it put Raj and our team on the ground in a difficult and potentially dangerous situation. Our people were now vulnerable to official retribution—and it wasn't long in coming.

In the Dominican Republic, authorities had reacted to the presence of an authoritative exit poll in two stages: first, confused silence, then, reluctant acceptance. Chavez took a different approach. Word quickly came out of Caracas that the government was preparing to declare victory. Considering the results of our poll, we didn't know how it was possible: such a declaration would clearly be a brazen, outright theft of the election. With observers from both the OAS and the Carter Center—including former President Jimmy Carter himself—on the scene that seemed unlikely, and yet as I perused the wire reports that seemed to be precisely what Chavez was gearing up to do. Already, we were hearing reports that the government was attacking our polling numbers and asserting that the true election results were the exact opposite of what we were claiming.

That was bad enough, but what was happening on the ground in Venezuela was far worse. In Caracas, Chavez and his supporters seemed on the verge of unleashing a campaign of mass arrests and terror, and our employees were firmly in their sights. I was in constant contact with Raj, who was meeting with international observers at his hotel in Caracas. At about 2:00 AM, he and the rest of our team headed out to the Sumate headquarters. As they approached their destination, they caught a glimpse of a van stopped in the middle of the road—and four plain-clothed individuals standing in front holding machine guns. The driver did some nifty driving, heading the wrong way down one-way streets to get them past the gunmen to Sumate's main entrance.

What they found inside was almost as bad. The Sumate office was in full-throated panic. Reports were circulating that militias were forming to hunt down anti-Chavez forces. There was a general feeling among many Sumate members that they were in serious danger of being assassinated. The decision was made to abandon the office. After a frightening run through an adjoining shopping center and about an hour of uncertainty as to what was happening, people began to calm down, and Raj decided to head back to his hotel. As the focal point for election observers and Western journalists, it seemed to be the safest possible place in Caracas for an American pollster.

On the way there, one of the stranger and more frightening episodes I've heard of in more than thirty years as a political consultant occurred. About a block from the Sumate headquarters, Raj's car was pulled over by uniformed officers. Raj assumed this was just more election night opposition harassment and that, as an American he was not likely to be harmed. Then one of the officers began asking a series of pointed questions and started searching his bag. He removed Raj's laptop. At first, Raj thought the policeman was simply taken with his sleek new computer; however, it quickly became clear he wasn't interested in just taking a look at the latest technology. The policeman asked Raj to turn it on. Then he asked Raj to pull up every document that he had been working on recently. It quickly dawned on Raj that the officers were actually looking for the press release he had written that said Chavez had lost the election. In all my years, I'd never heard anything like it. You had a highly repressive regime in complete control of a nation's armed forces that was seemingly afraid of a press release. Fortunately for Raj, at that very moment a vanload of television journalists arrived on the scene, causing the police to make a speedy departure. Amid the confusion, Raj rushed back to the hotel. The next day he caught the first flight back to the United States.

Unfortunately, for Sumate and the people of Venezuela, their battle was only beginning.

As morning dawned and Raj boarded the flight that would bring him to safety, the National Electoral Council, headed by a dedicated Chavez supporter, anointed Chavez the winner, with a startling 59 percent to 41 percent victory, almost the exact opposite of our results. It was an absurd claim. We weren't the only entity to have conducted exit polls on the referendum: the Primero Justicia party had done polling as well and come up with results that were almost identical to ours. It's one thing for an exit poll to be off by a few points, but the idea that two *independently conducted* polls were off by *twenty points*—it was absurdly implausible. Never in the history of modern survey research had an exit poll been so far off. The Sumate volunteers had conducted approximately twenty thousand exit interviews at 267 well-chosen voting stations, a huge sample for a country of 25 million people. (In contrast, national exit polls in the United States, a country of nearly 300 million people, typically sampled about twelve thousand people.) It was exactly the sort of methodology we had used in hundreds of polls both domestic and international. In the end, our poll had a razor-thin 1 percent margin of error. There was no doubt in my mind that we were correct—and that Chavez had been soundly defeated.

But, while we had the numbers on our side, we, and the opposition, were losing the public relations battle. In the days before the vote, Chavez's people had been talking to the international media claiming that his numbers were turning around and that new polls showed him taking the lead. As a result, many of the reporters were already preparing themselves for the possibility of a Chavez victory.

Far more debilitating was the support of international election observers, led by former President, Jimmy Carter, whose internationally recognized Carter Center was responsible for much of the international election monitoring. The night before his departure,

Raj had managed to meet briefly with both Jennifer McCoy, the head of the Carter Center's Venezuela mission, and Cesar Gaviria, the secretary-general of the OAS, to tell them more about the results and methodology of our exit poll. In my view, the evidence of fraud or at the very least irregularities was suspicious enough to warrant a wait-and-see attitude from Carter and other international observers, but neither Gaviria nor McCoy were open to bad news. "There's not going to be any fraud in this election," Gaviria told Raj after their quick meeting—a comment that seemed completely at odds with the situation on the ground. The next morning, Carter went on television and proclaimed the election "free and fair."

I was stunned. Two scrupulously conducted independent exit polls had captured an election result that was dramatically different from the official vote tally. Considering the huge discrepancy between the exit polls and the final result, as well as the voluminous reports of voter fraud that both Sumate and we were receiving, it was rash for international observers to so quickly accept the results. Carter went even a step further and proclaimed the election an "historic performance" and asserted that it was now, "the responsibility of all Venezuelans to accept the results and work together for the future." Chavez delivered his own piece of hyperbole calling the plebiscite a "democratic fiesta."

Meanwhile, back in the United States, we were assailed from all sides—by both the media and our competitors. "U.S. Poll Firm in Hot Water in Venezuela," reported the Associated Press. We were accused of producing a highly flawed poll that relied far too heavily on the assistance of an organization committed to getting rid of Chavez.

While our work was being vilified in the media, doubts on the ground were growing. Votes in the recall election had been cast on electronic voting machines, which were provided by a previously obscure company reputed to have close ties with the Chavez government.[1] Disturbing details about the way the vote had been tabulated and about discrepancies in the data were beginning to percolate throughout the country—even among the staff of the Carter Center. Election officials promised Sumate that they would conduct a small "hot audit" of 1 percent of the votes cast the evening of the election to verify the official results. They had also promised to provide access to both the opposition and international observers. Both promises had been ignored. Instead, the audit was pushed back to the following morning and contrary to the government's promises, only seventy-eight of the 192 boxes that were supposed to be counted were actually reexamined. The opposition was allowed to attend only twenty-eight of those counts. International observers saw a mere twenty.

But the worst was still yet to come.

Observers soon learned that, contrary to earlier assurances, the computer server used to tabulate results at election headquarters had not simply received reports from field stations but had also sent communications to them, which suggested that the central computer might have actually been transmitting vote tallies to the precincts. No OAS or Carter Center observers were allowed to observe operations at the computer hub. By the end of the day, it apparently became clear even to the credulous Carter Center that there were serious concerns about the legitimacy of the election. Two days after the election, both the Carter Center and the OAS approached the Chavez government with a demand that was both entirely appropriate and seemingly tough: the government needed to conduct a full audit that compared the supposed electronic results with actual paper records in a large number of districts. After some initial hesitation, Chavez agreed.

Of course, there were still problems. The government refused to allow a third party to generate the random selection; it also refused to seal selected boxes and deliver them to a secure central process. That was certainly a bad sign. Still, I didn't want to believe that the international community would stand by and allow fraud to occur in such an important country. Standing down now would unleash a dangerous dynamic—not only in America's backyard but also in a country that was our leading oil supplier.

The recount began on Wednesday morning, yet even as it began, President Carter and the OAS seemed intent on downplaying its importance. "We have no reason to doubt the integrity of the electoral system or the accuracy of the referendum results," President Carter told the press at a news conference on Tuesday. "There is no evidence of fraud, and any allegations of fraud are completely unwarranted."

The former president even went out of his way to criticize *us,* telling reporters, that Sumate "deliberately distributed this erroneous exit poll data in order to build up, not only the expectation of victory, but also to influence the people still standing in line."

I was puzzled by such comments. Of course there were doubts; that was why there needed to be an audit. Given the issue at stake, I couldn't imagine why my party's former president was championing Hugo Chavez. Still, I hoped for the best. The exit polling was too clear, the suspicions too numerous. Surely the audit would reveal at least an inkling of how Chavez had stolen the election.

By the end of the day, the audit was complete. At the end of the week, the OAS announced its results. "The audit is completed," declared Secretary-General César Gaviria. "The results we have obtained with this check are totally compatible with results of the electoral council." In short, Chavez had won.

I was speechless. Despite the disturbing connections between the company that had provided the electronic voting machines and the Chavez government, I had hoped that a rigorous audit of the paper records would expose the government's skullduggery. Instead, it had apparently confirmed the results. A handful of media outlets recognized that an unprecedented act of fraud had somehow been committed, among them the *Wall Street Journal*. An August 26 editorial noted that "As 'audits' go . . . this was akin to Arthur Andersen scrubbing Enron." The *Journal* then went on to list some of the problems afflicting the recount:

> The sample for the audit was selected by the National Electoral Council (CNE), which is controlled by Mr. Chavez, and was too small to be considered statistically reliable. On referendum day, there was no open audit at the polling stations to reconcile the paper ballots to the electronic voting machines, as the opposition requested, because Mr. Chavez would not allow it. There was also no closed-door audit with all of the National Electoral Council members present because the Chavez-controlled Council did not allow it. There was no inspection of the electronic voting machines immediately after the vote because Mr. Chavez would not allow it. And there was no impartial impounding of the election data paper or digital because . . . you get the idea.[2]

US News and World Report correspondent Michael Barone, author of the authoritative *Almanac of American Politics,* likewise defended our work in Venezuela:

> [A]bout 90 percent of voters [in Venezuela] approached by exit pollsters agree to participate. That is almost double the rate in the United States. Moreover, exit pollers work in teams; there would have to be massive collusion for them to produce fraudulent results. The Penn Schoen exit poll was conducted at about 200 polling places and produced more than 20,000 responses. Changing those results from something like 42–58 (the Chavez announced figure) to 59–41 would be quite a feat. The firm employed supervisors to make sure the polling was done right. And its results by precinct can be checked against the official results reported for that precinct.[3]

Barone went on to note, presciently, that "it would be far easier, given the touch-screen voting method and central tabulation used in Venezuela, for the central counting center to falsify the results."

Yet these were the exceptions. Most media outlets leapt to the conclusion that we'd gotten things wrong. It was unthinkable to me that two, independent exit polls could have been so wrong, yet the mystery of what happened—and why the audit, despite its problems, had seemingly vindicated the official election results—remained.

One month later in September 2004, two Boston-based academics, Ricardo Hausmann of Harvard University and Roberto Rigoban of the Massachusetts Institute of Technology, took a close look at the official results of the Venezuelan election. Their methodology was very sophisticated—too sophisticated, really for much of the press and the public—and so were their conclusions. The two men made no claims about finding a smoking gun or even any direct proof of fraud. What the pair did find, though, was a statistical anomaly so striking and so unlikely that fraud is by far the most logical explanation.

Hausmann and Rigobon began by rejecting the most common fraud hypothesis—the idea that individual machines had been tampered with. Early on outside observers were struck by the fact that a large number of precincts had come in with identical vote counts to oust Chavez, a circumstance that many suspected meant that some kind of electronic cap had been put in place that limited the maximum number of yes votes (i.e., votes to oust Chavez). After reviewing the odds, they concluded that this seemingly unusual convergence had almost certainly occurred as a matter of chance. They also rejected the idea that some machines but not others were tampered with at the precinct level after concluding that variation patterns across precincts were normal.

However, one possibility stood out—the idea that *every* machine in certain precincts had been slightly tampered with. To assess this possibility, the two researchers looked at two data sources. First, they examined the number of registered voters in each precinct who had signed petitions to recall Chavez. Then they looked at our sample of yes votes (votes in favor of recalling Chavez). The researchers found that the precincts where our results were most at odds with the official results were also the areas where the number of petitioners suggested that there should have been a much higher anti-Chavez vote. In other words, the recall tallies seemed lowest in the areas where it would have been reasonable to expect them to be much higher. It was as if some uniform percentage of the anti-Chavez vote had been systematically shaved away.

Hausmann and Rigobon then turned their attention to the precincts audited three days after the election by the Carter Center and OAS. There they found something startling: the incidence of yes votes—votes to oust Chavez—was 10.5 percent higher in audited precincts than in unaudited precincts. What was the chance that that kind of variation would occur naturally? To test this proposition, the authors set out to determine how often this pattern occurred in nonaudited precincts. After taking a thousand random samples of nonaudited precincts, they concluded that this level of variation "occurs with a frequency lower than one percent." In other words, it was extremely implausible that this voting pattern had occurred naturally.

Suddenly, certain aspects of the election looked much more suspicious. The central election commission had promised that its server would only receive election results; in fact, we had since learned that these computers also initiated communications with precinct computers. Second, election observers had been barred from monitoring the server computer. In short, there had been no safeguard in place to prevent the central election commission server from simply communicating to certain election precincts what results to print.

But didn't the Carter Center's August 18 audit rule out such interference? Wouldn't it have detected such a problem? Not necessarily, for here too Hausmann and Rigobon found serious problems. The biggest had to do with the so-called random audit of voting machines that occurred after protests were raised about the result. For an audit of machines to be successful, it was essential that the tested machines be chosen in a way that was truly random. The two researchers found that this had not occurred. Instead of allowing outside observers to generate a random selection of machines, the Carter Center instead allowed the Chavez-controlled electoral commission to use its own computer program on its own device to generate the supposedly random audit selection—twenty-four hours in advance of the actual count. Nor were those boxes moved to a central, secured location in advance of the recount, an omission that had led Sumate to boycott the audit process. In short, it was far from clear that the "random" recount was truly random or even secure. Chavez, it appears, orchestrated a widespread campaign of tampering (say in three thousand precincts), but in a rather subtle way—through transmissions from the central election computers. Then, he directed international observers to those precincts where no fraud occurred.

A handful of media outlets recognized the significance of Chavez's fraud. A September 9 *Wall Street Journal* editorial highlighted Hausmann and Rigobon's finding that there was a 99 percent chance of fraud and emphasized how that the act of fraud probably hinged on computer manipulation:

> Voting machines were supposed to print tallies before communicating by Internet with the CNE center. But the CNE changed that rule, arranging to have totals sent to the center first and only later

printing tally sheets. This increases the potential for fraud because the Smartmatic voting machines suddenly had two-way communication capacity that they weren't supposed to have. The economists say this means the CNE center could have sent messages back to polling stations to alter the totals.[4]

If this is indeed what did occur, the Chavez "affirmation" (as the government described it) marks the most sophisticated example of election fraud in modern history. Did the Chavezistas have the skills to pull it off? As I've thought about this question, I've turned time and time again to Raj's encounters with the Venezuelan secret police and in particular to the way in which they honed in on his computer and demanded to see all of his recent documents. These are sophisticated people.

Interestingly, more evidence of fraud in the referendum has recently become available thanks to an article by Maria Febres Cordero and Bernardo Marquez in the International Statistical Review. That study confirmed the conclusions of the Hausmann and Rigobon research, showing that up to one quarter of the votes were tampered with and almost 20 percent of the voting centers showed irregular voting patterns. In all, over 2 million votes were altered, and according to estimates by Cordero and Marquez, Chavez actually lost the referendum 56 percent to 41 percent—very close to the results of our exit poll. Sadly, exactly how this fraud was actually executed may never fully be revealed. One of the things I've learned in the course of working with democrats against authoritarian incumbents is that rulers who cheat will go to great pains to conceal evidence of fraud. One of the last things Serbian strongman Slobodan Milošević did as crowds prepared to sweep him from power in 2000 was order his subordinates to destroy all the evidence of how the regime manipulated the 1992 presidential campaign and the 1996 municipal elections.

Ominously, the story of Chavez's election fraud does not end in Venezuela. In 2005, the company at the center of the dubious recall election—Smartmatic—purchased a leading American e-voting company, Sequoia, in part by disingenuously presenting itself as a Florida-based company. (It has a small office in Boca Raton.) Reporters later revealed that Smartmatic's partners in the purchase were actually shell companies controlled by the Venezuelan government.[5] Earlier this year, a Carnegie Mellon University computer science professor concluded that Sequoia's system could in fact be manipulated, a finding that caused the government of Allegheny County, Pennsylvania, to scrap plans to buy the machines.[6] It's unlikely—though not impossible—that Chavez would attempt to tamper with U.S. elections, but in other parts of the world an apparently respectable and apparently North American e-voting company bent on fraud could do real damage. Chavez has made no secret of his desire to influence elections elsewhere, yet so far, his moves have gone largely unchecked.

The widespread acceptance of the fraudulent election results in Venezuela came as a demoralizing surprise to me. Just as the outgoing Bush (I) administration and the incoming Clinton administrations were unable to do anything about Milošević's stolen election in 1992, now the international community and the Bush II administration were again standing idly by while the hemisphere's most sinister politician—a man who also happened to be the United States's most prominent oil supplier—did the same thing. For me, the experience in Venezuela served to underscore the lessons of Serbia: exit polls on their own often weren't enough. In order to ensure that democracy would be upheld, opposition parties needed to organize campaigns of civil disobedience and secure the political backing of the United States and regional organizations such as the EU or the OAS.

Even if an exit poll can't topple a devious and determined dictator on its own, I continue to believe that exit polls can have a powerful role in contested elections. Perhaps the best example occurred in 2004 during Ukraine's so-called Orange Revolution, which I watched from an unusual vantage point.

In 2000, a wealthy Ukrainian businessman named Victor Pinchuk approached me with a compelling request: he wanted me to help him forge closer relations between Ukraine and the West, particularly the United States. It was an assignment I was happy to take on. Like Pinchuk, I believed that both the United States and the Ukraine had an interest in building strong ties with the other. As former national security director Zbigniew Brzezinski has noted (along with many others), as long as Ukraine is independent, Russia can never threaten the West in the way the Soviet Union once had. The fact that Pinchuk—who in addition to being one of the country's most successful businessman was also a member of parliament, the owner of a major Ukrainian television network, and President Leonid Kuchma's son-in-law—wanted my help in bringing about this alignment seemed like a win-win proposition. Yet by the spring of 2004, a major test was fast approaching in the form of presidential elections that pitted President Kuchma's chosen successor, Prime Minister Viktor Yanukovich, against a charismatic Ukrainian nationalist, Viktor Yushchenko.[7]

By the summer of 2004, the presidential election was in full swing—and getting ugly. Yushchenko was unwelcome in the Russophone east and Yanukovich was similarly unwelcome in the nationalist west, a substantial portion of which had been incorporated by Stalin into the Soviet Union after World War II and was irrevocably hostile to any Russian speaker. The increasingly heated campaign threatened to split the country in two.

And that was the nice part of the campaign. On September 6, Ukrainian doctors diagnosed Yushchenko with a severe case of food poisoning. However, as disfiguring sores spread across their candidate's face, the Yushchenko camp began to suspect something more nefarious. At first, the Yanukovich camp pooh-poohed this claim. Later that month, however, Yushchenko flew to Vienna for specialized treatment where he received a shocking diagnosis: dioxin poisoning. The face the candidate now presented to the world was shocking too. In the course of a few weeks, the handsome, young politician who had once been called Ukraine's John F. Kennedy looked like a horribly disfigured old man.

Although suspicions swirled around the intelligence services in the region, there was no clear evidence of who was responsible for the poisoning. Yushchenko continued the campaign even though he was in so much pain that he had a morphine drip attached directly to his spine. The first round of voting took place on November 1. Most experts predicted a win by Yushchenko—but not a big enough win to avoid a runoff election three weeks later. However, when official results came in, they showed *Yanukovich* with a narrow lead. The opposition immediately cried foul. It was common knowledge that the Yanukovich's camp's control of the mechanisms of govern-ment—"administrative resources," as it was called in Ukraine—would be good for up to 5 percent of the vote. "Honest graft," they would have called it back in East Harlem. The decisive vote would come on November 21. Every one knew that on that date Yushchenko would need a clear win in order to claim victory.

The three weeks that followed were a whirlwind of campaign stops, accusations, intimidation, and, occasionally, violence. I was in Kiev for part of the time, and even as an outsider, it was clear that passions were reaching the boiling point. Election Day actually came as something of a relief to me, and I dearly hoped that it would settle the issue of who should govern the country decisively. To the dismay of many, when polls closed at 8:00 PM, the official re-sults showed Yanukovich ahead by about 3 percent. It was immedi-ately clear that these results were unlikely to stand up. An exit poll funded by Western embassies had shown a very different result: it had Yushchenko ahead by an 11 percent margin, more than enough to offset Yanukovich's "administrative resources." The question now was what would happen next? Would Yushchenko fold, or would Yushchenko's supporters take to the streets, as the democratic resistance in Serbia had in 2000?

Yushchenko's camp had scheduled a massive rock concert—a massive *televised* rock concert—in Independence Square for the day of the vote. By the next morning, however, it had become clear that the rock concert was not just a gesture toward his young supporters. It was actually the first phase of a plan to mobilize tens of thousands of Ukrainians in the heart of Kiev. The day after the election hun-dreds of portable toilets and a food-serving operation that could feed thousands had suddenly materialized along Khreshchatik Av-enue, the capital's main thoroughfare. The goal was to get so many supporters into the core of the city that the roughly fifteen thousand troops in the region would be unable to forcibly evict them without major bloodshed.

The contest now entered a dangerous phase. By noon of the fol-lowing day, some thirty thousand pro-Yushchenko demonstrators had filled downtown Kiev, essentially seizing control of the center city. Yushchenko's team was quick to drive home the message. As the crowd swelled to an estimated two hundred thousand people, Yushchenko formally declared himself the victor and was "sworn in" during a mock ceremony. In Washington, Secretary of State Colin Powell served notice that the United States would not accept the "official" results, while Russia lashed out at "unprecedented interference" by the United States. With the country teetering on the edge of chaos, Ukraine's highest court called for delaying the certification of election results.

Pinchuk played a calming role in the conflict. Although he had supported Yanukovich, he sympathized with the protestors. He and his young daughter even mingled with the protestors, a brave move for a figure many Yushchenko supporters distrusted. Even more important was his unofficial role as a back channel between the U.S. government and the Yanukovich camp. According to the *New York Times,* just days after the November 21 vote, Yanukovich allies in the powerful Interior Ministry actually mobilized troops for a Tiananmen Square–style crackdown. However, news of the operation leaked, and at the behest of the U.S. Ambassador to Ukraine, Pinchuk made a frantic call to the president's chief of staff in an attempt to find out what was happening and to stop the government from making a rash decision. Several hours later, the operation—government officials would later describe it as a mere exercise—was called off. There would be no use of force. Instead, the government and the opposition reached a peaceful agreement to hold a third round of voting. The two sides agreed that the winner, whoever that might be, would accept a series of changes to the constitution that would reduce the powers of the presidency and move the country toward a more parliamentary system of government. Consensus reached, the high court ordered a new—and final—election for December 26.

This time I wouldn't be watching as an observer. Pinchuk asked me, along with Republican pollster Frank Luntz, to conduct a definitive exit poll for ICTV, the Pinchuk-owned private television station—something that would demonstrate with absolute confidence who had won. In a press conference two weeks before the election, he announced that ICTV would call the election soon after the polls closed. It was a brave stance. Most observers expected Yushchenko to win and saw the Yanukovich camp as complicit in electoral irregularities. Pinchuk—a Yanukovich supporter—felt it was important to produce an unimpeachable assessment of the country's preferences, whatever the consequences.

Given the stakes, it was imperative that this exit poll be correct. There could be no inaccurate predictions as was the case in Bush-Gore 2000. In order to guarantee an accurate count, ICTV agreed to conduct an extraordinary number of interviews on Election Day—three separate rounds of interviews that would query a total of roughly fifty thousand people. To better convey just what a massive sample that was, pollsters in the United States—a country with a population six times larger than that of Ukraine—typically conduct only ten thousand to twelve thousand exit interviews for a national election. In Ukraine, every effort would be made to avoid or hopefully eliminate statistical errors.

The day before the election—Christmas Day, 2004—I flew into Kiev to oversee this tremendous exit-polling operation. There in the Ukrainian capital, I found an extraordinary scene. Despite the bone-chilling weather, Yushchenko supporters of all ages continued to occupy the center of the city. The mood in their tent city was both festive and fatalistic; though people hoped for change—indeed, they were willing to risk their lives for it—they also feared the worse. If soldiers and tanks had suddenly appeared at the end of Khreshchatik, no one would have been surprised.

December 26 was Election Day and Round Three. My greatest fear was that our canvassers might have difficulty gaining access to polling places in certain regions of the country. By mid-morning, however, it was clear that such worries were unwarranted. Everything was proceeding smoothly. By noon, we had our first numbers, and as expected, they showed Viktor Yushchenko making a strong showing. By 2:00 PM, we had enough responses to predict with confidence that Yushchenko would be the clear winner. Yet as the afternoon passed, Yushchenko's margins began to decline. By 6:00 PM, our polls showed him ahead by ten or eleven points—a decisive margin but one that was nonetheless worrisome in light of the allegations of fraud that surrounded the previous rounds of voting.

As the polls closed at 8:00 PM, Luntz and I headed across town to brief Pinchuk on our results. I was curious to see how he would react to his candidate's defeat. Victor was the picture of sangfroid. He was intensely interested in the results, particularly in how people in his native region of Dnieperpetrovsk had voted, but there was never any question about the results' validity or about our next step. After about half an hour, he said simply, "It's time for you to go to the studio." We were soon back in the car and on our way to ICTV.

At 9:00 PM, we went live with ICTV's news anchor, Dmitry Kiselyov. Kiselyov's introduction made it clear that nothing would be withheld from his viewers. "The polls have now closed," he began (as best I could tell from the simultaneous translation that was taking place), "and while election officials will be counting ballots through the night, ICTV is now ready to tell you who won the election. We have as our guests . . ." Then we were on, live.

I had expected a brief segment—perhaps five minutes. Instead, Kiselyov quizzed us on our results for a full half-hour. His questions were good ones: Exit polls in America had gotten the 2000 elections wrong, he noted. Why should viewers accept these results? I responded by explaining the extraordinary care we had taken in conducting this survey. Our final numbers showed Yushchenko ahead by double digits; we were certain they were right. Ten minutes later, we repeated our results on Fox, essentially revealing the outcome of the election to the rest of the world as well.

My appearance with Luntz on ICTV marked an important turning point. By commissioning us to conduct an exit poll and then fearlessly authorizing its release, Pinchuk was sending an important signal that the incumbent Kuchma administration would adhere to the will of the people. Within hours, the state media was also publicizing poll results that echoed our findings: Yushchenko had won. While Yanukovich refused at first to acknowledge his defeat, official entities like the central election commission and the high court declined to support his claims of election fraud and improprieties. On December 31, parliament forced Yanukovich to resign as prime minister. The writing was on the wall. On January 23, 2005, Viktor Yushchenko was sworn in as Ukraine's third elected president.

From Mexico to Serbia, from the Dominican Republic to Ukraine—the record is clear: in contested elections, exit polls can be the decisive difference between an election that is fraudulent and an election that is free and fair. Unfortunately, the past decade has also revealed another, less hopeful truth: time and time again, our government has failed to respect and support public opinion in countries abroad, often with disastrous consequences for both U.S. foreign policy and for the world. If America had stood with Milan Panic against Serbian strongman Slobodan Milošević in 1992, horrors like the massacre at Srbrenica (where Serb forces killed thousands), the ethnic cleansing in Bosnia-Herzegovina, and the war in Kosovo might never have happened. Likewise, during a pivotal election in South Korea that would determine whether that country would continue in its historic alliance with the United States or instead elect a "see-no-evil," North Korea-courting regime, the (second) Bush administration would display virtually no interest in Korean public opinion, responding to a tragic accident involving the U.S. military with a *pro forma* apology that contributed greatly to the loss of the pro-American Lee Hoi Chang.

Exit polls are not in and of themselves a panacea. While exit polls alone were enough to focus international opinion and ensure that the will of the electorate prevailed in Mexico, the Dominican Republic, and (ultimately) Ukraine, in some cases—Serbia in 1992 and Venezuela in 2004—their impact was more disappointing. In my experience, countries where exit polling failed share a number of characteristics. One is that exit poll credibility often suffered from its association with the opposition. The referendum in Venezuela in 2004, is clearly the most dramatic example of this, and the fact that Sumate was involved in fielding the poll made it all too easy for other governments to resist evidence of fraud. Clearly, a nonpartisan, objective vantage point is essential for exit polling in contested, high-stakes elections.

The other critically important variable is the reaction of the international community or rather, the part of the international community that really matters (be it the United States, the European Union, or, in the case of Zimbabwe, South Africa). Exit polls can only be successful as a policing mechanism when the international community takes them seriously and makes it clear that flouting the will of the electorate will be viewed as a grave offense. In the case of Serbia, the West was so eager to strike a deal with Serbian strongman Slobodan Milošević that it was prepared to indulge cheating—at least in 1992, as the prospect of short-term gain led other countries to ignore their long-term interest in having a truly democratic neighbor.

Regrettably, there's no ready solution to the problem of misguided perceptions of national self-interest, but there is a straightforward institutional solution to the problem of exit polls being seen as partisan. Very simply, international organizations should strongly consider sponsoring multiple exit polls in closely contested elections where partisanship is high and emotions are raw. Think how different the situation in Venezuela would have been if there had been two or three officially sanctioned exit polls conducted by local professional pollsters that would not have had to rely on a supposedly biased group like Sumate as we did. This inability to recruit local partners would have been a red flag—and an international issue—in its own right. An OAS-funded poll would also have been much harder for the international community to ignore. And if, despite all the evidence to the contrary, one believes that Chavez actually won, the imprimatur of an international organization would have given the winner the uncontested legitimacy they crave.

Indeed, during the 2006 presidential election in Venezuela, there were multiple exit polls, including one by the National Electoral Commission, which did seem to confirm that Hugo Chavez had, in the words of Mary Anastasia O'Grady, "gotten more votes" than his opponent Manuel Rosales. While initially the election appeared to be more transparent than the 2004 referendum, it was hardly a step toward a constitutional democracy.

A number of additional exit polls completed in 2006 showed a narrower margin than the one the National Electoral Council reported. Subsequent investigation has shown that there may well have been large scale fraud and padding of the voter rolls.

According to a multidisciplinary team of professors from the Universidad Simon Boliver and the Universidad Metropolitana in

Caracas, Chavez may well have manipulated more than 4 million votes, thus casting his victory into doubt and suggesting, at the very least, that if he did win, it was by a far smaller margin than was reported in either the exit polls or the final count.

As Chavez's subsequent move toward authoritarian rule demonstrates, the clear losers here are Venezuelan people.[8] To be sure, exit polls should not replace elections, but when well-done exit polls produce results different from the vote tally—as happened in Venezuela in both 2004 and 2006—it should trigger heightened scrutiny of the whole electoral process.

Anyone remember Bush versus Gore?

The 2000 debacle in Florida and the constitutional crisis that followed, the feverish speculation surrounding Ohio in 2004—all could have been avoided or at least mitigated if multiple exit polls had been fielded. The United States was better off when the various networks conducted their own exit polls rather than relying on a single polling consortium. Even North America, I believe, would benefit from a return to multiple exit polls.

In the end, exit polls that are utilized strategically and in concert with the backing and support of the international community can play an important role in promoting democracy and preventing electoral fraud. This may seem like a strange notion after exit poll debacles during recent U.S. elections, but those bad experiences should not permanently taint a necessary tool in the fight for good governance. When I first began as a pollster thirty years ago, the notion of using exit polling to promote democracy would have been practically incomprehensible. Nevertheless, as communication technology has spread and the goal of promoting democracy and ensuring free, fair elections has taken root across the globe, we found in our hands a marvelous tool to protect the will of the people. It's an instrument we would be foolish to neglect.

CHAPTER TWELVE

PITFALLS ON THE PATH TO 2008

When it comes to political campaigning and communications, democracies (and aspiring democracies) abroad continue to look to the United States for inspiration and guidance. However, the condition of our own democracy—and of the Democratic Party—is far from healthy. Even as they swept congressional Republicans from power in 2006's midterm elections, 45 percent of voters said that Democrats "do not share their values." (Republicans, not surprisingly, fared even worse: almost 55 percent of voters say the same of the GOP.) Seventy-three percent of independent voters—by far, the single largest segment of the electorate—now say that they could support an alternative third political party. Understanding the causes of this discontent is (or should be) one of the most pressing challenges of our time. The party that focuses on and responds to this dissatisfaction first is the party that will win the presidency in 2008.

Democrats begin with certain advantages. The disastrous presidency of George W. Bush has energized the Democratic Party to an extent not seen since the dark days of Watergate. This broiling anger, together with the growth of websites like MoveOn.org and blogs like The Daily Kos have given rise to something new—the "net roots" voter. Highly active, generally liberal, and nonreligious, these voters have emerged as a significant force in the Democratic Party, championing Howard Dean in 2004 and defeating Senator Joseph Lieberman in the Connecticut Senate primary in 2006. These activists style themselves as, in Dean's words, the "Democratic wing of the Democratic Party"—a designation that of course excludes moderates and outright insults conservatives. I see them as something else—the George McGovern/Eugene McCarthy wing of the party.

It is important that Democrats not misunderstand the present moment. The lesson of Connecticut Democratic senatorial nominee Ned Lamont's defeat last year is clear: Activists may express their anger in the primary, but mainstream voters want sensible, centrist, bipartisan candidates in office. The state's Democratic activists may have denied Joe Lieberman the nomination, but they couldn't deny him a victory on Election Night, even in an overwhelmingly Democratic antiwar state. Voters in Connecticut value the bipartisan role Lieberman has carved out for himself in the Senate and distrust the pugnacious, partisan left wing of the Democratic Party. In this, the voters of Connecticut are not alone. Even after six years of Republican misrule, the electorate remains overwhelmingly hostile to core liberal ideas like higher taxes and the redistribution of wealth. The facts are clear. From North Carolina to Indiana, from Missouri to Arizona, the Democrats who won were, in the words of Democratic Leadership Council head Bruce Reed, "straight out of centrist casting." Moderate Democratic candidates like Robert Casey in Pennsylvania, James Webb in Virginia, and John Tester were the key to the Democratic victory. Their stances on issues like abortion and gun control allowed them to neutralize traditional Republican values attacks and capitalize on public outrage over the GOP's domestic and international failures.

The 2006 midterm elections should not be misinterpreted: they were a rebuke of President George W. Bush, not an embrace of the Democratic Party. If the 2008 presidential election comes down to a choice between a Republican advocating smaller, more limited government, and lower taxes and a Democrat championing a more activist government and "fairer" taxes—in other words, if Democratic activists saddle the party with a candidate who takes Howard Dean-like positions—the Democrat will lose, plain and simple.

For Democrats, this is a moment of opportunity. In many ways, it resembles the period immediately after President Bill Clinton's re-election in 1996—a moment when the Democratic Party had an opportunity to cement its hold over centrist voters, to break down the Red-Blue political divide, and bring a significant number of conservatives into the Democratic Party. In order to understand what Democrats need to do next, it's worth reconsidering what did—and did not—happen then. It's also worth touching on the electoral disaster that was the 2000 presidential race, when Democratic presidential nominee Al Gore—who was my client for several critical years—would inexplicably squander another opportunity to regain majority status for the Democratic Party.

It's a memory no one who was there will ever forget: Election Night, 1996. I was one of perhaps two-dozen people circulating through the Presidential Suite of Little Rock's Excelsior Hotel. The president was in fine spirits: confident in his own reelection, he graciously complimented our efforts on his behalf. Election Night is almost always a moment of high excitement but that evening at the Excelsior, the mood was one of quiet satisfaction. There was none of the uncertainty that usually marks Election Day. The only real question of the evening was would the president's popularity be enough to sweep a handful of contested Senate seats into the Democratic column. It was a remarkable turnaround for a president who two years earlier had faced the very real prospect of being swept away by a historic Republican tide.

Eight o'clock rolled around, and the networks began awarding states to Clinton, just as we had expected. Yet as the evening progressed, my thoughts took on a more serious cast. Despite our largely successful efforts to reposition the president as a fiscally prudent moderate who shared mainstream American values and despite the obvious contrast between the youthful and vigorous incumbent and his elderly and sardonic challenger, I soon realized that Clinton was not going to win a landslide victory. Democrats would not add to their total number of seats in the Senate (in fact, they would suffer a net loss of two), and President Clinton would not win a majority of votes cast (though with 49.2 percent of votes cast, he would come very close). Victories in critical swing states like Tennessee and Kentucky may well have had more to do with Texas billionaire Ross Perot's presence in the race than to the crossover appeal of Democratic ideas. It was clear that we still had a lot of work to do.

Later that evening, I found myself in the elevator with George Stephanopoulos. Stephanopoulos had resisted our admission to the Clinton camp two years earlier, but that night he was gracious in conceding that our embrace of fiscal discipline and mainstream, centrist values had worked—this time. I remember his look of surprise when I told him, "This isn't as big as it looks."

Stephanopoulos looked at me quizzically.

"Look," I continued, "it's reasonable to estimate that two-thirds of the Perot vote would have gone to Dole if Perot hadn't been in the race. That means that under normal circumstances we would be looking at a 52 percent to 48 percent win. That's not exactly a landslide in a race where we did everything right and the Republicans did everything wrong." Far from revealing that the electorate had embraced the Democratic Party, election results actually offered further evidence of just how conservative the American electorate had become.

Still, for both Mark and myself, Clinton's reelection was a heady moment. Just two years earlier, Republicans had taken control of Congress, leaving Democrats in Washington stunned and demoralized. In response to our polling, Clinton had embraced a balanced budget, "ended welfare as we knew it," and stopped the Republican "revolution" in its tracks. We were slowly erasing the old tax-and-spend stigma that Republicans had emblazoned on the Democratic Party, replacing it with a new catchphrase Clinton had first rolled out during his period heading the centrist Democratic Leadership Council before he launched his Presidential candidacy: opportunity *and* responsibility. In the process, we were making the Democratic Party the party of the American dream, the party that helped people who worked hard and behaved responsibly get ahead.

In retrospect, it's now clear that there was one important vulnerability we failed to neutralize: the Republican advantage on national security issues. Since the early 1970s, Republicans had enjoyed an advantage on foreign policy and national security. They claimed not only to be "tougher" but also more competent. (Who could forget Bob Dole's infelicitous exclamation in the 1976 vice presidential debates that the conflicts of the twentieth century were "Democratic wars"?) Voters seemingly agreed.

At the time of President Clinton's reelection, however, this weakness seemed less important than it does today. The Cold War, after all, was over. In war zones like Bosnia, the Clinton administration wanted to use the U.S. military to prevent genocide and promote democracy, not the GOP. On the contrary, many Republicans struck frankly isolationist notes in their hysterical attacks on the president. (When the Clinton administration launched air strikes against Serbia in an effort to stop that country's ethnic cleansing of Kosovo in the spring of 1999, then-Representative Tom DeLay of Texas compared the president to the Unabomber.) In short, it seemed that while the Democratic Party of Bill Clinton was embracing the mainstream and shedding its post-Vietnam distaste for military power, the GOP was abandoning the legacies of Abraham Lincoln and Teddy Roosevelt for the ideas of arch-isolationist Robert Taft and Newt Gingrich.

As the second term began, Mark and I began working on a proposal that would simultaneously address the country's most difficult long-term financial problem and make Democrats the party of bold initiatives—Social Security reform. President Bill Clinton was keenly aware of the grim prognosis facing Social Security. Social Security's surpluses would soon turn to deficits, and, as the population aged and the number of workers supporting retirees shrank, it would become ever more difficult for the federal government to make good on its obligations to seniors. Putting Social Security's finances on a safe foundation would be an historic achievement. It would demonstrate the president's zeal for nonpartisan reform and also appeal to young voters, who recognized that investing in equities was a better way to build wealth than government bonds. It would clearly have been a hard sell—the liberal wing of the party was adamantly opposed to this idea; however, the surging stock market of the 1990s at least gave us a chance.

But it was not to be. Instead of moving forward with an ambitious effort to place a new governing philosophy at the heart of the Democratic Party, the effort to remake the Democratic Party stalled, thanks in large part to the fallout that resulted from revelations of the president's relationship with Monica Lewinsky. To be sure, worries about the vagaries of stock market investing played a role as well. Still, it is hard to escape the view that privatization was seriously impacted by the clamor for the president's impeachment.

Special Prosecutor Kenneth Starr's singularly destructive investigation and the House of Representatives subsequent decision to impeach the president threw the White House into crisis mode. Although the president's popularity remained high throughout a period of nearly continuous harmful revelations (many of them seemingly leaked from sources close to the Starr camp), the need to stave off impeachment forced the president to protect his left flank and be solicitous of Congressional Democrats them in ways that most Presidents with approval ratings approaching 60 percent do not typically have to be. As a result, Clinton was unable to build on the "end of big government" agenda that had defined his reelection campaign. The push for a plan that would add personal accounts to Social Security was quietly abandoned. Rather than being in a position to push innovative ideas on liberal Democrats in Congress, Clinton was now at their mercy.

I remember Clinton calling me in the summer of 1998 soon after I had returned to New York from an overseas trip. Mark was unavailable, and the president was eager to add a question to an upcoming poll. It had just been announced that the president would testify before the grand jury convened by special prosecutor Kenneth Starr to investigate the president. As an attorney, I knew that by doing so the president risked putting himself in further legal jeopardy. When I asked the president why he had agreed to testify, he replied, wearily, that even his supporters wanted him to explain himself. "I need to protect my base in the House," he said with a sigh.

Luckily, Clinton was more fortunate in his enemies than in his friends. In particular, it was the president's immense good fortune to have House Speaker Newt Gingrich as a foe. The House Speaker and his colleagues clearly overplayed their hand on impeachment, and misreading the mood of the American people, appeared too partisan, but their problems went beyond impeachment. If Gingrich had negotiated constructively with Clinton and emphasized his willingness to make policy in a bipartisan fashion, he could reasonably have claimed credit for the responsible policies the president had advanced. But instead of sensible compromise, the GOP confused voters' dislike of *big* government with their own dislike of government per se. Well over two-thirds of the electorate believed that a fiscally prudent government could still provide decent medical care for senior citizens and children, a high-quality education system, and protection for the environment. Ignoring this reservoir of support for limited but effective government led congressional Republicans to their ruinous decision to shut down the government—twice. By 1999, Newt Gingrich was gone, and President Clinton's political dominance was undisputed.

The impeachment scandal notwithstanding, Democrats had a clear advantage going into the 2000 presidential elections. Values voters had already started returning to the Democratic Party, which had won a significant number of them in the 1998 midterm elections. These voters were motivated at least in part by the very creative positioning Mark Penn had developed for the 1998 midterm election—progress not partisanship—which helped consolidate support among the swing I and swing II voters who had moved toward Clinton and the Democrats in the 1996 contest. There was every reason to believe that Gore could overcome whatever "problems" lingered from the Lewinsky scandal by linking himself directly to the extraordinary accomplishments of the Clinton-Gore administration—low unemployment, low crime, low inflation, record economic growth, welfare reform, middle class tax cuts, a record budget surplus, the list went on and on. It was a record any politician would have been happy to run on.

But not, apparently, Vice President Al Gore.

I knew Gore well. For nearly three years during the mid-1990s, I did most of his polling and met with him and his core advisers at least once a month in the vice presidential residence at the Naval Observatory. I publicly defended him on national television during a mini-scandal involving allegations of improper fundraising in the vice president's office. Gore could be oddly detached; he was not a people person. During a lull in one of our early meetings, I remember being alone in his office with him. First came an awkward silence; then he started surfing the Web. When I attempted to make conversation by asking him a question about Serbia, he looked at me as if to say, "Why would you want to talk about that?"

Gore's core political principles also seemed oddly unsettled. He was constantly looking for new approaches, new answers to the question of what he should espouse. In spite of this restless drive for answers, he rarely took the suggestions Mark and I made. In early 1999, Mark did an elaborate message poll for Gore that was intended to help him develop themes for his upcoming presidential run. He urged Gore to stay in the center and even advocated a faith-based initiative as a cornerstone of his campaign. Gore made one speech on faith-based initiatives and then dropped the idea—because, he said, of criticism from the Left.

Still, as long as Clinton's popularity was climbing, Gore was with Clinton. In early 1999, Gore switched over to working mainly with Mark, who had moved to Washington and become one of the Clintons' closest advisers. Mark didn't last long either. After the news of Monica Lewinsky broke, Gore began distancing himself from so-called "Clinton people." By this time, Mark was working on the nascent New York senatorial campaign of Hillary Clinton, whom the Gore camp distrusted and saw as a potential rival. As a result, Mark and I became suspect.

Mark made the specific error of defending Bill Clinton directly to the vice president. At a time when some of Gore's advisers were recommending that the he steer clear of the president, Penn suggested that the vice president embrace Clintonism and expand on it. Bob Shrum, who had been working on the Gore campaign, later told me that Mark's honesty with Gore led the vice president to fire him, instructing him and his top aides to replace Mark with someone who had a different view of public opinion and the way it was affecting his candidacy. I believe this decision by Gore was wrongheaded and may well have cost him the election.

To be sure, there were good reasons for Gore to move to the Left. The vice president was deeply concerned by the insurgent primary campaign of former New Jersey Senator Bill Bradley, which centered on an innovative proposal for providing universal health insurance. He also worried that liberals who remembered the conservative positions he had espoused during his first run for the presidency in 1988 might abandon his candidacy. Of course, they did not. Gore edged Bradley out in New Hampshire and went on to win decisively in the primary contests that followed. But instead of responding by moving back to the center, Gore embraced his newest incarnation: he was now a feisty, left-wing populist. It would have been easy for Gore to decry the president's personal conduct—which clearly had shocked and distressed him—and still associate himself with the Clinton/Gore extraordinary record of accomplishment, for which the vice president indeed deserved considerable credit. Instead, Gore distanced himself from Clinton and from his own achievements. The Gore campaign also kept Clinton out of other races where his presence might have helped, such as the Senate race in New Jersey. By the last week of the race, polls in New Jersey had tightened to single digits, and Clinton was eager to help. I remember him calling to ask if I felt he could help Corzine and Gore in New Jersey. It was clear Clinton was surprised and hurt that his old pollster, Stanley Greenberg, would have given Gore such advice; he wanted my opinion on whether he could have helped. I told him the truth: he could have. Corzine wouldn't make the same mistake again. Five years later when Corzine ran for governor, Clinton publicly endorsed him at a rally in New Jersey, and Corzine insisted that a clip of that endorsement be prominently featured in a heavily run campaign commercial.

Even worse than Gore's decision to distance himself from Clinton was his decision to move away from the political strategy that saved the president in 1996. Instead of embracing mainstream political values and the philosophy of opportunity and responsibility, he took a populist tack—"the people against the powerful"—that was more William Jennings Bryan than Bill Clinton. Unfortunately, in 2000 (as opposed to 1892) such slogans were a sure-fire loser with Southern whites. Gore's unfortunate rhetoric ultimately played a major role in the loss of his home state of Tennessee and the election.

To my mind, Gore's class-based appeals were counterproductive. As early as 1982, Mark and I had argued in the pages of the *New York Times* that in order to win back the support of working-class voters, Democrats needed to "abandon their appeal·to class resentment and economic fairness" and focus instead "on issues that cut across class and income lines." Populism didn't work then and was even less resonant now, after the Democratic boom of the mid-1990s. Yet instead of reacting with alarm to Gore's strange metamorphosis from New Democrat technocrat to circuit-riding populist, many Democratic activists thrilled to the promise of renewed government activism.

I, however, was not one of them. The polling data is absolutely clear on this point: most Americans do not want a large, activist government. Like it or dislike it, this is a bedrock belief among the American electorate. That President Clinton understood this and reacted accordingly was to his credit. His vision of a limited but still crucial role for government was one that a majority of Americans embraced. Unfortunately, many Democrats saw this as apostasy. As the 2000 presidential election race got under way, I realized with dismay that the vice president appeared to be among them. Instead of attempting to offset Republican advantages on say, values, by embracing something like faith-based initiatives, the Gore camp preferred to present Americans with a straight-up choice between a classically liberal Democratic platform and compassionate conservatism. While Gore showed no ability to address his weaknesses, Texas Governor George W. Bush did, with a platform that cleverly sought to neutralize Democratic advantages. It was excruciating to watch from the sidelines while Bush used a strategy we had pioneered against my own party's candidate.

The whole world knows how the 2000 election turned out. Losing was bad enough, but the way many Democrats reacted was even worse. For many, the election was not lost; it was stolen. By insisting that victory had been snatched from Gore's hands by the Supreme

Court, they failed to grasp the political advantages and acumen of their opponents. Indeed, Democrats' obsession with Florida masked troubling evidence of serious weaknesses: a drop in party identification, a lack of support in America's growing suburbs and exurbs, and a failure to express a more appealing political philosophy.

These vulnerabilities would become painfully obvious during Democrat John Kerry's run against President Bush in 2004. Bush could hardly have been more vulnerable. No weapons of mass destruction had been found in Iraq; Americans had not been greeted with open arms but with roadside bombs. The Republican-controlled federal government had gone on a huge spending binge, appeared initially to bungle a new prescription drug benefit for seniors and slashing taxes for the wealthiest Americans while consigning mountains of debt to future generations.

Yet sadly, the very magnitude of the GOP's failures contributed to Senator Kerry's ultimate undoing. President Bush, who as a candidate had promised to be a "compassionate conservative" who would conduct a "humble" foreign policy while working with Democrats to bridge "the partisan divide," had betrayed his promises so flagrantly and governed so recklessly that the case for change seemed to many Democrats to be self-evident. Unfortunately, this attitude was not one shared by a majority of Americans in 2004. In order to win, Democrats needed a compelling message. Instead, the Party went searching for a resume. Vietnam veteran and war hero, John Kerry received the nod for being the most "electable" and ultimately won the nomination.

This is not to say that Kerry was an entirely bad choice. Like any candidate, he had negatives: he alternated between indecisiveness and arrogance; his personal style was somewhat cold; his complicated position on the war allowed Republicans to portray him as a flip-flopper. However, these weaknesses were offset by considerable strengths. John Kerry is a smart, articulate person. Those strengths showed. He won three debates, raised an extraordinary amount of money, and received more votes than any Democratic candidate for president in American history. He won 54 percent of moderate voters and 15 percent of conservative voters—and still lost the election by three points.

It was a distressing loss on many levels. Kerry was a decorated war hero; Bush had been a National Guardsman with a history of unexplained absences. On paper, the Democratic agenda trumped the GOP agenda on almost every major issue. Of the one in five voters who said the economy and jobs was the most important issue, 80 percent voted for John Kerry. Similar ratios existed for those who said health care and education were most important to them. When bread-and-butter issues that affect the middle class are on the table, Democrats have a clear advantage. Yet in 2004, this advantage did not translate into results the way it had in the 1990s. Instead, the embattled middle-class started voting Republican.

Anne Kim and Jim Kessler of the Third Way project define "middle class" voters as those earning between thirty thousand dollars and seventy-five thousand dollars. In a startling and important study entitled "Unrequited Love: Middle Class Voters Reject Democrats at the Ballot Box" that came out in the summer of 2005, Kim and Kessler note that that Kerry lost this segment by six points. His record with middle-class white voters, who represent one-third of the electorate and three-fourths of the middle class, was even worse: he lost those voters by twenty-two points. That's considerably worse than Al Gore's mediocre showing in 2000, when he lost the middle-class vote by 2 percent and white middle-class voters by 15 percent. Here's how serious Kerry and the Democratic Party's problem with white voters is: once white voters start earning over $23,700—a figure just over the poverty line—they start voting Republican.

This represents a historic—and profoundly worrying—shift. In 1996, President Clinton had an eight-point advantage over Senator Bob Dole among voters who made between thirty thousand dollars and fifty thousand dollars. For Kerry, it was 50 percent to 49 percent—a statistical tie. President Clinton had enjoyed a slight advantage among earners in the fifty thousand dollar to seventy-five thousand dollar bracket (which is the largest financial grouping in the country at 23 percent). Kerry lost them by thirteen points.

The Democratic nominee even failed to win the support of many low-income voters. As Al From and Bruce Reed from the Democratic Leadership Council noted, "Of the twenty-eight states with the lowest per-capita incomes, Bush carried twenty-six. An administration whose overriding motive has been to protect the rich was just given a second term by the very people who will suffer the most for it."

There was worse: Democratic support among female voters also slipped. In 1996, female voters supported President Clinton by 54 percent to 38 percent. In 2004, John Kerry won among women by only three points. Among non-college-educated female voters, Bush increased his advantage by nine points.

Kerry's loss in 2004 was nothing less than a repudiation of the Democratic Party. Kerry won 85 percent of liberals and a majority of moderate voters (54 percent to Bush's 45 percent), but still lost the popular vote by more than three million votes. The problem is that only about 20 percent of voters identify themselves as liberals. In contrast, 35 percent say they're conservatives. Fundamentally, Bush beat Kerry because he won 84 percent of conservative voters, a group that makes up 35 percent of the electorate. Republicans also won large majorities among their two strongest groups: voters primarily concerned with national security, Iraq, and terror and voters most concerned about values issues. To win national elections, Democrats must win an even larger share of moderate voters and/or make at least some inroads with conservatives. That means abandoning its fixation on redistribution, moving to the middle, addressing values issues, and taking strong stances on national security issues.

Unfortunately, instead of responding with a realistic appraisal of what had gone wrong, many Democrats preferred to focus on tactics. House Democrats like Nancy Pelosi have embraced UC Berkeley linguist George Lakoff's argument (expressed most fully in his book, *Don't Think of an Elephant!: Know Your Values and Frame the Debate—The Essential Guide for Progressives*)[1] that Democrats were losing because Republicans were able to frame the political discourse in ways that hurt Democrats. According to this diagnosis, Republicans were winning because they used clever catchphrases like "tax relief," "the death tax," and so forth. When Republicans started to rail against trial lawyers and "frivolous law suits," Democrats needed to respond by describing them as "public protection attorneys." Taxes should be recast as "membership fees" or "investments" in America. Words are undoubtedly important—and

Lakoff writes savvily about the ways in which Republicans have shrewdly linked their preferred policies to common values—but as a fundamental critique of what is wrong with the Democratic Party, Lakoff's critique is laughable. Only a linguist would think that the Democratic Party's problems are fundamentally linguistic.

Yet variants of this argument abound. Democratic intellectuals like the author Thomas Frank claim, "the reason the Right has prevailed is its army of journalists and public intellectuals"—and that in order to win, Democratic intellectuals need more plush think tanks too. Others say that Democrats needed to be tougher and more ruthless, like Republicans are. This argument was little more than a meaningless tautology: not being "tough enough" becomes a catchall explanation for Democratic failure. In a display of truly marvelous historical myopia, some Democrats argue that the party is losing elections because it has lost touch with its populist roots. Rather than seeing Al Gore's populist, "the-people-versus-the-powerful" stance in 2000 as a mistake, this camp seems to believe that if only the Democratic Party was even more populist, then lower- and middle-class voters would return to the Democratic fold. During a stint last summer as a guest columnist for the *New York Times,* Frank wrote, "Everything I have written about in this space points to the same conclusion: Democratic leaders must learn to talk about class issues again."

It's understandable why Democrats like Frank would think this way. Over the course of the past three decades, the gap between rich Americans and everyone else has widened to an alarming degree. A recent study by economists Thomas Piketty and Emmanuel Saez found that since 1973, real income for the top 1 percent of earners have risen by 148 percent while incomes for the bottom 90 percent of Americans have declined by 7 percent.

This fact that a growing number of hard-working Americans are struggling to get by is a serious problem, one that is tailor-made for the Democratic Party to address. Our party's calling is (or at least should be) to help hard-working people who are under increasing economic stress. However, it should not be redistributing wealth to combat inequality per se. Nevertheless, the redistribution of wealth is precisely what many Democrats seem eager to do. Whatever its merits as policy, as a political strategy, this is seriously misguided. Only 26 percent of voters believe that government should pursue policies that redistribute wealth from the richest to the middle class and poorest. As the Third Way's Kim and Kessler have noted, "Voters who are solidly middle class had voting patterns nearly identical to wealthy voters, not low income voters, and thereby gave large middle class electoral margins to Republican candidates." Populism simply does not animate the average American voter. There are simply not enough liberals to elect a candidate who espouses such ideas. John Kerry did a great job of energizing the liberal base in 2004, winning 85 percent of liberals (versus, say, Carter's 72 percent in 1976)—and still lost. Yet among self-styled "progressive" Democrats what Elaine Kamarck and Bill Galston have dubbed "the myth of mobilization" lives on. Unfortunately, Democratic victories in the 2006 midterm elections have rekindled liberal hopes that Americans have started to "see through" Republican deception and are coming around to old-fashioned Democratic ideas. In fact, nothing could be further from the truth.

To avoid defeat in 2008, Democrats must face a hard truth: we do not have the natural majority coalition in American politics. When they're not completely bungling fundamental issues like the war in Iraq, Republicans can typically build an organic coalition of conservatives, whites, upper-income earners, Southerners, and evangelicals that has the potential to be almost as formidable as the old New Deal coalition. The long-term historical trends are clear: In 1978, Democrats held thirty-seven statehouses; by 2004, the number had dropped to twenty-two. Although Democrats regained a majority in 2006, it would be a mistake to interpret this as a fundamental realignment. In many ways, the gains Democrats experienced are akin to the Democratic surge immediately after Watergate in 1974—a short-term reaction to outrageous Republican misconduct. The elements for a GOP majority coalition are still there.

Despite heartening signs of health last year, the long-term story of the Democratic Party is still one of decline. In 1980, Democrats enjoyed a fifteen-point lead in party identification. Today, that edge has been reduced to a few percentage points. And while it may have bumped up somewhat more after the 2006 midterm election, it will most likely drop back toward parity, assuming (as is virtually certain) that the Republicans nominate a competitive candidate in 2008.

The decline of the Democratic Party has important implications for the way political campaigns are conducted. The fact that more than a third of voters self-identify as conservatives means that to a considerable extent Republicans can run campaigns directed at their base. Democrats have no such luxury: the percentage of liberal voters in the electorate is much smaller, about 20 percent.

Some analysts have argued that demographics will restore the Democratic Party's position. In their book *The Emerging Democratic Majority*,[2] John Judis and Ruy Teixeira argued that the rising proportion of Hispanic (Democratic) Americans and the comparative decline of white (Republican) Americans will propel the Democratic Party to victory—in a decade or so. In time, Judis and Teixera may well be proven correct; however, short-term demographic trends don't favor Democrats. A recent analysis by the Progressive Policy Institute (PPI) found that the voting age population in large, Republican-leaning counties (those where George Bush won more than 55 percent of the vote in 2004) grew 9.8 percent between 2000 and 2004 while the population of Democratic-leaning counties grew by just 2.2 percent. Ninety-seven of the one hundred fastest growing counties voted Republican. "All told," noted PPI's Ed Kilgore, "the Republican Party's base areas—especially suburbs, exurbs, and small towns—have a greater combined population and are growing faster than the Democratic Party's current base areas in cities and inner-suburbs."

The need to reach these voters should be obvious, yet there's a sense in the Democratic Party that any movement away from the Left somehow betrays the party's roots. Advocates of this viewpoint tend to think less about what's wrong with the Democratic Party and more about what's wrong with the American voter. Their line of thinking goes something like the following: "If only the electorate was more progressive, we would be a better country with a progressive/liberal/Democratic party in power doing more redistributing of wealth."

I completely reject that notion. To me the Democratic Party is at its core a broad-based, centrist party that attempts to promote the common good to a much greater degree than the business- and upper-class-inflected GOP does. Wishing for a mass conversion is not a political strategy. Neither is rallying the base. Quite simply, for Democrats to win elections they must come up with a compelling, broad-based vision that reaches beyond the party's core interest groups. It must be inclusive, it must bring a disparate group of voters together, and it must be moderate—even conservative—in tone. If Democrats can't win a majority of moderate voters and make significant inroads into the GOP's conservative base, they simply can't win a national election. In 1996, Clinton received 57 percent of moderate votes and 20 percent of the conservative votes. Had Kerry received the same level of support he almost certainly would have won.

Democrats' 2006 victories should not obscure this fact. This is a crisis of survival and viability. Unless the party learns the lessons of 2006 and continues to move back to the center, it will face serious electoral challenges going forward.

In order to win elections, Democrats must surmount three challenges. First, they must bridge the chasm with conservative voters in general and religious voters in particular. This is not simply a matter of using the right "code words," as President Bush so masterfully has. Rather, the Democratic Party must sincerely embrace mainstream views on values and religion—positions that unfortunately are anathema to a vocal minority of secular Democrats. Second, they must address their debilitating perceived weakness on the war on terror. Last, they must develop a domestic agenda that respects voters' core values—their beliefs' in equal opportunity, community, and responsibility.

Values matter. In the 2004 presidential elections, 22 percent cited it as their top issue, trumping even the war on terror. Needless to say, 80 percent of these voters opted for George W. Bush while only 18 percent voted for John Kerry. Values continue to be a formidable Republican advantage. In the 2006 midterm elections, "values/morality" was also the top issue among voters, second only to the war in Iraq. (Twenty-two percent of voters cited

Iraq as their top issue; fourteen percent cited "values/morality.") Sixty-four percent of these values voters cast their ballots for Republican candidates.

Things need not work this way. Candidates like John Tester in Montana, Jim Webb in Virginia, and Robert Casey (as well as Harold Ford in Tennessee, who ran a remarkably strong if ultimately unsuccessful campaign) showed how Democratic candidates can defang Republican attacks by taking divisive issues like gun control off the table. Adopting "Republican" stances on social issues is not the only way to win on values issues. Democrats can also win by trumpeting three basic values that our polling has consistently shown as unifying principles for the American people: opportunity (giving every American a chance to succeed), responsibility (the duties we owe to each other), and community (preserving and promoting families). The genius of Bill Clinton's approach was that he couched a wide array of positions and initiatives in values-based terms that connected with voters more deeply than the social issues used by Republicans. For instance, Clinton sought to expand the Head Start program for young children not because it would benefit the national economy but because giving all children the chance to get an education and go to college was the right thing to do. Environmental initiatives were embraced not on dry, scientific grounds but because leaving a better world to our children and grandchildren is the right thing to do.

Clinton also knew how to demonstrate his commitment to traditional social values. Take, for example, his deft handling of Hollywood, a key source of campaign funds for the Democratic Party but one that irritates many Americans. During the 1996 campaign, President Clinton came out in support of the V-Chip, a tool parents could use to keep objectionable content on television away from their children. President Clinton was as close to Hollywood as possibly any president in history, but with his support for the V-Chip, he was able to demonstrate that he understood the fears of families who were concerned about the programs that their children were seeing on television.

Conversely, during the 2004 election, John Kerry attended a fundraiser at Radio City Music Hall in which some performers, including Whoopi Goldberg, made off-color jokes about the president. It could have been a Sister Souljah moment: Kerry should have defended the dignity of the presidency while criticizing the performance of this particular president. Instead, he publicly embraced the raunchy spectacle, declaring, "Every single performer . . . conveyed to you the heart and soul of our country." Such comments are disastrous to Democrats seeking to establish a rapport with middle America. For Kerry, they made an already difficult race that much harder to win. The Bush campaign understood this instantly. To mock the Kerry campaign and demonstrate who really spoke to "the Heart and Soul of America," they adopted it as their slogan, printing up banners with the catchphrase and proudly displaying it at Bush campaign events.

In short, values are important—far more important than many Democrats realize. Yet instead of observing that values consistently rank as one of the most important issues to voters and responding accordingly, too many Democrats persist in believing that voters are delusional. Books like Thomas Frank's *What's the Matter with Kansas?*[3] illustrate the problem. Frank notes, correctly, that many middle-America voters (Kansans are his particular interest) consistently vote for Republicans and against their own economic self-interests. This is undoubtedly true, but what explains it? To Frank the answer is simple: the GOP has created an artificial cultural crisis that has tricked voters into supporting politicians who "may talk Christ but walk Republican." It's an enticing theory, but it's just not true. Republicans do play cultural issues more deftly than Democrats, but that doesn't mean that they've duped large numbers of voters into supporting them who would otherwise be Democrats. As Princeton political scientist Larry Bartels has noted, outside of the South, lower-income whites have not in fact moved over to the Republican column *en masse,* and lower-income voters are in fact no more concerned about values than they were thirty years ago.[4] The groups that have become more concerned about values are the middle- and upper-class and Southern whites. To win elections in places like Kansas (a state that, incidentally, has long been overwhelmingly Republican despite a Democratic breakthrough in 2006), Democrats need to address the moral concerns of these voters.

This does not mean that Democrats should abandon their principles on contentious social issues. Rather they should recognize that many hold divergent views, and understand that those views are worthy of respect and deference—even if they diverge from traditional Democratic party orthodoxy. Here again, President Clinton—and more recently, Hillary Rodham Clinton, can serve as useful models: both are unabashedly pro-choice yet both acknowledge the moral qualms surrounding abortion by saying it should be "safe, legal, and rare." Senator Clinton struck exactly the right note in a January 2005 speech when she declared, "I, for one, respect those who believe with all their hearts and conscience that there are no circumstances under which any abortion should ever be available." She then went on to warn of a situation where "government substitutes its opinion for an individual's," a formulation that resonates strongly with pro-choice voters but that pro-lifers who are suspicious of government meddling can also understand.

Often going hand in hand with the concept of values voting is the role that religion has come to play. Kerry's failure to win religious voters was a major contributor to his defeat. In 1996, President Clinton won the Catholic vote by a sixteen-point margin. John Kerry, a practicing Catholic, lost the same voting bloc by five points. Among regular churchgoers, 64 percent supported President Bush while only 36 percent voted for John Kerry.

The growing identification of religious belief with Republicanism is a disaster in the making for the Democratic Party. Already Republicans hold an extraordinary advantage over Democrats among religious voters. According to a 2005 Pew Forum poll on Religion and Public Life, by a wide margin—51 percent to 28 percent—the Republican Party is seen as the party that is "most concerned with protecting religious voters." While 55 percent of the electorate sees the GOP as friendly to people of faith, only 29 percent of voters feel the same about Democrats, a figure that is down an alarming thirteen points from only two years earlier. Among white evangelicals and conservative Republicans, the numbers are even worse: nine out of ten see liberals as having "gone too far" in imposing their nonreligious views. Even many Democrats agree: more than half of all

Democrats believe liberals have pushed too far in trying to keep religion out of public schools and government.

During the 2006 elections, Democratic candidates in Ohio and Pennsylvania showed that Democrats could make significant inroads among religious voters. However, on the whole, the Republican Party held on to its striking advantage with one-quarter of the electorate who consider themselves born-again or evangelical voters, winning 70 percent of those voters, essentially the same ratio they won two years earlier. Winning a larger percentage of this vote should be a major Democratic priority. Some Democrats insist that the Republican identification with the Religious Right alienates as many voters as it attracts. This is just not the case: some voters are repelled by the Religious Right, but an equal number believe that Democrats are too beholden to "nonreligious liberals." Democrats need to embrace mainstream values and recognize that religion should play a larger role in public life, not a smaller one. Here again, Democrats would do well to consider the lessons of the 2006 midterm elections: both Jim Webb in Virginia and Claire McCaskill in Missouri won 33 percent of the evangelical voters, a number that proved critical to their victories. Democrats need to go after these voters at every opportunity. A good place to start would be with faith-based initiatives. More than two-thirds of voters believe that churches, synagogues, and mosques should be able to apply for government funding to provide social services. More than 60 percent of Americans believe the government should be involved in supporting traditional institutions like the church, the family, and marriage. Democrats need to demonstrate that they understand and accept this. That means, for instance, that it would be a serious mistake for Democrats to push for gay marriage (as opposed to civil unions and civil rights for gays, which a majority of voters now support). To do otherwise undermines Democratic chances for victory, as the 2004 election cycle demonstrates. Thirteen states voted on same-sex marriage initiatives that year, and, while the details varied from state to state, the conservative position prevailed in every instance. Gay Americans weren't the only people to lose out. So did Democratic presidential nominee John Kerry. By driving conservatives to the polls, Republicans gave President Bush a critical boost in battle-ground states like Ohio.

No doubt some liberals will object to my proposal that Democrats need to come to terms with majority beliefs about the role of religion. However, Democrats who care about social policy need to realize that institutions of faith are not a distraction but rather only venues for social action that a clear majority of Americans support. This is an issue that Democrats still have time to claim. While President Bush made faith-based programs a focal point of his 2000 campaign, the lack of follow-through during his presidency suggests that his interest was more in the political saliency of the message then in actual accomplishments.

The current governor of Virginia, Tim Kaine, provides an interesting model for how Democrats can use religion to turn the tables in contentious moral debates. Kaine's campaign was orchestrated by Penn, Schoen & Berland alumnus Peter Brodnitz, who worked closely with both Mark and myself and in fact was engaged to handle the campaign while still with our firm. Kaine won in 2005 not by running from religion but instead by making it the focal point of his campaign. A devout Catholic and former missionary, Kaine frequently talked about his religious values, even running a television ad in which he said, "The Bible teaches we can accomplish great things when we work together." When attacked by his opponent for his opposition to capital punishment, he responded with an ad that explained his opposition in religious terms. At the same time, he also made clear that as governor he would abide by the law and enforce it. Virginians may have disagreed with him, but they were willing to give him a pass as long as they thought his views were rooted in religious faith. By taking his most obvious vulnerability off the table, Kaine was then able to focus on the issues that best favored Democrats, such as education, fiscal leadership, and in this case, Virginia's strong economy under the Democratic incumbent, Mark Warner. It was a pitch perfect example of inoculation.

None of these prescriptions should be new or unfamiliar to Democrats. John Kennedy spoke candidly about his religious views (because of—and despite—scurrilous attacks on his Catholic faith).

Jimmy Carter was, famously, a born-again Christian. Bill Clinton is likewise a deeply religious man who felt comfortable using the language of faith, particularly during his fight to prevent impeachment where he openly spoke of having "sinned" in his relationship with Monica Lewinsky. Polls show that religious voters are increasingly open to Democratic overtures to people of faith. By mid-2006, only 54 percent of evangelicals approved of President Bush's job performance; among regular churchgoers, a majority now disapprove of his performance.

This opportunity for Democrats is real. Unfortunately, the skills and understanding needed to win them over is still lacking. In 1996, President Bill Clinton ran ads on Christian radio. Today, Fox News is the country's most popular cable news outlet yet they often have difficulty attracting Democratic politicians as guests. Democrats on venues like conservative talk radio are even rarer. This is shortsighted in the extreme. In an atmosphere where the word "liberal" is often uttered in the same tone of voice as "terrorist," Democrats win simply by showing up and appearing to be moderately reasonable. It also wouldn't hurt for Democrats to have more firsthand exposure to their opponents so they can learn just how smart the other side is. The fact that today's Democratic Party is better at catering to the needs of obscure groups of like-minded activists than to evangelicals has probably cost Democrats the past two presidential elections.

While religion plays a crucial role in Democrats' ability to attract new voters to their tent, few issues bedevil them as much as national security. During the darkest days of the Vietnam War, in 1972, Richard Nixon was able to thrash George McGovern *because of* McGovern's dovish stance on the war. Then came the presidency of Jimmy Carter, the Iranian hostage crisis, the identification of Ronald Reagan, the GOP, and patriotism, and Michael Dukakis on the tank. This series of Republican triumphs and Democratic mishaps created the impression that Democrats were not tough enough to confront external threats. Then came the miracle that transformed the world and remade the American political landscape—the end of the Cold War and the collapse of the Soviet Union.

These events made it possible for a charismatic young governor from Arkansas to defeat an experienced foreign policy hand and war hero. Once in office, Bill Clinton began the slow and difficult process of demonstrating that Democrats could safeguard America's interest. To be sure, the Clinton administration had its challenges—the "Black Hawk down" incident in Somalia, terrorist attacks on the Khobar towers, the *U.S.S. Cole,* and the genocide in Rwanda. However, it also had successes. NATO went to war for the first time in the alliance's history to stop Serbian genocide in Kosovo; Slobodan Milošević was (belatedly) toppled; and most important of all, Clinton preserved peace in the world. Then came the attacks of 9/11 and the transformation of George W. Bush from hapless minority to president to vigorous wartime leader.

Of course, recent events have badly dented that image. However, President Bush's blunders in Iraq should not obscure the fact that the war on terror is real and that the American public wants a tough, resolute president to lead it. If not addressed effectively, the perception of Democratic weakness on national security threatens to return the party to the 1980s, when "weak" Democratic presidential candidates lost election after election to "strong" Republicans. Staking out strong and clear positions on national security should be at the top of the Democratic political agenda.

Redefining the party's stance on these issues is by no means an unmanageable task. There was once a time when Democrats were trusted by the electorate as the stronger of the two parties on the use of force. In 1948, Harry Truman established the modern Democratic Party as a party of Cold War, anti-Communist hawks. Later, in 1960, John F. Kennedy "out-hawked" Richard Nixon on the issue of the supposed "missile gap" between the United States and the Soviet Union and the defense of two long forgotten islands off the coast of China, Quemoy and Mattsu. Today, the American electorate views Michael Moore, MoveOn.org, and Howard Dean as the faces of the Democratic Party on national security issues. That must change. Until it does, national security will continue to be one of the GOP's single most powerful rhetorical weapons.

The Democratic Party must find a way to show Americans that they are serious about national security, that they were willing to use unilateral force, and that they are not beholden to the United Nations or our international allies when it comes to defending American's national interests. Democrats need a strong plan and cannot win a national election if they are unable to neutralize this issue.

This does not mean that Democrats need to support costly misadventures like the Bush administration's war in Iraq. The party's presidential nominee should not hesitate to embrace a timetable for withdrawing U.S. forces from the region. Such a position is probably necessary to win the Democratic presidential nomination; however, it's also a winner with the public at large. The American public does not want to see a precipitous withdrawal from Iraq; a majority now clearly supports a timetable for a phased withdrawal. Democrats need to provide such a plan. A majority also believes that the war has wasted American lives and resources and increased the risk of terrorism, without advancing U.S. interests. Democrats should not hesitate to pound away on these points. The war began under false pretences, was prosecuted under incorrect assumptions about the number of troops needed to secure the peace, and was horribly botched in its execution during the postinvasion occupation. Democrats should say so.

That being said, the Democrats should not adopt anything that can be interpreted as a "cut-and-run" strategy for both substantive and political reasons. This means the party cannot simply move to precipitously cut off funding for the war as a means of demonstrating the failure of the Bush policies. That would be irresponsible, and the American people would rightly judge Democrats harshly for such an action. Instead, Democrats need a plan for Iraq that sets clear goals and objectives for withdrawal and seeks to work more constructively with other nations both inside and outside of the region. In short, voters need to sense that they have a strategy to at least try to stabilize Iraq and which also ties into larger foreign policy goals like winning the war on terror.

These foreign policy mishaps have lessened the impact of the Republican national security advantage, which should allow Democrats to exploit their traditional edge on domestic social issues. As was the case in the 1980s, Americans prefer Democratic positions on health care, education, the environment, and economic stewardship to Republican positions. On paper, the domestic Democratic agenda generally trumps its GOP counterpart. In practice, however, negative perceptions of Democrats on issues like the role of government, cultural issues, and national security all too often outweigh the positives while the public's positive perceptions of Republican stands on national security and cultural values largely outweigh the party's negatives on bread-and-butter issues. Democrats win when they neutralize Republican advantages—as Clinton did in 1996 by cracking down on crime and embracing welfare reform, family-friendly policies, and balanced budgets—and fight their political battles on friendlier turf.

George W. Bush's fecklessness has given Democrats control of Congress. However, House Speaker Nancy Pelosi should not overestimate her strength. In many ways, her position is strikingly similar to that of former House Speaker Newt Gingrich after the Republican "revolution" of 1994. Both delivered victory to their parties; both are beloved of their activist wing; both have a political profile that is at odds with that of the majority of the country. Gingrich governed as a Far Right Republican all the way to his early resignation. If Pelosi governs as a liberal, she will be following suit. However, if Pelosi instead seizes the center and stays there, she has the rare opportunity to change the Democratic Party's image in the mind of the public. Pelosi needs to anchor herself to the political center. One way to do that is to publicly and unwaveringly embrace a commitment to balanced budgets and fiscal responsibility.

Even Republicans recognize that Democrats have an opening on this issue. In his recent book, *Impostor: How George W. Bush Bankrupted America and Betrayed the Reagan Legacy,*[5] conservative economist and former Reagan and George H. W. Bush adviser Bruce Bartlett rebukes George W. Bush for recklessly expanding federal spending and the scope of government. Bartlett, a dyed-in-the-wool conservative, contrasts the GOP's record unfavorably with, of all people, Bill Clinton, whom he praises for his adherence to fiscal conservatism. Clinton was not alone. Since the days of John F. Kennedy, Democratic presidents have been far more responsible stewards of the federal budget then Republican presidents. Even the free-spending Lyndon B. Johnson insisted on balancing the budget. If Democrats in Congress can make a clear break with President Bush and the GOP's profligate spending, they will go a long way toward persuading the general public to recognize this fact as well.

Democrats should also recognize the danger that corruption poses to the party in power. In 2006, the scandals surrounding Republican lobbyist Jack Abramoff and former Congressman Mark Foley played an important role in tipping key races toward Democrats. Voters will expect a reform agenda that limits lobbying and tightens rules on campaign donations by special interests. Democrats need to deliver it.

Though Bush's blunders have laid the groundwork for the Democratic presidential campaign in 2008, his failures alone will not win back the White House for Democrats. Democrats should resist the temptation to challenge President Bush at every turn. Instead, Democrats need to swallow hard, move past the years of venomous Republican attacks, and actively look for opportunities to work with President Bush (who, like President Clinton after 1994, will in turn be under pressure to work with the new majority). The polling data is absolutely clear on this point: eighty-five percent of voters say it is critically important that both parties be bipartisan in moving forward. Americans are ready to move past partisanship to an era of cooperation. Some 68 percent want the two political parties to work with corporate America and organized labor to formulate policy. Any party that disregards this strong and clear desire will suffer. Democrats in Congress need to propound a sweeping, unifying vision of America that provides a big tent for Americans of all stripes to gather under. Domestically it means offering the middle class in America a bargain: if they play by the rules, their government will be on their side. Every Democratic policy proposal should reflect the bedrock values of opportunity and responsibility. Internationally, the party needs to be clear that it will aggressively protect Americans, promote American values, and hunt down and kill those who would threaten us.

There is no shortage of innovative Democratic ideas. Former Clinton aide Matt Miller has argued that with just two cents of the national dollar, the United States could provide every American with access to health care, ensure that every full-time worker received the minimum wage, and put good teachers in fixed-up schools. Senator Hillary Rodham Clinton has championed five hundred dollar "baby bonds," to be issued for every child at birth and then again at age ten, funds that could be for college or vocational training, buying a first home, even retirement. Gene Sperling has presented well-thought out proposals to cushion the impact of globalism and create a more dynamic workforce. Rahm Emanuel and Bruce Reed's book, *The Plan,*[6] proposes sensible ways to reduce health insurance premiums and simplify the tax code in a way that encourages retirement and college savings while reducing taxes for people earning less than a hundred thousand dollars a year; it also strikes a smart stance on national security issues, calling for an additional hundred thousand soldiers in the U.S. military, and a domestic antiterrorism agency for the United States modeled on Great Britain's highly effective MI5.

These ideas are worthy of serious consideration. However, from the viewpoint of a political strategist, it is not the individual policies that are most important; it is the ability to understand and connect with the electorate. In this era of constant polling and omnipresent focus groups, this might seem like an easy task. In fact, I believe that most of my colleagues are missing something big. Recent polling has convinced me that there is a huge vacuum in American politics today, one that centers on a growing crisis that neither party is addressing. I call it the *crisis of affordability*. It might also be termed *a crisis of the American Dream*.

Looking at this data closely a common theme stands out—affordability. For many Americans, health care has become unaffordable; a college education is increasingly out of reach; pensions for retirement are vanishing; utilities, child care, housing, and gasoline have become much more expensive. Ninety percent of survey respondents say that retirement is less secure than ever before. Eighty percent believe that a middle-class life has become unaffordable for most people. Eighty-four percent think that even if you keep up your skill set, you may still lose your job to outsourcing. In short, most Americans are profoundly anxious and pessimistic about their economic futures.

Such sentiments may come as a surprise to many politicians, pundits, journalists, and, yes, pollsters—high earners all who have benefited from the economy's relatively strong performance through most of the presidency of George W. Bush. As conservatives never tire of noting, gross domestic product grew smartly throughout most of the Bush presidency, rising by 2.5 percent in 2003, 3.9 percent in 2004, and 3.2 percent in 2005. Yet Americans have consistently faulted the president for his handling of the economy. Frustrated Republicans have responded by insisting that, in President Bush's words, "Things are good for American workers." In fact, the public's dissatisfaction and anxiety reflects very real economic trends. The benefits of the Bush boom have gone almost entirely to corporations and the wealthy: as Steven Greenhouse and David Leonhardt of the *New York Times* have noted, the economic expansion of the Bush years was the first since World War I not to lift real wages.

"[W]ages and salaries now make up the lowest share of the nation's gross domestic product since the government began recording the data in 1947, while corporate profits have climbed to their highest share since the 1960s," note Greenhouse and Leonhard.[7] In fact, wages have declined since 2003.

Not surprisingly, Americans are stressed. Eighty-one percent of Americans still believe that this is the land of opportunity but faced with the rising cost of health care and education, as well as stagnating wages and increased pension costs, deep-seated doubts are developing. This is particularly true of the middle class. Only 49 percent of voters now believe that if you work hard and play by the rules, you can lead a solid middle-class life. Sixty-one percent of Americans say they are not living the American dream; an equal number believe they never will. The old rules by which people lived have gone by the wayside: the electorate no longer has any sense of security about their jobs, their health care, or their pensions. No wonder that nine out of ten people say it is harder to achieve the American Dream than ever before.

Currently, neither party is addressing this profound anxiety. Republicans focus on tax credits and the estate tax, issues that are hardly central to most Americans. Democrats talk about the minimum wage and higher taxes for the wealthy, another set of ideas that are not central to the way most people live and work. Democrats need to aggressively address the concerns of these voters with a platform that emphasizes security—health security, retirement security, job security, security against terrorists, and security against Internet pornography and sexual predators.

(If you're scoffing at this last prescription, you're probably part of the Democratic Party's problem. According to the Third Way, there are 420 million individual pornographic web pages today, and children are its biggest consumers. A 25 percent "smut tax" on the $12 billion porn industry is one tax Democrats should support, with proceeds used to combat sexual predators and unscrupulous website operators who imbed children's words like Teletubbies and Pokemon into websites as "meta-tags" to drive kids to their sites.)

It is critically important that they address these problems with optimistic solutions. Even if voters themselves are pessimistic, pessimism is not a trait Americans like in their political leaders. Voters will respond positively to upbeat, intelligent proposals like Rahm Emanuel and Bruce Reed's ideas for universal college savings, universal retirement plans, and a simpler tax code (with three brackets rather than six), and a 10 percent tax rate for those making less than a hundred thousand dollars a year—as long as Democrats can show that these proposals will be enacted in a fiscally responsible way that won't impose substantially higher taxes on them. They will not respond positively to attacks on the wealthy or calls for a greater distribution of wealth. It's a point that bears repeating: only 26 percent of voters believe that government should pursue policies that redistribute wealth from the richest to the middle class and poorest. This is true among even the poorest Americans: only 39 percent of people earning less than twenty thousand dollars a year want the government to pursue redistributionist policies. In contrast, 74 percent say government should reject redistribution and instead pursue policies that grow the economy for everyone. Moreover, Americans across the political spectrum want a new style of politics that emphasizes consensus and cooperation and brings the American people together.

In crafting their platform for 2008, Democrats need to bear in mind some simple statistics. Liberals today make up 20 percent of the electorate; moderates account for 35 percent; and conservatives make up another 35 percent. In order to win national elections, Democrats have to find a way to attract a third of conservative voters, 60 percent of moderates, and virtually all liberals. The only way to do that is to develop a broad-based, centrist philosophy that promotes security while avoiding class warfare, is fiscally prudent, and reflects core American values. If Democrats can do that, they have the chance to regain the political dominance that seemed within reach during President Clinton's second term. If they don't—and if the GOP continues to self-destruct—American politics could change in fundamental ways.

I s the Democratic Party likely to change? I hesitate to offer a prediction: there's no tested methodology for prophecy. Nevertheless, there are some hopeful signs. In selecting candidates for the 2006 midterm elections, the national Democratic Party acted on many of the ideas that we have long championed. Instead of shunning pro-life Democrats like Robert Casey of Pennsylvania, the party invited them back into the tent—and won Pennsylvania as a result. Jim Webb of Virginia was a former political appointee of President Ronald Reagan. John Tester of Montana is an ardent supporter of tax cuts and gun rights. Claire McCaskill of Missouri is an unrepentant centrist who worked hard to reach rural voters. High profile Senators like Evan Bayh of Indiana and Hillary Rodham Clinton of New York have sought to reach across party lines and have championed the kind of mainstream ideas (such as better protecting this country's borders) that Democrats once ignored.

These centrist, occasionally even conservative, figures followed the same playbook Mark and I used with President Clinton: they neutralized Republican strengths on contentious social issues and that (along with the backlash to President Bush, of course) allowed them to win on Democratic strengths. Their success vindicates the positions we have advocated for most of the past two decades. It also clearly strengthens centrist potential presidential candidates like Republican Senator John McCain of Arizona and Senator Hillary Rodham Clinton of New York, who it should be noted won 20 percent of Republican voters on her way to a huge 67 percent win in New York.

Both McCain and Clinton will face challenges should they choose to run—McCain from either former Mayor Rudy Guiliani or a right-wing candidate buoyed by extremist Republican primary voters, and Hillary from Senator Barack Obama or John Edwards (who has a strong base in Iowa and South Carolina). Nonetheless, McCain and Clinton are still the clear front-runners and the likely nominees of their respective parties at this point.

There is a deep hunger for nonpartisan, cooperative leadership in this country. On contentious issues, such as health care, our research shows that voters want their elected officials to reach across the aisle and work with members of the other party. They also want their elected officials to work with—not fight—industries like the pharmaceutical industry. My own experience working with New York Mayor Mike Bloomberg—a former Democratic-turned-Republican whose attitude is perhaps best described as one of pragmatic centrism—has shown me just how liberating it can be to operate with an eye toward competence and results instead of encrusted political constituencies and special interests.

Americans desperately want this kind of leadership. If the two political parties don't deliver it—if, for instance, the dominant parties were to nominate right- or left-wing ideologues or otherwise fail to cooperate in a bipartisan fashion to advance the common good—American politics might well change in unpredictable ways. The desire for an alternative to the two-party status quo is clearly there.

As for me, I'm proud of what I've accomplished in politics. We as a country are better off for electing politicians like Ed Koch, Jay Rockefeller, Evan Bayh, Bill Clinton, Jon Corzine, and Mike Bloomberg—and for defeating the likes of Slobodan Milošević. I feel passionately that the business I helped create and refine is, on the whole, a force for good. It has made politicians more responsive to the very real needs and desires of the electorate, while forcing electoral systems to be more transparent. In short, it is a profession I am proud to have been associated with.

Although I clearly have strong opinions on the 2008 Presidential election, I plan to observe this election. I'll still be analyzing the polls and talking strategy, but this time as a commentator for Fox News. My goal for the next election cycle is to move beyond facile "horse-race" political analysis to a more nuanced understanding of politics, one that demystifies and explains what political consultants without devaluing the difficult and often courageous things that politicians both here and abroad continue to do. It's the same goal that motivated this book. At a time when few people put much trust in politics, it may be a long shot—but it's the kind of long shot I like. I've won against longer odds before.

NOTES

Chapter 1

1. Clayton Knowles, "Recount Due in Assembly Race Won Uptown by 6-Vote Margin," *New York Times*, November 16, 1970.

2. Thomas Meehan, "The Carter Burden Question: Can a Rich, Handsome Young Member of the Jet Set from Dry Dock County Find Happiness as a Reform Machine Poll," *New York Times Magazine*, November 7, 1971.

3. "Primary Choices," *New York Times*, September 9, 1971.

4. See note 2.

Chapter 2

1. Kevin P. Phillips, *The Emerging Republican Majority* (New Rochelle, NY: Arlington House, 1969).

2. Richard M. Scammon and Ben J. Wattenberg, *The Real Majority* (New York, Coward-McCann, 1970).

3. Arthur Browne, Dan Collins, and Michael Goodwin, *I, Koch: A Decidedly Unauthorized Biography of the Mayor of New York City, Edward I. Koch* (New York: Dodd, Mead, & Company, 1985).

4. Ultimately, a compromise was found between the neighborhood and the Lindsay administration: the towers were okayed, but at half the original size. The mediator of the conflict was a young

Italian American lawyer from Queens named Mario Cuomo. Cuomo's role in settling the Forest Hills fiasco positioned him as hero of liberal Democrats and a champion of middle-class ethnic white voters, marking the beginning of Cuomo's political career.

5. Nicholas Lehman, "The Other Underclass," *Atlantic Monthly*, December 1991.

Chapter 3

1. Arthur Browne, Dan Collins, and Michael Goodwin, *I, Koch: A Decidedly Unauthorized Biography of the Mayor of New York City, Edward I. Koch* (New York: Dodd, Mead & Company, 1985).

2. Jack Newfield and Wayne Barrett, *City for Sale: Ed Koch and the Betrayal of New York* (New York: Harper and Row, 1977), 126.

3. See note 2, 129–130.

4. Adam Clymer, *Edward M. Kennedy: A Biography* (New York: HarperCollins, 1999), 286–287.

Chapter 4

1. David Shipler, "Autonomy Plan for Palestinians Is Fading Away," *New York Times*, March 8, 1981, A8.

2. David Shipler, "Begin, Invigorated by Campaign, Has the Look of a Front-Runner," *New York Times*, May 9, 1981, A6.

3. Howard Rae Penniman and Daniel Judah Elazar, *Israel at the Polls, 1981: A study of Knesset Elections, Washington: American Enterprise Institute* (Bloomington: Indiana University Press, 1986), 89.

4. William E. Farrell, "Begin's New Coalition Wins Vote in Parliament in a Bare Majority," *New York Times*, August 6, 1981.

5. "Begin Attacks Schmidt on Role Under Hitler," *New York Times*, May 7, 1981.

6. David Shipler, "Israelis Worry Over What Some View as a Tendency to 'Growth of Fascism,'" *New York Times,* June 25, 1981.

7. "The Heavy Voters Are Big Winners in Israeli Vote," *New York Times,* July 5, 1981.

8. Serge Schemann, "The Trials of a Peace Seeker: A Special Report; Terrorism Forces Peres From Brink of Triumph," *New York Times,* March 10, 1996, A1.

9. Gitelman and Goldstein, "The Russian Revolution in Israeli Politics," in *The Elections in Israel, 1999,* ed. Alan Arian and Michal Shamir (Albany: State University of New York Press, 2002), 143.

10. Deborah Sontag, "Israeli Race for Prime Minister Is Down to the Wire, with Little Cheering," *New York Times,* May 12, 1999.

11. Mendilow, "The Likud's Campaign and the Headwaters of Defeat," in *Israel at the Polls, 1999,* ed. Daniel Judah Elazar and M. Benjamin Mollov (Portland, OR: F. Cass, 2001), 216.

12. Deborah Sontag, "The Israeli Vote: The Overview; Israelis Choose a New Leader and Remake their Parliament," *New York Times,* May 18, 1999.

Chapter 5

1. *After Milošević: A Practical Agenda for Lasting Balkans Peace* (Brussels, Belgium: International Crisis Group, 2001), 3–4.

2. The conflict in Yugoslavia first flared under the watch of President George H. W. Bush; however, his administration resolutely ignored the conflict and the killing, declaring in the infamous words of then-Secretary of State James Baker, "We don't have a dog in this fight." Soon after President Clinton took office, he reportedly read Robert Kaplan's book *Balkan Ghosts*—a book that described the Balkans as a hotbed of ancient ethnic hatred—and decided against intervening.

3. Stephen Kinzer, "Yugoslav-American in Belgrade Leads Serbs Who Won't Follow," *New York Times*, August 24, 1992, A1.

4. Interviews with David Calef, Milan Panic's press secretary, March 3, 2003, and June 11, 2003.

5. The Yugoslav economy was experiencing hyper-inflation, but rather than address the root causes of the problem, the government attempted to force residents and visitors to exchange their currencies for the dinar at an artificially low rate. The result was a huge black market, used by citizens and visitors alike.

6. The difference between our approach in Serbia and more recent U.S. efforts in Iraq is striking. Four years after the invasion of Iraq, Americans still did not understand the most basic questions about the virulent insurgency or the partisan militias there. (See James Risen, "The Struggle for Iraq; Sunni and Shia Insurgents Remain Mystery to U.S., Iraq Report Charges," *New York Times*, December 11, 2006.) An intensive effort to understand the varieties of Iraqi public opinion, using the full arsenal of anthropological and public opinion research techniques, would almost certainly have resulted in far more effective U.S. policy.

Chapter 6

1. In South Korea, the president is limited to a single five-year term.

2. "Ruling Party Candidate in Fight Back," *Financial Times*, November 27, 1997, 6.

3. Philip Gourevitch, "Alone in the Dark: Kim Jong Il Plays a Canny Game with South Korea and the U.S.," *New Yorker*, September 8, 2003.

4. Yonhap news agency, Seoul, "Cyber Fan Club Takes Credit for Success of South Korean Presidential Nominee," May 2, 2002 (supplied by BBC Worldwide Monitoring).

5. "Kwang-Chul Denies 'Unworthy' Hyundai/Hynix Stock Rigging Claims," AFX News Limited Wire, October 28, 2002.

Chapter 7

1. Amy Shapiro, *Millicent Fenwick: Her Way* (New Brunswick: Rutgers University Press, 2003), 2002.

2. Albert R. Hunt, "Millicent Fenwick Is Rich, Feisty, Candid and Smokes a Pipe," *Wall Street Journal*, May 21, 1982.

3. Mark Penn and Doug Schoen, "Reagan's Revolution Ended?" *New York Times*, November 9, 1986.

4. Dennis Farney, "Indiana Race for Governor Finds GOP Stumped by Democrat with a Fresh Face, Legendary Name," *Wall Street Journal*, October 24, 1988.

Chapter 8

1. "On Target: Bill Clinton and the Radical Republicans under Newt Gingrich; Victory March: The Inside Story Inside," *Newsweek*, November 18, 1996.

2. Adam Clymer, "Whether Friend or Foe, Most Think Clinton Is Playing Politics on the Budget," *New York Times*, June 16, 1995.

3. During the 2004 election, I had dinner with George Soros, who was organizing a media campaign against President Bush under the banner of America Coming Together (ACT). I told him ACT's advertisements were by no means as effective as they could have been. Soros said he was no political expert and had in fact delegated design and creation of those advertisements to the man who had, in his words, "created and initiated" Clinton's brilliant early-1996 advertising campaign—Harold Ickes. Needless to say, this was news to me.

4. See Ron Fournier, *Douglas B. Sosnik and Matthew J. Dowd, Applebee's America: How Successful Political, Business, and Religious Leaders Connect with the New American Community* (New York: Simon & Schuster, 2006) for a detailed discussion of the campaign's approach to swing voters.

5. Democrats also needed to address the so-called values gap. The issue of tobacco provides an interesting insight into just how effective such issues can be. In the summer and fall of 1995, I was doing polling for Paul Patton, who was running for governor in Kentucky. The tobacco issues had taken on new prominence after a decision by the Food and Drug Administration to declare nicotine addictive. Patton was afraid the issue would sink his campaign, but my polling consistently showed that even in tobacco states, as long as you are talking about tobacco in terms of protecting children from cigarettes, you would have strong voter support even from those who smoked or had ties to the tobacco industry. Patton won, and my research contributed to the president's decision to focus on curtailing cigarette advertising aimed at children.

Chapter 9

1. How do you poll the appeal of a hypothetical candidate? Basically, by presenting voters with a description of the candidate, typically a brief biography and a list of his positions, and then asking them to compare him with his likely opponent, in this case Jim Florio. In order to gauge how a campaign will play out, pollsters typically lead respondents through a series of statements, attacks, and rebuttals, measuring how additional information affects voters' perceptions of the candidate at each stage of the process. This process is as much an art as a science, and in this case, it rested on the assumption that our candidate would in fact be able to define himself on his own terms.

2. James Dao, "A Vast Fortune, Liberal Views, a Political Dream," *New York Times,* September 3, 1999, B1.

Chapter 10

1. Bloomberg spent $74 million during his 2001 campaign and another $85 million during his reelection campaign.

Chapter 11

1. See Richard Brand, "Forget Dubai: Worry about Smartmatic Instead," *Miami Herald,* March 27, 2006; and Gary Walsh, "Alderman Sees a Plot in Voting Machines: Burke Connects Dots to Venezuela Leader," *Chicago Tribune,* April 8, 2006, for details.

2. "Hugo, Jimmy, and Colin," *Wall Street Journal,* August 26, 2004.

3. Michael Barone, "Exit Polls in Venezuela," *USNews and World Report,* August 20, 2004.

4. "Conned in Caracas," *Wall Street Journal,* September 9, 2004.

5. Richard Brand, "Forget Dubai: Worry about Smartmatic Instead," *Miami Herald,* March 27, 2006.

6. Gary Washburn, "Alderman Sees a Plot in Voting Machines: Burke Connects Dots to Venezuela Leader," *Chicago Tribune,* April 8, 2006.

7. In the final months of 2003, Pinchuk set aside what arguably would have been in his own personal self-interest and helped persuade his father-in-law not to run for a third term.

8. Interestingly, in the 2006 Venezuela election, there were four or five exit polls, including one done by Penn, Schoen and Berland, which consistently showed President Hugo Chavez leading his challenger, Zulia State Governor Manuel Rosales, by a wide margin, which we estimated at about 16 percent. Whatever the precise margin—the Central Election Commission reported a larger gap—it is clear that a large part of that lead reflected the considerable advantage Chavez enjoyed as an autocratic incumbent. His monopolization of the state-run media, his use of state resources to subsidize the life of the poor and the middle class, the ongoing intimidation of his political opponents—all of these things made organizing a successful campaign in the slightly more than three months Rosales had extremely difficult. These advantages may also explain why in the final days the 10 percent of the electorate that

had been undecided appeared to break for Chavez, providing his larger-than-expected margin of victory. It remains to be seen if fraud played any direct role in his victory.

Chapter 12

1. George Lakoff, *Don't Think of an Elephant!: Know Your Values and Frame the Debate: The Essential Guide for Progressives* (White River Junction, VT: Chelsea Green Publishing, 2004).

2. John B. Judis and Ruy Teixeira, *The Emerging Democratic Majority* (New York: Scribner, 2002).

3. Thomas Frank, *What's the Matter with Kansas? How Conservatives Won the Heart of America* (New York: Metropolitan Books, 2004).

4. Larry Bartels, "What's the Matter with *What's the Matter with Kansas?*" *Quarterly Journal of Political Science,* 2006, vol. 1, 201–226.

5. Bruce Bartlett, *Impostor: How George W. Bush Bankrupted America and Betrayed the Reagan Legacy* (New York: Doubleday, 2006).

6. Rahm Emanuel and Bruce Reed, *The Plan: Big Ideas for America* (New York: Public Affairs, 2006).

7. Steven Greenhouse and David Leonhardt, "Real Wages Fail to Match Rise in Productivity," *New York Times,* August 28, 2006.

ACKNOWLEDGMENTS

I t had to be the headiest day of my life.

I was on a receiving line in late October 2006 in New York's Beacon Theatre waiting to be introduced to the Rolling Stones by former President Bill Clinton. When my turn came, Clinton, in a fit of exuberance, gave me a big hug and told Mick Jagger and Keith Richards that I was the "genius" who had gotten him reelected.

And while that was certainly a very generous and obviously over-stated characterization of my role, it did make me feel incredibly good when Mick Jagger and Keith Richards, in deep bass voices both re-peated "You the Man" and high-fived me, as we posed for a group picture before a fund-raising event for President Clinton's Foundation.

Were that it had always been that way.

In 1976, when Mark Penn and I first began to work for Ed Koch, there was real doubt whether the "kid pollsters" as our com-petitor Peter Hart called us, were up to the task. Koch's own treas-urer, Bernie Rome, called us "nickel-and-dime" researchers, and while we dismissed that appellation angrily, it did accurately de-scribe the capital we had to invest in our first polling office.

There are many people I need to thank for giving me a helping hand on the journey from local New York political activist to one of America's and the world's most visible practitioner of his craft.

First, I have to thank my mother for encouraging me to abandon competitive sports for politics—a transition former U.N. Ambassador Richard Holbrooke later told me he made himself. To my early mentors, Dick Morris, Dick Gottfried, and Jerry Nadler, I can only thank you for sending me to graduate school in Lincoln Towers.

To the late Tony Olivieri and the late Carter Burden, I can only thank them posthumously for giving me a chance to run campaigns at a high level at a very early age. And to E.J. Dionne, world-renowned columnist and author for sharing much of my early work in East Harlem and beyond. From Appellate Court Judge Eugene Nardelli I came to understand integrity—both intellectual and the street version.

The late Pat Cunningham and David Garth gave me the chance to step up to prominent roles in citywide and statewide campaigns before my twenty-first birthday. State and Nina Solarz were there for me from the beginning and proved that you could build deep and lasting friendships in this business. My father and mother always encouraged me in my pursuit of my unusual career.

Academically, three people made me who I am: Martin Kilson, Bill Schneider, and R.W. (Bill) Johnson. Kilson taught me about real politique; Schneider taught me how to analyze and collect data; and Johnson taught me how to think critically about politics and indeed, life. I am profoundly grateful to them all. I have likewise benefited from my many conversations with John McIntyre from Real Clear Politics and David DesRossiers from the Manhattan Institute, two friends from whom I hope to continue to learn.

Al From told me when we began work in the White House that we had to appreciate his role in creating both the New Democrat movement and Bill Clinton's candidacy if I was ever going to really understand the mission we had set out to pursue. And he was absolutely right and I remain deeply in his debt.

A number of people have assisted me greatly in helping me gather the history and background necessary to produce this manuscript. Michael Cohen and John Buntin did a prodigious amount of research for the project and helped me develop as compelling an account of each episode of my career as was possible. I am profoundly grateful to them both. As a fact-checker, Edith Honan caught many errors. Those that remain are certainly my own. Finally, a particular thank you to my long-time associate Bradley Honan for his great contributions to many of the campaigns and projects described herein.

Specific individuals helped with sections on particular countries: Laura Silber for Serbia, Zev Furst for Israel, and Diego Arria for Venezuela. Needless to say their efforts made the manuscript much better than it already was and I greatly benefited from their insights and judgments. People like Bernardo Vega in the Dominican Republic and Victor and Elena Pinchuk in the Ukraine, have become great friends as well as clients. I am convinced that I have learned as much from them as they hopefully have gotten from me.

But the whole task of turning my life into a compelling story is the result of detailed discussions over many years with two people: Judith Regan and Cal Morgan.

It was Judith who first suggested that I write this book and spent long periods of time suggesting questions to answer and issues to explore, resulting in a narrative that I believe is certainly more compelling than it otherwise would have been. Her attention to detail and the attention she and her colleagues have shown to my work has made this an infinitely better work than it might ever have been. Judith, in particular, could not have been more of an advocate for me and this book and I will always be profoundly grateful to her for her efforts.

Cal Morgan picked up where Judith left off and shaped the manuscript editorially, forcing me to see what was most important in the story and working to refine each chapter, painstakingly and yet elegantly. He and his wife Cassie Jones once again managed to not only shape the final product but to produce it on an exacting schedule. Finally, Matt Harper's detailed editorial suggestions made the manuscript immeasurably better and his knowledge of politics, his writing skill, and his judgment were a welcome benefit to me as I completed the book.

Over the past two years I have served as a Fox News Contributor and have found the experience very useful intellectually and substantively. I refined a number of the arguments I have made herein while commentating on Fox and for that I must thank Roger Ailes, his number two Bill Shine, and the head of all guest booking, Laurie Luhn. All three made it clear in their own way that I was a voice worth hearing and their unceasing efforts to give me increased visibility has been an enormously positive opportunity for me. Also my dear friend Susan Estrich has been a consistent source of good counsel and advice and has made me better at all I do.

A final professional word for the two people whose influence has been most profound. Mark Penn, whose work is described expansively herein, has one of the great minds in American politics and had a profound impact on my own intellectual development as well as putting up with the fact that I have always lacked his technological acuity. Mike Berland has been a wonderful friend, a source of profound new approaches to American politics, and has demonstrated that if you think long enough and work hard enough, you will ultimately prevail. I would also like to thank Sir Martin Sorrell and Howard Paster from WPP who acquired our firm in 2001 and demonstrated very quickly why WPP is the most sophisticated and successfully integrated marketing services company in the world.

I am convinced myself, Mark Penn, and Mike Berland formed the most thoughtful, innovative, and yes, successful political consulting firm in recent American political history, a firm that continues to grow as part of WPP. This work is intended to be a testament to that thirty-year collaboration.

It is left for me to thank the most important person in my life for all that he has done with me and for me, my son Josh. It is to him that his book is most gratefully dedicated.

INDEX